ADVANCES IN CHEMICAL PHYSICS

VOLUME LXV

EDITORIAL BOARD

Advances in
CHEMICAL PHYSICS

EDITED BY

I. PRIGOGINE

University of Brussels
Brussels, Belgium
and
University of Texas
Austin, Texas

AND

STUART A. RICE

Department of Chemistry
and
The James Franck Institute
The University of Chicago
Chicago, Illinois

VOLUME LXV

AN INTERSCIENCE® PUBLICATION
JOHN WILEY & SONS
NEW YORK · CHICHESTER · BRISBANE · TORONTO · SINGAPORE

An Interscience® Publication

Library of Congress Cataloging Number: 58-9935

ISBN 0-471-83800-4

Printed in the United States of America

10 9 8 7 6 5 4 3 2 1

CONTRIBUTORS TO VOLUME LXV

L. S. CEDERBAUM, Lehrstuhl für Theoretische Chemie, Institut für Physikalische Chemie, Universität Heidelberg, Heidelberg, West Germany

H. TED DAVIS, Department of Chemical Engineering and Materials Science, University of Minnesota, Minneapolis, Minnesota

J. B. DELOS, Department of Physics, College of William and Mary, Williamsburg, Virginia

W. DOMCKE, Lehrstuhl für Theoretische Chemie, Institut für Physikalische Chemie, Universität Heidelberg, Heidelberg, West Germany

D. HSU, Department of Chemistry, Columbia University, New York, New York

JOHN KERINS, Department of Chemical Engineering and Materials Science, University of Minnesota, Minneapolis, Minnesota

J. NAGHIZADEH, Department of Chemistry, University of Tennessee, Knoxville, Tennessee

J. SCHIRMER, Lehrstuhl für Theoretische Chemie, Institut für Physikalische Chemie, Universität Heidelberg, Heidelberg, West Germany

L. E. SCRIVEN, Department of Chemical Engineering and Materials Science, University of Minnesota, Minneapolis, Minnesota

J. L. SKINNER, Department of Chemistry, Columbia University, New York, New York

W. VON NIESSEN, Lehrstuhl für Theoretische Chemie, Institut für Physikalische Chemie, Universität Heidelberg, Heidelberg, West Germany

INTRODUCTION

Few of us can any longer keep up with the flood of scientific literature, even in specialized subfields. Any attempt to do more and be broadly educated with respect to a large domain of science has the appearance of tilting at windmills. Yet the synthesis of ideas drawn from different subjects into new, powerful, general concepts is as valuable as ever, and the desire to remain educated persists in all scientists. This series, *Advances in Chemical Physics*, is devoted to helping the reader obtain general information about a wide variety of topics in chemical physics, which field we interpret very broadly. Our intent is to have experts present comprehensive analyses of subjects of interest and to encourage the expression of individual points of view. We hope that this approach to the presentation of an overview of a subject will both stimulate new research and serve as a personalized learning text for beginners in a field.

I. PRIGOGINE
STUART A. RICE

CONTENTS

Optical Dephasing of Ions and Molecules in Crystals 1
By J. L. Skinner and D. Hsu

Quasi-Two-Dimensional Phase Transitions in Paraffins 45
By J. Naghizadeh

Correlation Effects in the Ionization of Molecules: Breakdown of the Molecular Orbital Picture 115
By L. S. Cederbaum; W. Domcke, J. Schirmer, and W. von Niessen

Semiclassical Calculation of Quantum Mechanical Wavefunctions 161
By J. B. Delos

Correlation Functions in Subcritical Fluid 215
By John Kerins, L. E. Scriven, and H. Ted Davis

Author index 281

Subject index 293

ADVANCES IN CHEMICAL PHYSICS

VOLUME LXV

OPTICAL DEPHASING OF IONS AND MOLECULES IN CRYSTALS

J. L. SKINNER AND D. HSU

Department of Chemistry, Columbia University, New York, New York 10027

CONTENTS

I. Introduction . . . 1
II. Experimental Overview . . . 3
III. Theoretical Overview: A Historical Perspective . . . 4
IV. Nonperturbative Theory . . . 6
V. Coupling to Acoustic Phonons . . . 14
VI. Coupling to Optical Phonons . . . 19
VII. Coupling to Pseudolocal Phonons . . . 23
VIII. Comparison With Experiment: Acoustic Phonons . . . 31
A. 1,3-Diazaazulene in Naphthalene . . . 31
B. Cr^{3+} in Al_2O_3 (Ruby) . . . 34
C. Additional Remarks . . . 36
IX. Comparison With Experiment: Pseudolocal Phonons . . . 37
A. 3,4,6,7-Dibenzopyrene in *n*-Octane . . . 37
B. Pentacene in Benzoic Acid . . . 39
C. Additional Remarks . . . 40
X. Conclusion . . . 40
References . . . 41

I. INTRODUCTION

An electronic excitation on a molecule or ion in a crystal can suffer a wide variety of fates. Due to multipolar or exchange interactions it can transfer to a nearby molecule or ion. In a neat molecular crystal or stoichiometric inorganic crystal this interaction is responsible for the formation of delocalized Frenkel exciton states.[1] In a mixed crystal, this interaction does not necessarily produce delocalized eigenstates but nonetheless should be experimentally observable.[2–6] An excitation can also be transferred to a nearby molecule or ion with an accompanying absorption or emission of a phonon.[7] This process is called *phonon-assisted transfer.* An excitation can decay radiatively or nonradiatively to a lower electronic level in the same moledule or ion.[8] Both of these possibilities involve population relaxation and are called lifetime or T_1 processes. This is in contrast to a *pure dephasing* or T_2 process, where no population is transferred.

Pure dephasing can be viewed semiclassically as being the result of the dynamic modulation by the environment of the energy difference between two levels of an ion or molecule. Imagine creating a linear superposition of, for example, the ground state and a particular electronic excited state. In the absence of any relaxation processes, there will always be a well-defined phase relation between the amplitudes of the linear superposition state. If, however, the energy difference between the levels is randomly modulated, then the phase relation will gradually disappear, or *dephase.* This is called *pure* dephasing to distinguish it from dephasing due to population relaxation. For an ion or molecule in a crystal this energy modulation is produced by either the hyperfine interaction or the electron–phonon interaction. In the former case, the electronic energies depend on nuclear spin states. A given nuclear spin can interact with the nuclear spins of neighbors, and this interaction can lead to a modulation of the electronic energy.[9] The electronic energy levels also depend sensitively on the relative position and orientation of neighboring molecules or ions. Phonons modulate these relative coordinates, which in turn produces a modulation of the electronic energy levels.

Here, we are concerned only with the pure dephasing of optical transitions by phonons. For earlier reviews on this subject see refs. 10–17. In order to study this process independently, one must exclude all of the other relaxation channels mentioned above. For example, consider a mixed crystal composed of a low-concentration species (the guest or impurity) and the host. If one is interested in a particular electronic transition of the guest, and if the difference between this transition energy and all host transition energies is large, then energy transfer between guests and hosts will be negligible. If the guest concentration is sufficiently small, then guest–guest energy transfer will be negligible. If the guest transition of interest is between levels that are well separated from other guest levels, then nonradiative decay may be unimportant. Radiative decay channels are always present, but because the radiative lifetime can easily be measured by fluorescence experiments and is not temperature dependent, radiative lifetime effects can be separated from pure dephasing. Finally, by looking at a guest with a small or zero nuclear spin moment, one need not be concerned with nuclear spin flip dephasing. Thus, in fact, it is relatively easy to find systems where pure dephasing by phonons can be studied without the interference of competing processes.

At $T = 0$ K there are no phonon modes populated, and thus there is no pure dephasing.[18] As the temperature is increased, the population of phonon modes increases, and the pure dephasing rate increases. It is the temperature dependence of the pure dephasing rate constant that is of fundamental interest. This (often dramatic) temperature dependence provides useful information about the coupling of the optical excitation to phonons.

Primarily, this review is concerned with the theory of the temperature dependence of the pure dephasing of optical transitions by phonons and discusses our own work in this field. In the next section, however, we provide the experimental context for the theory, and in Section III we briefly discuss previous theoretical approaches. In Sections IV–VII we present the main results of the theory, and in Sections VIII and IX we compare our results with experiment.

II. EXPERIMENTAL OVERVIEW

Traditionally, one measures pure dephasing of a particular optical transition by the homogeneous width of the absorption line shape. Assuming that the optical Bloch equations are valid,[19] it is easy to show that the homogeneous line shape is Lorentzian, with a FWHM given by $\Delta\nu = 1/\pi T_2$, where T_2 is the total dephasing time.[10,11] One finds that $1/T_2$ has both lifetime and pure dephasing contributions: $1/T_2 = 1/T_2' + 1/2T_1$, where T_2' is the pure dephasing time and T_1 is the excited-state lifetime. As mentioned earlier, since it is easy to obtain T_1 from independent experiments, from the observed linewidth one can determine $1/T_2'$. We might note that the usual derivation of the Bloch equations assumes weak electron–phonon coupling. Recently, however, Skinner[20] showed that the Bloch equations are valid even for strong electron–phonon coupling, as long as $1/T_2'$ is calculated to all orders in the electron–phonon interaction.

Unfortunately, it is often not possible to obtain homogeneous linewidths of optical transitions in crystals because of inhomogeneous broadening. Both homogeneous and inhomogeneous broadening can be viewed as arising from the modulation of the transition energy that was discussed above. The distinction between the two is simply a matter of time scales; that is, homogeneous broadening or pure dephasing occurs if the modulation time scale is much shorter than the experimental time scale (T_2), while inhomogeneous broadening occurs if the modulation time scale is much longer than the experimental time scale. As discussed earlier, homogeneous broadening due to phonons is very temperature dependent and vanishes at $T = 0$ K. Inhomogeneous broadening does not vanish at $T = 0$ K and is temperature independent at low temperatures. It is thought to arise from crystal strains and defects. Because the time scale for defect relaxation is so long at low temperatures, to a first approximation one can consider the defects to be frozen. Thus one often speaks of inhomogeneous broadening as static and of homogeneous broadening as dynamic.

Several ingenious techniques have been invented to circumvent the problem of inhomogeneous broadening. For example, in the photon echo method,[19] one subjects the sample to two short laser pulses separated by a time t and measures the intensity of the "echo" generated at a time t later. Theoretically, one finds that if the Bloch equations are valid, then the echo decays exponentially with a decay rate $1/T_2$. Thus with this technique one can directly measure the homogeneous linewidth. Another useful technique is that of hole burning. In this nonlinear method, one depletes a region of the inhomogeneous spectrum with a strong but narrowband laser and then probes the hole with a weak laser. One finds[21] that the hole width is given by $\Delta\nu_{\mathrm{hb}} = 2/T_2$, so this method can also be used to measure directly the homogeneous linewidth.

The optical spectrum for dilute impurities in crystals is characterized at low temperatures by two features: a sharp zero-phonon line (ZPL) and a broad phonon sideband to the blue of the ZPL.[14–16] The ZPL results from a purely electronic transition, while the sideband results from an electronic transition with an accompanying creation of one or more phonons. The electron–phonon coupling produces a broadening of the ZPL, and it is the temperature dependence of this broadening that is measured by absorption, photon echo, and hole burning experiments.

At relatively high temperatures (typically high enough so that pure dephasing produces a homogeneous linewidth that is much larger than the inhomogeneous linewidth, and thus can be obtained directly from the absorption spectrum), one finds that the ZPL has a very steep temperature dependence. This dependence has been attributed to the coupling of the optical excitation to acoustic phonons. At low temperatures (typically so low that the homogeneous linewidth is dominated by the inhomogeneous broadening, and therefore must be obtained by photon echo or hole burning techniques), one finds that over a limited range of temperature, the temperature dependence of the ZPL is Arrhenius. This dependence has been attributed to the coupling of the electronic excitation to pseudolocal phonons.

III. THEORETICAL OVERVIEW: A HISTORICAL PERSPECTIVE

Phonons cause dephasing of the ZPL because the electron–phonon coupling produces a difference in the ground and excited electronic state

Born–Oppenheimer potential surfaces. One can perform a Taylor series expansion of this difference, which has terms that are linear and quadratic in the ground-state phonon coordinates (to second order). Silsbee[22] was the first to recognize that the linear coupling produces no dephasing, and that the quadratic coupling is responsible for the ZPL width. Using the method of moments, for the Debye model of acoustic phonons he predicted the now famous T^7 temperature dependence at low temperatures for the ZPL width. Shortly thereafter, McCumber and Sturge[23] (see also McCumber[24] and Krivoglaz[25]) provided a more quantitative calculation of the broadening, treating the quadratic electron–phonon interaction with second-order perturbation theory. The McCumber–Sturge analysis provides the standard result in the field of optical dephasing and has been rederived with different methods several times.[8,26,27]

Several years later Small[28] realized that the McCumber–Sturge result could be modified to account for the coupling to pseudolocal phonons, which are often produced near an impurity site in a crystal, and he showed that this leads to an Arrhenius temperature dependence of the ZPL width. This result was also found by Jones and Zewail.[26] Harris[29] and deBree and Wiersma[30] took a somewhat different approach to this problem. They noted that in many systems the pseudolocal peaks in the phonon sideband are very prominent, indicating that perhaps the pseudolocal phonons should be treated on "equal footing" with the optical excitation.[30] Harris[29] followed the exchange ideas of Kubo and Tomita[31] and Anderson[32] and found that the ZPL width is Arrhenius. deBree and Wiersma[30] took a more microscopic approach using Redfield theory[33,34] (second-order perturbation theory in the *system-bath* interaction). They found that under certain conditions their results reduced to exchange theory, but, under other conditions, their results predicted a bi-Arrhenius temperature dependence of the ZPL width.

The linear electron–phonon coupling term is primarily responsible for the creation of the phonon sideband.[14] For many systems, the Huang–Rhys factor, which determines the ratio of the ZPL to sideband intensities, shows that the linear coupling must be considered strong. If this is the case, there is no reason to assume that the quadratic coupling is weak, which is what is required by the perturbative theory of McCumber and Sturge.[23] (In fact, even if the linear coupling is weak, it is not correct to assume that the quadratic coupling is also weak.) Perhaps this fact motivated two researchers to perform a nonperturbative calculation of the ZPL width. Both Abram[35] and Osad'ko[36] considered linear and quadratic coupling to harmonic phonons, although the models are slightly different in that Abram made the rotating-wave approximation in the quadratic term. Both workers found that the linear coupling does not produce a temperature-dependent broadening of the line, but the results they obtained for the quadratic term are not in agreement. In particular, Abram found that the width is an even function of the coupling strength, while Osad'ko obtained no such simplification.

We were motivated to perform a nonperturbative calculation of the ZPL width in part to resolve the discrepancy between Osad'ko[36] and Abram[35] and in part to understand the relationship among the various approaches[28–30] to dephasing by pseudolocal phonons. The body of this review summarizes our work[37–40] to this end. Although our method is quite different from that of Osad'ko,[36] we obtain complete agreement with him for our general results[37] in Section IV. We then further analyze our results for different models of coupling to phonons in Sections V–VII.

IV. NONPERTURBATIVE THEORY

We consider[37] two electronic levels of an isolated impurity in a crystal. The adiabatic nuclear Hamiltonians for the two electronic states are

$$H_0 = T + V_0 \tag{4.1}$$

and

$$H_1 = T + V_1 \,. \tag{4.2}$$

Here T is the nuclear kinetic energy operator (for all the nuclei in the crystal) and V_0 and V_1 are the Born–Oppenheimer ground- and excited-state potential surfaces. Within the harmonic approximation, the ground-state Hamiltonian is

$$H_0 = \sum_k \hbar\omega_k (b_k^\dagger b_k + \tfrac{1}{2}) \,, \tag{4.3}$$

where $b_k^\dagger$ and b_k are the creation and annihilation operators for the kth normal mode, and ω_k are the normal mode frequencies. It is important to emphasize that these are the normal modes of the crystal *plus* (ground-state) impurity system. The excited-state Hamiltonian can clearly be written

$$H_1 = H_0 + (V_1 - V_0) \,. \tag{4.4}$$

The difference between the potential energies can be expanded in a

Taylor series in the *ground*-state normal mode coordinates. Keeping only up to quadratic terms (to be consistent with our harmonic ground-state surface), we obtain

$$V_1 - V_0 = \hbar\omega_0 + \sum_k g_k(b_k^\dagger + b_k) + \frac{1}{2}\sum_{kq} g_{kq}(b_k^\dagger + b_k)(b_q^\dagger + b_q)\,, \qquad (4.5)$$

where ω_0 is a constant and g_k and g_{kq} are the linear and quadratic expansion coefficients.

In what follows we actually adopt a less general form for the potential difference. We assume that it can be expanded in a single collective coordinate, ϕ, which is linear in the normal mode coordinates,

$$V_1 - V_0 = \hbar\omega_0 + a\phi + \frac{W}{2}\phi^2 \qquad (4.6)$$

and

$$\phi = \sum_k h_k(b_k^\dagger + b_k)\,, \qquad (4.7)$$

where a and W are constants. We choose W to be dimensionless; thus h_k and a have units of the square root of energy. When coupling to acoustic phonons, ϕ is the strain field, while for coupling to a pseudolocal phonon, ϕ is the *local* mode coordinate. More generally, when coupling to more than one type of collective mode is important, we have

$$V_1 - V_0 = \hbar\omega_0 + \sum_i a_i\phi_i + \sum_{ij}\frac{W_{ij}}{2}\phi_i\phi_j \qquad (4.8)$$

and

$$\phi_i = \sum_k h_{ik}(b_k^\dagger + b_k)\,. \qquad (4.9)$$

This more general situation is considered elsewhere.[41]

Within the Condon approximation (the transition dipole moment is independent of phonon coordinates) from linear response theory one finds that the optical absorption lineshape is given by[42,43]

$$I(\omega) \propto \int_{-\infty}^{\infty} dt\, e^{i\omega t}\langle e^{iH_0t/\hbar}e^{-iH_1t/\hbar}\rangle\,, \qquad (4.10)$$

where

$$\langle \cdots \rangle = \mathrm{Tr}\,\frac{[e^{-\beta H_0}\cdots]}{\mathrm{Tr}[e^{-\beta H_0}]}\,. \tag{4.11}$$

Here the brackets indicate a thermal phonon average with the ground-state phonon Hamiltonian. Phenomenologically introducing the excited-state lifetime, T_1, and using Eqs. (4.4) and (4.6), this becomes

$$I(\omega) \propto \int_{-\infty}^{\infty} dt\, e^{i(\omega-\omega_0)t} e^{-|t|/2T_1} \langle F(t) \rangle\,, \tag{4.12}$$

where

$$F(t) = e^{iH_0 t/\hbar} e^{-i(\Delta + H_0)t/\hbar} \tag{4.13}$$

and

$$\Delta = a\phi + \frac{W}{2}\phi^2\,. \tag{4.14}$$

$F(t)$ can also be written[44]

$$F(t) = \exp_T\left[\frac{-i}{\hbar}\int_0^t d\tau\, \Delta(\tau)\right], \tag{4.15}$$

where

$$\Delta(\tau) = e^{iH_0\tau/\hbar}\Delta e^{-iH_0\tau/\hbar} \tag{4.16}$$

and T is the chronological operator that orders the $\Delta(\tau)$'s in order of increasing time from right to left for $t > 0$, and from left to right for $t < 0$.

After performing a cumulant expansion[45] of $\langle F(t) \rangle$, we find[37] that under certain conditions the line shape is Lorentzian with a width $1/T_2$ and a shift δ given by

$$\frac{1}{T_2} = \frac{1}{T_2'} + \frac{1}{2T_1}\,, \tag{4.17}$$

$$\frac{1}{T_2'} \equiv \gamma = -\mathrm{Re}\{K\}\,, \tag{4.18}$$

and

$$\delta = -\mathrm{Im}\{K\}\,, \tag{4.19}$$

where

$$K = \sum_{m=1}^{\infty} K_m \tag{4.20}$$

and

$$K_m = \left(-\frac{i}{\hbar}\right)^m \int_0^\infty d\tau_1 \int_0^\infty d\tau_2 \cdots \int_0^\infty d\tau_{m-1} \times \langle T\Delta(\tau_1 + \tau_2 + \cdots + \tau_{m-1}) \cdots \Delta(\tau_1)\Delta(0)\rangle_c . \tag{4.21}$$

In the above $\langle \cdots \rangle_c$ denote a cumulant average.[45] Thus the line shift, δ, and the pure dephasing contribution to the linewidth, γ, are the imaginary and real parts of an infinite sum of cumulants. (To get the line shift in hertz, one must divide δ by 2π). We note that the usual weak coupling results[23–26] can be obtained by truncating Eq. (4.20) at $m = 2$. The condition under which Eqs. (4.17)–(4.21) are valid[20,37] is that the separation of time scales $T_2 \gg \tau_c$ must be satisfied, where τ_c is the phonon correlation time (the time in which $\langle\Delta(t)\Delta(0)\rangle$ decays). At the end of the calculation we verify that this condition is indeed satisfied.

In what follows, we express each term in Eq. (4.20) as a sum of diagrams. A judicious resummation of the resulting infinite series gives a closed form analytic expression. In formulating the diagrammatic analysis,[37] we make use of the finite-temperature version of Wick's theorem,[44] which, along with the cumulant average,[45] dictates that all diagrams must be connected. The diagrams fall naturally into two classes—chain and loop diagrams—which are shown in Figures 1 and 2. The rules for evaluating these diagrams are as follows:[37]

1. A line joining two points (a point is a dot or circle) represents a factor $C(t)$.
2. With each point is associated a time argument: 0 with the left most point, t_1 with the next point to the right, $t_1 + t_2$ with the next point to the right, and so on.
3. The time argument of a factor $C(t)$ represented by a line joining two points is the time associated with the right point minus the time associated with the left point.
4. For a diagram with m points, there are $m - 1$ time integrals of the form

$$\int_0^\infty dt_1 \int_0^\infty dt_2 \cdots \int_0^\infty dt_{m-1} .$$

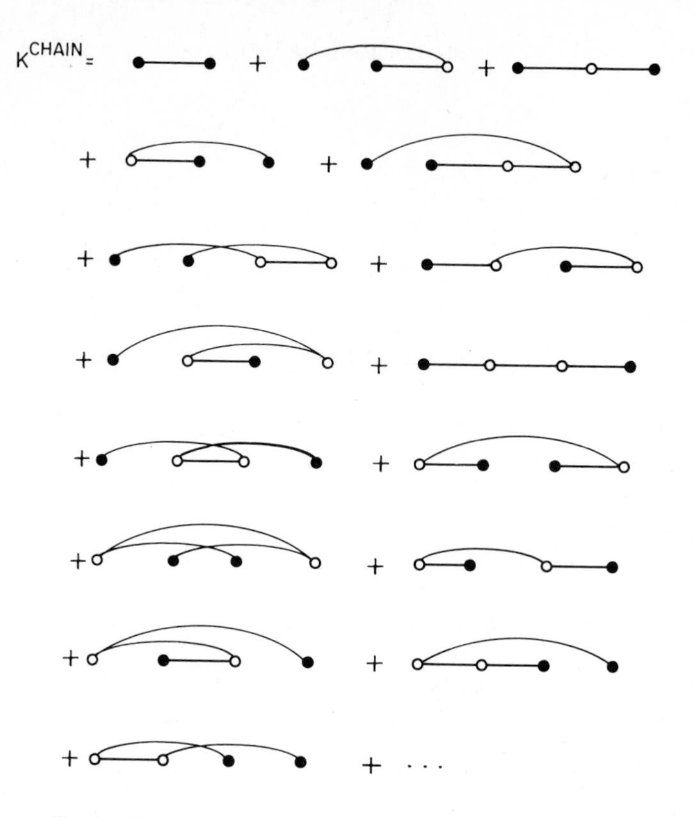

Fig. 1. Chain diagrams that contribute to the cumulant sum.

$K^{LOOP} =$

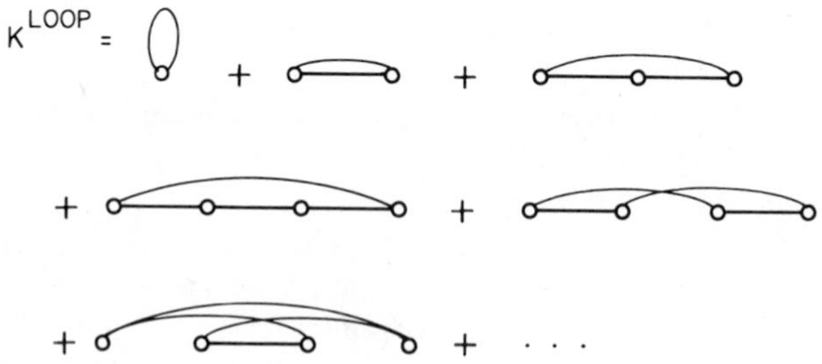

Fig. 2. Loop diagrams that contribute to the cumulant sum.

5. With each dot is associated a factor $-ia/\hbar$.
6. With each circle is associated a factor $-iW/\hbar$.
7. If the two lines from each circle do not go to two different points, there is a factor of $\frac{1}{2}$.

The factor $C(t)$ in the above is the time-ordered correlation function (Green's function),

$$C(t) = \langle T\phi(t)\phi(0)\rangle . \tag{4.22}$$

Using Eq. (4.7) this becomes

$$C(t) = \sum_k h_k^2[(n_k + 1)e^{-i\omega_k|t|} + n_k e^{i\omega_k|t|}] , \tag{4.23}$$

where

$$n_k = n(\omega_k) = \frac{1}{\exp(\beta\hbar\omega_k) - 1} . \tag{4.24}$$

For future use we need the Fourier transform of $C(t)$, which is defined by

$$\hat{C}(\omega) = \int_{-\infty}^{\infty} dt\, e^{i\omega t} C(t) . \tag{4.25}$$

$\hat{C}(\omega)$ can be computed explicitly, with the result

$$\hat{C}(\omega) = [2n(\omega) + 1]\hbar\Gamma_0(\omega) + [2n(-\omega) + 1]\hbar\Gamma_0(-\omega) + i\hbar\Omega_0(\omega) , \tag{4.26}$$

where

$$\Gamma_0(\omega) = \frac{\pi}{\hbar}\sum_k h_k^2\delta(\omega - \omega_k) \tag{4.27}$$

and

$$\Omega_0(\omega) = \frac{1}{\hbar}\sum_k h_k^2\left[P\left(\frac{1}{\omega - \omega_k}\right) - \left(\frac{1}{\omega + \omega_k}\right)\right] . \tag{4.28}$$

This can also be written

$$\Omega_0(\omega) = \int_0^{\infty} \frac{d\omega'}{\pi}\Gamma_0(\omega')2\omega' P\left(\frac{1}{\omega^2 - \omega'^2}\right) . \tag{4.29}$$

In the above, $\Gamma_0(\omega)$ is the dimensionless weighted phonon density of states. That is, it is the density of states weighted by the extent to which each normal mode contributes to the collective coordinate. Since all $\omega_k > 0$, it follows that $\Gamma_0(-\omega) = 0$ for $\omega \geq 0$.

By defining certain diagram classes and performing the appropriate variable changes in the time integrations, it is possible to sum exactly both the infinite series for the chain and loop diagrams. The results are[37]

$$K^{\text{chain}} = \left(\frac{-i}{\hbar}\right) \frac{a^2\Omega_0(0)}{2} \frac{1}{1 - W\Omega_0(0)} \tag{4.30}$$

and

$$K^{\text{loop}} = -\int_0^\infty \frac{d\omega}{2\pi} \ln\left[1 + iW\frac{\hat{C}(\omega)}{\hbar}\right]. \tag{4.31}$$

Using Eqs. (4.18) and (4.19) we now take the real and imaginary parts of $K = K^{\text{loop}} + K^{\text{chain}}$ to find[37]

$$\delta = \delta_0 + \delta_T, \tag{4.32}$$

$$\delta_0 = \frac{a^2\Omega_0(0)}{2\hbar} \frac{1}{1 - W\Omega_0(0)} + \int_0^\infty \frac{d\omega}{2\pi} \arctan\left(\frac{W\Gamma_0(\omega)}{1 - W\Omega_0(\omega)}\right), \tag{4.33}$$

$$\delta_T = \int_0^\infty \frac{d\omega}{2\pi} \arctan\left(\frac{2n(\omega)W\Gamma_0(\omega)[1 - W\Omega_0(\omega)]}{[1 - W\Omega_0(\omega)]^2 + [2n(\omega) + 1]W^2\Gamma_0(\omega)^2}\right), \tag{4.34}$$

and

$$\gamma = \int_0^\infty \frac{d\omega}{4\pi} \ln\left(1 + \frac{4n(\omega)[n(\omega) + 1]W^2\Gamma_0(\omega)^2}{[1 - W\Omega_0(\omega)]^2 + W^2\Gamma_0(\omega)^2}\right). \tag{4.35}$$

δ_0 and δ_T are the temperature-independent and thermal contributions to the line shift, respectively.

These results are in complete agreement with Osad'ko,[36] who used an integral equation approach. First, we see that the broadening and shift depend only on the weighted phonon density of states, $\Gamma_0(\omega)$, from Eq. (4.27), the linear and quadratic coupling constants a and W, and the temperature. [$\Omega_0(\omega)$ is determined by $\Gamma_0(\omega)$ from Eq. (4.29).] We also note that when $T \to 0$ K, $n(\omega) \to 0$, and thus $\gamma \to 0$; there is no pure dephasing at $T = 0\,\mathrm{K}$. This is in accord with more general arguments.[18] Second, we see that when $W = 0$ (no quadratic coupling), $\gamma = 0$; linear coupling alone will not produce line broadening.

We also showed[37] that the broadening and shift of the ZPL in the fluorescence spectrum are identical to the above; the ZPLs in absorption and emission coincide exactly. We can rewrite the linewidth formula to reflect this symmetry of the ground- and excited-state surfaces by introducing $\Gamma_1(\omega)$, the weighted normal mode density of states for the *excited*-state surface[37] (see also Osad'ko[15]):

$$\Gamma_1(\omega) = \frac{\Gamma_0(\omega)}{[1 - W\Omega_0(\omega)]^2 + W^2\Gamma_0(\omega)^2} . \tag{4.36}$$

With this Eq. (4.35) can be written

$$\gamma = \int_0^\infty \frac{d\omega}{4\pi} \ln[1 + 4n(\omega)(n(\omega) + 1)W^2\Gamma_0(\omega)\Gamma_1(\omega)] . \tag{4.37}$$

Not all values of the quadratic coupling constant, W, are physically acceptable. In particular, we require that the force constants of all the excited-state normal modes be positive. With the following normalization of the expansion coefficients (or equivalently, the weighted density of states),

$$\sum_k \frac{2}{\hbar\omega_k} h_k^2 = \frac{2}{\pi}\int_0^\infty \frac{d\omega}{\omega}\Gamma_0(\omega) = 1 , \tag{4.38}$$

the above requirement leads[38] to a lower bound for W of -1, that is, $-1 < W < \infty$.

We can make the connection with the more usual perturbative theories[23–26] of the ZPL width and shift by expanding our results for $|W\Gamma_0(\omega)| \ll 1$ to obtain

$$\gamma = \frac{W^2}{\pi}\int_0^\infty d\omega\, n(\omega)[n(\omega) + 1]\Gamma_0(\omega)^2 \tag{4.39}$$

and

$$\delta_T = \frac{W}{\pi}\int_0^\infty d\omega\, n(\omega)\Gamma_0(\omega) . \tag{4.40}$$

By substituting in Eq. (4.27), these can also be written

$$\gamma = \frac{W^2\pi}{\hbar^2} \sum_{kk'} h_k^2 h_{k'}^2 n(\omega_k)[n(\omega_k) + 1]\delta(\omega_k - \omega_{k'}) \tag{4.41}$$

and

$$\delta_T = \frac{W}{\hbar} \sum_k h_k^2 n(\omega_k) . \tag{4.42}$$

V. COUPLING TO ACOUSTIC PHONONS

In this section we consider the coupling of the optical excitation to acoustic phonons. As discussed previously, the difference between the impurity electronic levels is very sensitive to the local environment, which is perturbed by acoustic phonons. In Section IV we assumed that this difference could be expressed in terms of a single scalar collective coordinate, ϕ, which in this case defines the strain field of the acoustic phonons. The collective coordinate is then expanded in the normal modes of the system.

The introduction of an impurity into a crystal breaks the translational symmetry of the perfect crystal, and thus the normal modes of the system are no longer described by wavevectors. Nonetheless, because a single impurity produces such a minor perturbation on the long-wavelength eigenfunctions, we consider it a good approximation to treat these long-wavelength normal modes to be the normal modes of the perfect crystal. In this case the sum over normal mode indices, k, in Eq. (4.7) or (4.27) becomes a sum over the three acoustic phonon branches, $s = 1, 2, 3$, and a sum over all wavevectors, $\mathbf{q}$.

In the Debye model, all three branches of acoustic phonons are described by the dispersion relation $\omega_{\mathbf{q}s} = c|\mathbf{q}|$, where c is the sound speed. The sum over k then becomes

$$\sum_k = \sum_{\mathbf{q}s} \rightarrow \frac{3N}{\omega_D^3} \int_0^{\omega_D} d\omega\, \omega^2 , \tag{5.1}$$

where $\hbar\omega_D = kT_D$ and T_D is the Debye temperature. The coupling constants in Eq. (4.7), $h_{\mathbf{q}s}$, describe the relative importance of different phonons in producing the distortion of the excited-state potential. These coupling constants are in general not independent of phonon branch or wavevector orientation.[40] Thus, for example, because of the particular orientation of the impurity molecule in the crystal, a certain direction of a particular branch may not couple at all. However, in the simplest approximation (the long-wavelength approximation[8]) the *frequency* dependence of the coupling constants is the same for all modes: $h_{\mathbf{q}s} \sim (\omega_{\mathbf{q}s})^{1/2}$.

With this result, from Eqs. (4.27), (4.29), (4.38), and (5.1) we have[38]

$$\Gamma_0(\omega) = \begin{cases} \dfrac{3\pi}{2}\left(\dfrac{\omega}{\omega_{\mathrm{D}}}\right)^3, & 0 < \omega < \omega_{\mathrm{D}} \\ 0, & \omega > \omega_{\mathrm{D}} \end{cases} \tag{5.2}$$

and

$$\Omega_0(\omega) = -\left[1 + 3\left(\frac{\omega}{\omega_{\mathrm{D}}}\right)^2 + \frac{3}{2}\left(\frac{\omega}{\omega_{\mathrm{D}}}\right)^3 \ln\left(\frac{1-\omega/\omega_{\mathrm{D}}}{1+\omega/\omega_{\mathrm{D}}}\right)\right]. \tag{5.3}$$

Substituting Eqs. (5.2) and (5.3) into Eqs. (4.34) and (4.35) gives the nonperturbative expressions for the width and thermal shift, which in hertz are ($\Delta\nu = \gamma/\pi$, $\delta\nu = \delta_T/2\pi$):[38]

$$\Delta\nu = \frac{\omega_{\mathrm{D}}}{4\pi^2}\int_0^1 dx \ln\left\{1 + 9\pi^2 W^2 x^6 \frac{e^{xT_{\mathrm{D}}/T}}{(e^{xT_{\mathrm{D}}/T}-1)^2}\left[g(x)^2 + W^2\frac{9\pi^2}{4}x^6\right]^{-1}\right\}, \tag{5.4}$$

$$\delta\nu = \frac{\omega_{\mathrm{D}}}{4\pi^2}\int_0^1 dx \arctan\left\{\frac{3\pi W x^3 g(x)}{e^{xT_{\mathrm{D}}/T}-1}\left[g(x)^2 + \coth\left(\frac{xT_{\mathrm{D}}}{2T}\right)W^2\frac{9\pi^2}{4}x^6\right]^{-1}\right\}, \tag{5.5}$$

and

$$g(x) = 1 + W\left[1 + 3x^2 + \frac{3}{2}x^3 \ln\left(\frac{1-x}{1+x}\right)\right]. \tag{5.6}$$

Before we analyze the full temperature dependence of these results, it is instructive to consider first the weak coupling limit. Expanding Eqs. (5.4) and (5.5) for $|W| \ll 1$ we obtain the familiar weak coupling results:[8,23,26]

$$\Delta\nu = \omega_{\mathrm{D}}\frac{9W^2}{4}\left(\frac{T}{T_{\mathrm{D}}}\right)^7\int_0^{T_{\mathrm{D}}/T} dx \frac{x^6 e^x}{(e^x-1)^2} \tag{5.7}$$

and

$$\delta\nu = \omega_{\mathrm{D}} \frac{3W}{4\pi} \left(\frac{T}{T_{\mathrm{D}}}\right)^4 \int_0^{T_{\mathrm{D}}/T} dx \frac{x^3}{e^x - 1}. \tag{5.8}$$

These expressions could also have been obtained from Eqs. (4.39) and (4.40). For $T \ll T_{\mathrm{D}}$ they display T^7 and T^4 temperature dependencies, respectively. For $T \gg T_{\mathrm{D}}$, they go like T^2 and T, respectively. It should be noted, however, that for $T \gg T_{\mathrm{D}}$, it is only for *extraordinarily* weak coupling ($W \ll T_{\mathrm{D}}/T$) that the expansion of Eqs. (5.4) and (5.5) leading to the above results is justified.

It is also interesting to investigate the low-temperature limit of the exact results. Expanding Eqs. (5.4) and (5.5) for $T \ll T_{\mathrm{D}}$ we obtain:[38]

$$\Delta\nu = \omega_{\mathrm{D}} \frac{9W^2}{4} \int_0^1 dx \frac{e^{xT_{\mathrm{D}}/T} x^6}{(e^{xT_{\mathrm{D}}/T} - 1)^2} \left[g(x)^2 + W^2 \frac{9\pi^2}{4} x^6 \right]^{-1} \tag{5.9}$$

and

$$\delta\nu = \omega_{\mathrm{D}} \frac{3W}{4\pi} \int_0^1 dx \frac{x^3}{e^{xT_{\mathrm{D}}/T} - 1} g(x) \left[g(x)^2 + W^2 \frac{9\pi^2}{4} x^6 \right]^{-1}. \tag{5.10}$$

Because of the exponentials, when $T \ll T_{\mathrm{D}}$, the only contribution to the integrals will come for $x \ll 1$. Expanding the rest of the integrand we obtain for $W \neq -1$:

$$\Delta\nu = \omega_{\mathrm{D}} \frac{9}{4} \left(\frac{W}{1+W}\right)^2 \left(\frac{T}{T_{\mathrm{D}}}\right)^7 \int_0^{T_{\mathrm{D}}/T} dx \frac{x^6 e^x}{(e^x - 1)^2} \tag{5.11}$$

and

$$\delta\nu = \omega_{\mathrm{D}} \frac{3}{4\pi} \frac{W}{1+W} \left(\frac{T}{T_{\mathrm{D}}}\right)^4 \int_0^{T_{\mathrm{D}}/T} dx \frac{x^3}{e^x - 1}. \tag{5.12}$$

On the other hand, for $W = -1$, we obtain:

$$\Delta\nu = \frac{\omega_{\mathrm{D}}}{4} \left(\frac{T}{T_{\mathrm{D}}}\right)^3 \int_0^{T_{\mathrm{D}}/T} dx \frac{x^2 e^x}{(e^x - 1)^2} \tag{5.13}$$

and

$$\delta\nu = \frac{\omega_{\mathrm{D}}}{4\pi} \left(\frac{T}{T_{\mathrm{D}}}\right)^2 \int_0^{T_{\mathrm{D}}/T} dx \frac{x}{e^x - 1}. \tag{5.14}$$

Comparing Eqs. (5.7) and (5.8) and Eqs. (5.11) and (5.12) we see that for $W \neq -1$ and $T \ll T_D$, the perturbative results give the correct temperature dependencies although the overall coefficient is incorrect. However, for $W = -1$, comparing Eqs. (5.7) and (5.8) and Eqs. (5.13) and (5.14), we see that the weak coupling results predict the wrong power law for $T \ll T_D$; the exact dependencies are T^3 and T^2, respectively, for the width and shift, while the perturbative results give T^7 and T^4. In fact, the weak coupling result for the shift predicts the wrong sign!

Finally, we discuss the exact results for all temperatures, which are obtained by numerically integrating Eqs. (5.4) and (5.5). Here we focus only on the temperature behavior of the linewidth (see ref. 38 for a discussion of the line shift), in part because it is believed that to provide a useful comparison with the experimental line shift, one has to include the contribution from thermal expansion,[15,46] which is not considered in our theory. In Figs. 3 and 4 we show $\log_{10}(\Delta\nu/\omega_D)$ versus $\log_{10}(T/T_D)$ for various values of W between -1 and ∞. For comparison we also show the perturbative results, Eq. (5.7). Focusing first on positive W (Fig. 3), we

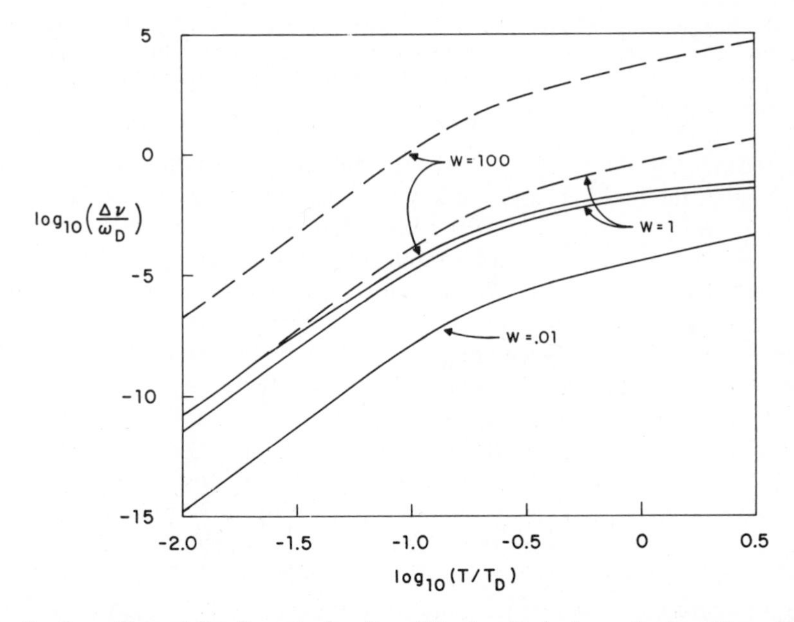

Fig. 3. Logarithm of the dimensionless linewidth, $\log_{10}(\Delta\nu/\omega_D)$, vs. the logarithm of the dimensionless temperature, $\log_{10}(T/T_D)$, for the Debye model of acoustic phonons and various positive values of the coupling constant W. The nonperturbative theory, Eq. (5.4), is given by the solid line and the perturbative theory, Eq. (5.7), is given by the dashed line. For $W = 0.01$ the two theories are indistinguishable.

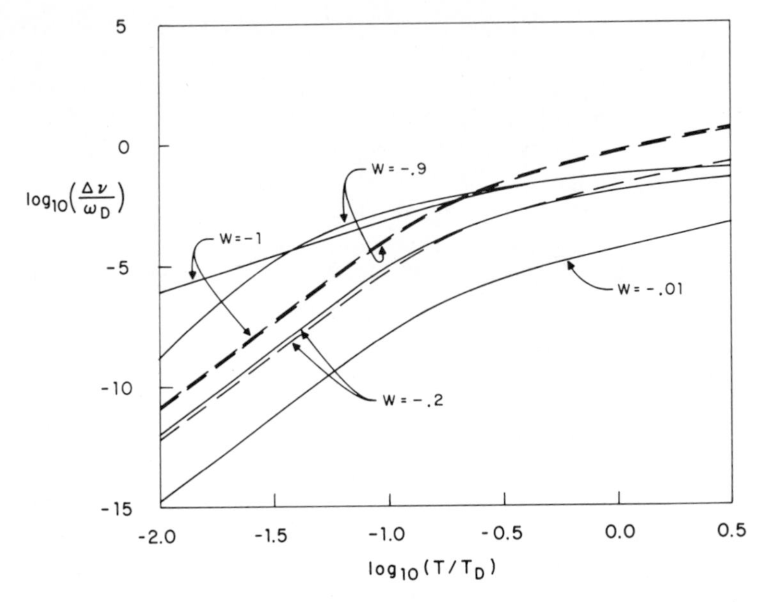

Fig. 4. $\mathrm{Log}_{10}(\Delta\nu/\omega_D)$ vs. $\log_{10}(T/T_D)$ for the Debye model of acoustic phonons and various negative values of W. The nonperturbative theory, Eq. (5.4), is given by the solid line and the perturbative theory, Eq. (5.7), is given the dashed line. For $W = -0.01$ the two theories are indistinguishable.

see that for $W = 0.01$, the perturbative and nonperturbative results agree. Moreover, the limiting slopes at low and high temperatures display the T^7 and T^2 power laws. For $W = 1$ and 100, the perturbative and nonperturbative results differ substantially (several orders of magnitude for $W = 100$), but all show the T^7 behavior at low temperatures, as discussed above. More importantly, however, the *temperature dependencies* of the two theories do not agree for $T > 0.03T_D$. In particular, as previously pointed out by Osad'ko,[36] the nonperturbative theory shows the property of saturation for large W, while the weak coupling results can increase without bound. Now considering negative W (Fig. 4), we see that for $W = -0.01$, the perturbative and nonperturbative results agree. As W approaches -1, at low T the nonperturbative result is greater by several orders of magnitude than the perturbative result, while at high T the perturbative result is greater by several orders of magnitude. Moreover, the temperature dependencies are very different. In particular, $W = -1$ shows a T^3 low-temperature dependence, in agreement with Eq. (5.13).

One of the central assumptions behind the cumulant expansion method, that leads to the nonperturbative results analyzed above, is the separation of time scales between the system and bath;[20,37] that is, the phonon correlation time should be much less than the dephasing time. The phonon correlation time is defined by the characteristic time over which the correlation function, $C(t)$, decays. From Eq. (4.23) it follows that for high temperatures $T > T_D$, the decay rate of this function is ω_D. For lower temperatures it decays with this rate and also the slower rate $kT/\hbar$. On the other hand, the dephasing rate is simply $\Delta\nu$. Thus the separation of time scale requirement is that

$$\frac{\Delta\nu}{\omega_D} \ll 1\,, \qquad T > T_D \tag{5.15}$$

and

$$\frac{\Delta\nu}{\omega_D} \ll \frac{T}{T_D}\,, \qquad T < T_D\,. \tag{5.16}$$

From Figs. 3 and 4 it is seen that these inequalities are satisfied, especially at the lower temperatures.

VI. COUPLING TO OPTICAL PHONONS

In this section we consider the coupling of the optical excitation to optical phonons. Even though the impurity breaks the translational symmetry of the lattice, as in the previous section, to a first approximation we take the normal modes of the impurity plus crystal system to be those of the perfect crystal. In the Einstein approximation, optical phonons are described by the trivial dispersion relation $\omega_k = \omega_0$, where ω_0 is the Einstein frequency [not to be confused with the electronic frequency in Eq. (4.5)]. The normalized weighted density of states is then [from Eqs. (4.27) and (4.38)]

$$\Gamma_0(\omega) = \frac{\pi\omega_0}{2}\,\delta(\omega - \omega_0)\,. \tag{6.1}$$

Below we show that this delta function prescription for $\Gamma_0(\omega)$ does *not* lead to broadening of the zero-phonon line. Therefore we have chosen[38] to modify Eq. (6.1) to incorporate a finite width:

$$\Gamma_0(\omega) = \begin{cases} \dfrac{3\pi}{8}\,\dfrac{\omega_0}{\bar{\omega}^3}\,[\bar{\omega}^2 - (\omega - \omega_0)^2]\,, & \omega_0 - \bar{\omega} < \omega < \omega_0 + \bar{\omega} \\ 0\,, & \omega < \omega_0 - \bar{\omega},\ \omega > \omega_0 + \bar{\omega}\,. \end{cases} \tag{6.2}$$

This is a parabolic density of states peaked at ω_0 with baseline width $2\bar{\omega}$. This density of states is actually a more realistic description than Eq. (6.1) since the finite width reflects the nontrivial dispersion relation for optical phonons. In what follows, we are concerned with narrow optical bands such that $\bar{\omega} \ll \omega_0$. In this case the normalization condition of Eq. (4.38) is approximately satisfied by Eq. (6.2), and thus -1 is still the lower bound for W. For later use we note that the delta function limit can be obtained from Eq. (6.2) by letting $\bar{\omega} \to 0$.

With Eqs. (6.2) and (4.29), we find $\Omega_0(\omega)$ to be[38]

$$\Omega_0(\omega) = -\frac{3}{8}\frac{\omega_0}{\bar{\omega}^3}\left\{[\bar{\omega}^2 - (\omega + \omega_0)^2] \ln\left(\frac{\bar{\omega} + \omega_0 + \omega}{\omega_0 + \omega - \bar{\omega}}\right)\right.$$

$$\left. + [\bar{\omega}^2 - (\omega - \omega_0)^2] \ln\left(\frac{\bar{\omega} + \omega_0 - \omega}{\bar{\omega} - \omega_0 + \omega}\right) + 4\bar{\omega}\omega_0\right\}. \tag{6.3}$$

As mentioned above, here we are concerned primarily with *sharp* densities of states. Taking the limit $\bar{\omega} \ll \omega_0$ gives

$$\Omega_0(\omega) = \frac{3}{8}\frac{\omega_0}{\bar{\omega}^3}\left\{2\bar{\omega}(\omega - \omega_0) + [\bar{\omega}^2 - (\omega - \omega_0)^2] \ln\left(\frac{\bar{\omega} - \omega_0 + \omega}{\bar{\omega} + \omega_0 - \omega}\right)\right\}. \tag{6.4}$$

Substituting Eqs. (6.2) and (6.4) into Eqs. (4.34) and (4.35) gives us the nonperturbative expressions for the broadening and shift of the electronic transition due to optical phonons:[38]

$$\Delta\nu = \omega_0 \frac{V}{4\pi^2}\int_{-1}^{1} dx \ln\left\{1 + \left(\frac{3\pi W}{4V}\right)^2 (1 - x^2)^2 \frac{e^{T_0(1+Vx)/T}}{(e^{T_0(1+Vx)/T} - 1)^2}\right.$$

$$\left. \times \left[h(x)^2 + \left(\frac{3\pi W}{8V}\right)^2 (1 - x^2)^2\right]^{-1}\right\}, \tag{6.5}$$

$$\delta\nu = \omega_0 \frac{V}{4\pi^2}\int_{-1}^{1} dx \arctan\left\{\left(\frac{3\pi W}{4V}\right)\frac{(1 - x^2)h(x)}{e^{T_0(1+Vx)/T} - 1}\right.$$

$$\left. \times \left[h(x)^2 + \coth\left(\frac{T_0(1 + xV)}{2T}\right)\left(\frac{3\pi W}{8V}\right)^2 (1 - x^2)^2\right]^{-1}\right\}, \tag{6.6}$$

and

$$h(x) = 1 - \frac{3W}{8V}\left[2x + (1 - x^2) \ln\left(\frac{1 + x}{1 - x}\right)\right], \tag{6.7}$$

where V is the dimensionless density of states width,

$$V = \frac{\bar{\omega}}{\omega_0}, \tag{6.8}$$

and

$$T_0 = \frac{\hbar\omega_0}{k}. \tag{6.9}$$

As in the case of acoustic phonons, it is instructive to take the weak coupling limit, $|W\Gamma_0(\omega)| \ll 1$, which in this case is $|W|/V \ll 1$. Expanding the above we obtain

$$\Delta\nu = \omega_0 \frac{W^2}{V}\frac{9}{64}\int_{-1}^{1} dx(1-x^2)^2 \frac{e^{T_0(1+Vx)/T}}{(e^{T_0(1+Vx)/T}-1)^2} \tag{6.10}$$

and

$$\delta\nu = \omega_0 W \frac{3}{16\pi}\int_{-1}^{1} dx \frac{(1-x^2)}{e^{T_0(1+Vx)/T}-1}. \tag{6.11}$$

In the case of $T/T_0 \gg V$ (which is not difficult to attain, since $V \ll 1$) these expressions can be simplified to give

$$\Delta\nu = \omega_0 \frac{W^2}{V}\frac{3}{20} n(\omega_0)[n(\omega_0)+1] \tag{6.12}$$

and

$$\delta\nu = \omega_0 \frac{W}{4\pi} n(\omega_0). \tag{6.13}$$

These perturbative results have been found before by several others.[11,26,28] For $T \ll T_0$ they lead to Arrhenius expressions for both the width and shift:

$$\Delta\nu = \omega_0 \frac{W^2}{V}\frac{3}{20} e^{-\hbar\omega_0/kT} \tag{6.14}$$

and

$$\delta\nu = \omega_0 \frac{W}{4\pi} e^{-\hbar\omega_0/kT} . \tag{6.15}$$

On the other hand, if $T \gg T_0$, then we find that the width and shift go like T^2 and T, respectively. However, in this case it is only for *extraordinarily* weak coupling ($W \ll V(T_0/T)$ that the expansions of Eqs. (6.5) and (6.6) are justified.

It is also important to examine the low-temperature limit of the *exact* results. For $T \ll T_0$ (but $T \gg VT_0$) we find that both the width and the shift are Arrhenius, and that the prefactors are complicated functions of W and V. One interesting result is that for the linewidth, the prefactor of the Arrhenius expression is an increasing function of $|W|/V$ until $|W|/V >$ 1.5, where it is a constant. This behavior differs dramatically from that of the weak coupling result, Eq. (6.14). Thus we find that the maximum linewidth is[38]

$$\Delta\nu_{\max} = 0.12\omega_0 V \exp\left(\frac{-\hbar\omega_0}{kT}\right) . \tag{6.16}$$

It is also interesting to examine what happens to the exact results in the limit of an infinitely narrow density of states—the Einstein limit. In this limit ($V \to 0$) the integrands of both Eqs. (6.5) and (6.6) become independent of V but remain finite. However, there is an overall factor of V in front of each integral. *Therefore in the Einstein limit, the broadening and shift of the zero-phonon line are both zero.* This interesting result, also found by Zhdanov and Osad'ko,[36] cannot be obtained from the perturbative expressions, Eqs. (6.10) and (6.11). In fact, when the $V \to 0$ limit is taken there, one finds that the broadening diverges while the shift is finite! [This behavior is perhaps seen most transparently by substituting the Einstein density of states, Eq. (6.1), into the weak coupling results, Eqs. (4.39) and (4.40).] The reason for this apparent discrepancy is that it is inconsistent to take first the weak coupling ($|W|/V \to 0$) and *then* the Einstein limits; to calculate the Einstein limit correctly one must work to all orders in perturbation theory. This issue has been confused by the erroneous claim of several workers that an Einstein density of states *does* lead to line broadening.[11,35]

On a final note, we should again verify that the separation of time scales required by the theory is indeed satisfied. The phonon correlation function decays on the time scale $1/\bar{\omega}$ for the optical phonon model, so that the separation of time scales requirement is

$$\frac{\Delta \nu}{\omega_0} \ll V. \tag{6.17}$$

At low temperatures one can verify[38] that this is certainly valid. The fastidious reader can verify from Eq. (6.5) that at least for low temperatures, even in the Einstein limit ($V \to 0$), the separation of time scales is satisfied. That is, although the phonon correlation time is diverging, the dephasing time is always still larger.

VII. COUPLING TO PSEUDOLOCAL PHONONS

With the substitution into a crystal of an impurity molecule (or ion) sufficiently unlike the molecules (ions) of the host crystal, new vibrational modes localized about the impurity are often created. If the local mode frequencies fall within one of the phonon bandwidths of the host crystal, the local modes can mix with the band modes and are called pseudolocal modes. We now consider the coupling of an electronic transition to a single such pseudolocal mode. In this case, the collective coordinate, ϕ, in Eq. (4.6), which produces the difference between the ground and excited electronic potential surfaces of the impurity, is simply the *local* mode coordinate.

In what follows, we are interested mainly in low-frequency modes, since these dominate the dephasing at very low temperatures. A low-frequency local mode will couple to phonons of the broad acoustic band of the crystal, producing the normal modes of the impurity plus crystal system. The local mode coordinate, ϕ, can be expanded in terms of the normal modes as in Eq. (4.7). One expects that the expansion coefficients, h_k, will only be appreciable when the normal mode frequencies are close to the local mode frequency. Thus we expect that the weighted density of states, $\Gamma_0(\omega)$ [see Eq. (4.27)], will be sharply peaked at the local mode frequency. Taking the local mode frequency to be ω_0 and keeping in mind the fact that there are no normal modes with zero frequency, we choose for $\Gamma_0(\omega)$ the following form:[39]

$$\Gamma_0(\omega) = \begin{cases} \dfrac{\omega/4\tau_0}{(\omega - \omega_0)^2 + (1/2\tau_0)^2}, & \omega \geq 0 \\ 0, & \omega < 0. \end{cases} \tag{7.1}$$

For $\Gamma_0(\omega)$ sharply peaked ($\omega_0 \tau_0 \gg 1$), the pseudolocal density of states is approximately normalized according to Eq. (4.38) and is also approximately Lorentzian with HWHM $1/2\tau_0$. τ_0 can be simply interpreted as the local mode lifetime.[39]

Defining the dimensionless variable $z = \omega/\omega_0$ and the dimensionless width $v = 1/2\omega_0\tau_0$, and calculating the integral in Eq. (4.29), we write for future reference the ground-state pseudolocal phonon density of states $\Gamma_0(\omega)$ and its associated function $\Omega_0(\omega)$:[39]

$$\Gamma_0(\omega) = \frac{vz/2}{(z-1)^2 + v^2}, \tag{7.2}$$

$$\Omega_0(\omega) = \frac{1}{4\pi}\left\{\frac{(z-1-v^2)[\pi + 2\,\mathrm{arctg}(1/v)] - vz\ln[(1+v^2)/z^2]}{(z-1)^2+v^2} - \frac{(z+1+v^2)[\pi + 2\,\mathrm{arctg}(1/v)] - vz\ln[(1+v^2)/z^2]}{(z+1)^2+v^2}\right\}. \tag{7.3}$$

Since we concentrate on sharply peaked pseudolocal densities of states ($v \ll 1$), we can approximate $\Omega_0(\omega)$ as:[47]

$$\Omega_0(\omega) = \frac{1}{2}\left[\frac{z-1-v^2}{(z-1)^2+v^2} - \frac{1}{z+1}\right]. \tag{7.4}$$

A convenient expression for the linewidth, Eq. (4.37), is written in terms of $\Gamma_0(\omega)$ and $\Gamma_1(\omega)$, the pseudolocal phonon weighted density of states in the *excited* electronic state. Thus it will be useful to derive approximate expressions for $\Gamma_1(\omega)$ in the limit of small v. Using Eq. (4.36) we find[39]

$$\Gamma_1(\omega) \simeq \frac{v_1/2z_1}{(z-z_1)^2 + v_1^2}, \quad z \text{ near } z_1, \tag{7.5}$$

where

$$z_1 = (1+W)^{1/2} \tag{7.6}$$

and

$$v_1 = v\left(\frac{z_1+1}{2}\right)^2. \tag{7.7}$$

We see from Eq. (7.5) that $\Gamma_1(\omega)$ near its maximum is Lorentzian with a peak at z_1, defined by Eq. (7.6), and HWHM v_1, defined by Eq. (7.7). From Eqs. (7.6) and (7.7) it follows that the excited-state local mode frequency and lifetime can be written

$$\omega_1 = \omega_0(1+W)^{1/2} \tag{7.8}$$

and

$$\frac{1}{\tau_1} = \frac{1}{\tau_0}\left[\frac{1+(1+W)^{1/2}}{2}\right]^2 . \tag{7.9}$$

Recalling that W is bounded from below by -1, we observe that at that lower bound there is a soft mode ($\omega_1 \simeq 0$).[47] For W positive, we also observe that the excited-state pseudolocal mode has a higher frequency than that of the ground state and that it has a shorter lifetime. For W negative, the converse is true.

It is easy to obtain expressions for the linewidth and shift in the weak coupling limit, which in this case is $|W|/v \ll 1$. From Eqs. (4.39), (4.40), and (7.1) we find

$$\gamma = \tfrac{1}{4} W^2 \omega_0^2 \tau_0 n(\omega_0)[n(\omega_0)+1] \tag{7.10}$$

and

$$\delta_T = \tfrac{1}{2} W \omega_0 n(\omega_0) . \tag{7.11}$$

These expressions are Arrhenius at low temperature and are often used in the analysis of experimental data.[15]

We turn now to the evaluation of our general line shape formulas, Eqs. (4.34) and (4.35) or (4.37), using as input the pseudolocal density of states Eq. (7.1). In general, the integrals can be easily determined numerically—we discuss these results in more detail at the end of this section. Here, however, we present some analytic results for the linewidth. For low temperatures such that $kT \ll \hbar\omega_0, \hbar\omega_1$, then $n(\omega) \ll 1$ and the logarithm in Eq. (4.37) can be expanded. If, in addition, the densities of states $\Gamma_0(\omega)$ and $\Gamma_1(\omega)$ are very sharply peaked ($1/\tau_0, 1/\tau_1 \ll kT/\hbar$), then we find for the linewidth[39]

$$\gamma = \frac{(\delta\omega\tau^*)^2}{1+(\delta\omega\tau^*)^2}\left(\frac{1}{2\tau_0} e^{-\hbar\omega_0/kT} + \frac{1}{2\tau_1} e^{-\hbar\omega_1/kT}\right), \tag{7.12}$$

where

$$\delta\omega = \omega_1 - \omega_0 , \tag{7.13}$$

$$\frac{1}{\tau^*} = \frac{1}{2}\left(\frac{1}{\tau_0} + \frac{1}{\tau_1}\right), \tag{7.14}$$

and the excited-state local mode frequency and lifetime, ω_1 and τ_1, are given in terms of W, ω_0, and τ_0 by Eqs. (7.8) and (7.9). Alternatively, one can consider ω_0, ω_1, and τ_0 to be the three independent parameters; τ_1 is then given by [see Eqs. (7.8) and (7.9)]

$$\frac{1}{\tau_1} = \frac{1}{\tau_0}\left(\frac{1 + \omega_1/\omega_0}{2}\right)^2 . \tag{7.15}$$

The width of the ZPL is thus given by the sum of two Arrhenius terms, the activation energies being the pseudolocal mode frequencies in the ground and excited electronic states.

There are several interesting limits of Eq. (7.12) which are most easily obtained by considering the dimensionless parameters W and $v = 1/2\omega_0\tau_0$. First, we consider the limit $|W| \ll 1$. In this case, $\tau_1 \simeq \tau_0$ and $|\delta\omega|/\omega_0 \ll 1$. If, in addition, $\hbar|\delta\omega|/kT \ll 1$ (recall that $\hbar\omega_0/kT \gg 1$), then in the exponent of Eq. (7.12) ω_1 may be replaced by ω_0 and we obtain

$$\gamma = \frac{1}{\tau_0}\frac{(\delta\omega\tau_0)^2}{1 + (\delta\omega\tau_0)^2}\exp\left(-\frac{\hbar\omega_0}{kT}\right). \tag{7.16}$$

This linewidth result has been previously derived by Harris[29] using exchange theory.

An interesting special case of this result is the fast exchange limit ($|\delta\omega|\tau_0 \ll 1$):

$$\gamma = (\delta\omega)^2\tau_0 \exp\left(-\frac{\hbar\omega_0}{kT}\right). \tag{7.17}$$

The limit $|\delta\omega|\tau_0 \ll 1$ also implies the weak coupling limit $|W|/v \ll 1$. Indeed, when Eq. (7.10) is evaluated at low temperatures we recover Eq. (7.17). [From Eq. (7.8) one sees that for $|W| \ll 1$, $\delta\omega = W\omega_0/2$.] This is in essence the approach of Small[28] and Jones and Zewail.[26] Thus the weak coupling results of these authors are essentially equivalent to the fast exchange limit of Harris.[29] This fact does not seem to have been appreciated in the literature.

In the opposite limit of slow exchange, $|\delta\omega|\tau_0 \gg 1$, one obtains the simple result

$$\gamma = \frac{1}{\tau_0} e^{-\hbar\omega_0/kT} . \tag{7.18}$$

The other interesting limit is when $|W|/v \gg 1$ (the strong coupling limit), which implies that $|\delta\omega|\tau^* \gg 1$. This limit is in general different from the slow exchange limit in that $|W|$ is not necessarily small compared to 1, and therefore τ_0 and τ_1 are not necessarily approximately equal, and $|\delta\omega|/\omega_0$ is not necessarily small compared to 1. From Eq. (7.12) we find

$$\gamma = \frac{1}{2}\left[\frac{1}{\tau_0}\exp\left(-\frac{\hbar\omega_0}{kT}\right) + \frac{1}{\tau_1}\exp\left(-\frac{\hbar\omega_1}{kT}\right)\right] . \tag{7.19}$$

This linewidth result has been previously derived by deBree and Wiersma.[30] We observe that when $\tau_1 \simeq \tau_0$ and $\hbar|\delta\omega|/kT \ll 1$, this strong coupling result agrees with that of the exchange result in the slow exchange limit ($|\delta\omega|\tau_0 \gg 1$).

Thus our quite general low-temperature linewidth formula, Eq. (7.12), reduces in various limits to the weak coupling result of Small[28] and Jones and Zewail,[26] the exchange theory of Harris,[29] and the "uncorrelated phonon scattering" result of deBree and Wiersma.[30] To assess the accuracy of our approximate low-temperature result, Eq. (7.12), and the validity of the exchange result, Eq. (7.16), we compare these results to the exact numerical ZPL width from Eq. (4.35). In what follows we take $v = 0.01$, and plot $\ln(\gamma/\omega_0)$ versus T_0/T, where $T_0 = \hbar\omega_0/k$. For $W = 0.01$ (Fig. 5) we see that the exchange expression works rather well at low temperatures. At $W = 0.1$ (Fig. 6) the exchange result deviates slightly, while for $W = 1$ (Fig. 7) the exchange result is significantly in error. In all three cases our approximate expression is accurate. In Fig. 8 we examine the ZPL width for $W = -0.8$. Our approximate expression works well in the low-temperature regime, but the exchange expression does not. Here we also see a dramatic change in the activation frequency in the low-temperature region. Whereas in previous graphs all the low-temperature linewidths exhibited activation frequencies of ω_0, here the width exhibits an activation frequency of $0.45\omega_0$. This is the frequency of the excited-state pseudolocal mode [see Eq. (7.8)]. Optical dephasing at low temperatures is determined, as one would expect, by the availability of the lowest-frequency phonons. In summary, we see that the exchange theory result is accurate in

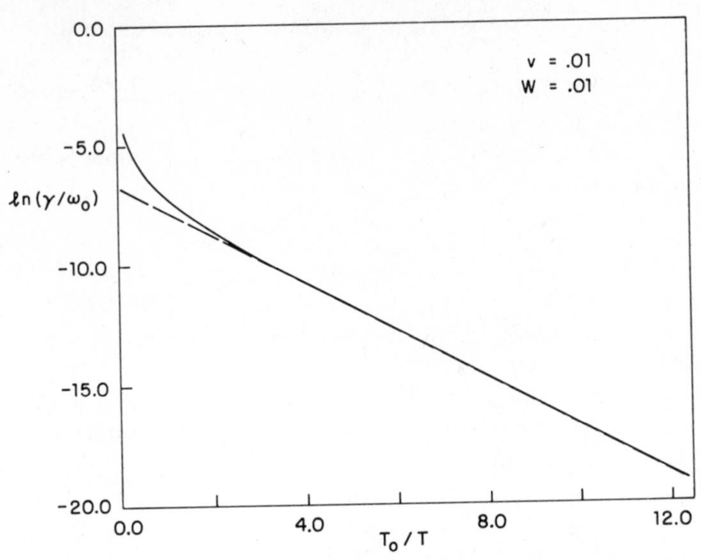

Fig. 5. Logarithm of the dimensionless linewidth, $\ln(\gamma/\omega_0)$, vs. the reciprocal dimensionless temperature, T_0/T, for the indicated values of v and W. The solid line is the exact result from Eq. (4.35), and the dashed line is the exchange result, Eq. (7.16), which is indistinguishable from our approximate expression, Eq. (7.12).

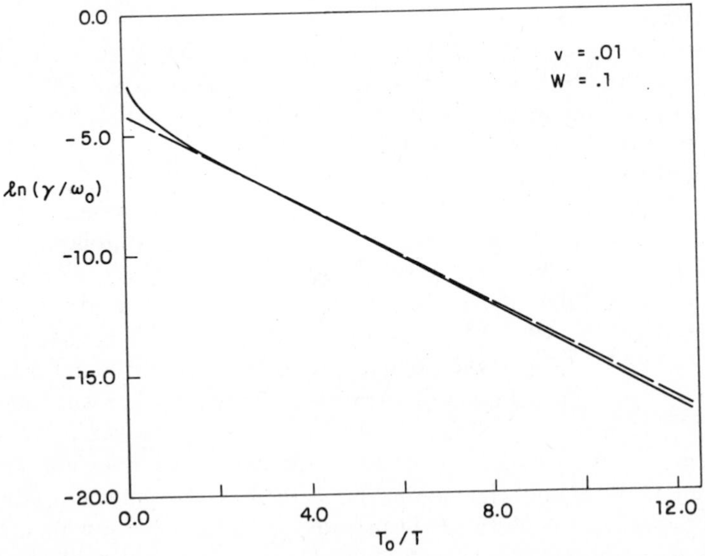

Fig. 6. Same as Fig. 5 except $W = 0.1$. Again the solid line is Eq. (4.35) and the dashed line is Eq. (7.16). Our approximate expression, Eq. (7.12), is indistinguishable from the exact result (—) for $T_0/T > 2$, and indistinguishable from the exchange result (– –) for $T_0/T < 2$.

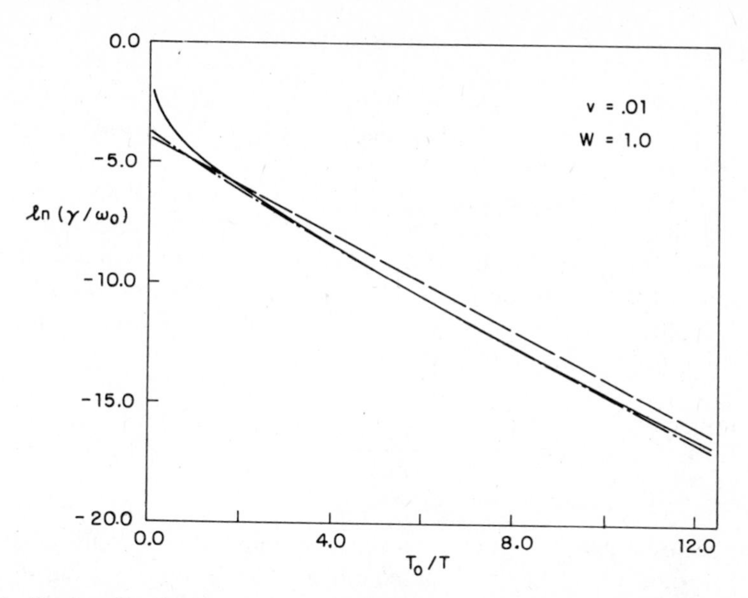

Fig. 7. Same as Fig. 1 except $W = 1.0$. As before the exact result (—) is from Eq. (4.35) and the exchange result (– –) is from Eq. (7.16). Here, our approximate result, Eq. (7.12), is given by (–·–).

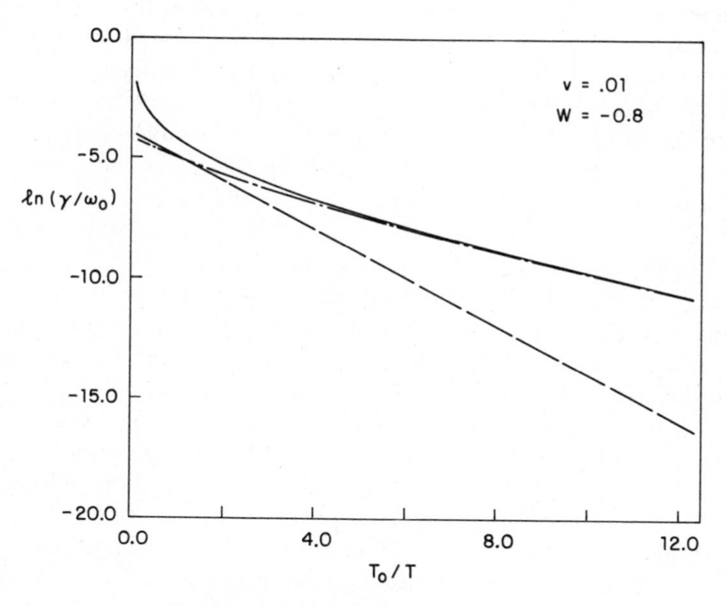

Fig. 8. Same as Fig. 7 except $W = -0.8$.

its expected regime of validity $|W| \ll 1$ (also $\hbar|\delta\omega|/kT \ll 1$) but is not accurate for larger W. On the other hand, our approximate expression seems to be reasonably accurate for all W.

We turn next to a discussion of the low-temperature line shift. We were not able to derive an Arrhenius expression for the line shift except in the weak coupling limit [see Eq. (7.11)]. In fact, numerical calculations from our exact expressions show[39] that the line shift is *not* Arrhenius, except in the weak coupling limit. We also showed that the exchange expression for the line shift,[29]

$$\delta_T = \frac{\delta\omega}{1 + (\delta\omega\tau_0)^2} e^{-\hbar\omega_0/kT} , \tag{7.20}$$

is not accurate except in the fast exchange limit, when it reduces to the low-temprature limit of the weak coupling result, Eq. (7.11). We suspect that exchange theory is not accurate for the line shift because it neglects a temperature-dependent renormalization of the optical frequency.[39]

As a final check on our results we must verify that the separation of time scales $T_2' \gg \tau_c$ is satisfied. In this case the condition is $\gamma \ll 1/\tau_0$, $1/\tau_1$. Inspection of the exact results in Figs. 5–8 shows that this separation of time scales is satisfied for all W at temperatures such that $kT/\hbar\omega_0 > 1/2$. Thus our theory is only rigorously valid in this low-temperature regime.

Before we leave this section it is important to remark that an expression for the linewidth that is identical to Eq. (7.12) can be easily derived from an intermediate result of deBree and Wiersma.[30] In fact, there is one difference: Eq. (7.12) was derived for a *harmonic* Hamiltonian, and therefore of the four parameters, ω_0, ω_1, τ_0, and τ_1, only three are independent—the fourth can be related to the other three, for example, by Eq. (7.15). In addition, because the Hamiltonian is harmonic, the lifetimes τ_0 and τ_1 are necessarily temperature independent. In contrast, in the work of deBree and Wiersma,[30] the Hamiltonian was not necessarily harmonic and therefore ω_0, ω_1, τ_0, and τ_1 may be independent, and τ_0 and τ_1 may be temperature dependent. The correspondence between these two results, derived by two very different methods, is important on two counts. First, for harmonic Hamiltonians, our derivation of Eq. (7.12) from the exact nonperturbative theory provides confirmation that the less rigorous approach of deBree and Wiersma is perfectly adequate. Second, the correspondence of the two results strongly suggests that Eq. (7.12) is more general than its (harmonic) derivation and is in fact valid for anharmonic systems as well.

VIII. COMPARISON WITH EXPERIMENT: ACOUSTIC PHONONS

The temperature dependence of the optical homogeneous linewidth of many impurity-crystal systems has been attributed to pure dephasing by acoustic phonons.[48–63] Because the temperature dependence of the linewidth due to this process is so rapidly increasing, one finds that dephasing by acoustic phonons tends to dominate at higher temperatures. Typically, these temperatures are high enough so that the homogeneous linewidth overcomes the inhomogeneous broadening and can be measured directly from the absorption spectrum.

All these systems have been analyzed with the "weak" coupling result of McCumber and Sturge.[23] The coupling is weak when the magnitude of the quadratic coupling constant as defined in Section V is much smaller than 1. We note that although the strength of the *linear* coupling constant is determined by the Huang–Rhys factor for the relative intensities of the phonon sideband and the zero-phonon line, the strength of the *quadratic* coupling constant cannot be determined simply from the spectrum. Therefore, even systems with a small sideband cannot be assumed to be in the weak coupling limit.

For some of the systems referenced above, this weak coupling approach appears to be valid. For others, however, there are indications that the weak coupling limit is not justified. In this section we reanalyze[40] data from one organic and one inorganic system with our nonperturbative theory as presented in Section V.

A. 1,3-Diazaazulene in Naphthalene

These experiments were performed by Burke and Small[48] on the 4500 Å $^1B_1 \leftarrow {}^1A_1(S_1 \leftarrow S_0)$ transition of 1,3-diazaazulene in a naphthalene host. They measured the temperature-dependent linewidths of the (0, 0) line and several vibronic lines. The temperature dependence of each vibronic linewidth was analyzed with the weak coupling result of McCumber and Sturge:[23]

$$\Delta\bar{\nu} = \bar{\alpha}\left(\frac{T}{T_D}\right)^7 \int_0^{T_D/T} dx \frac{x^6 e^x}{(e^x - 1)^2}, \tag{8.1}$$

where $\Delta\bar{\nu} = \Delta\nu/c$ is the linewidth in cm^{-1} (c is the speed of light), and $\bar{\alpha}$ and T_D are adjustable parameters. Making the identification

$$\bar{\alpha} = \frac{9}{2} W^2 \pi \bar{\nu}_D, \tag{8.2}$$

where $\tilde{\nu}_D = \omega_D/2\pi c$ is the Debye frequency in cm^{-1}, this is identical to Eq. (5.7).

The coupling constant, $|W|$, determined from the best-fit values of $\bar{\alpha}$ and Eq. (8.2) varies from 0.31 to 0.64, and the Debye temperature ranges from 50 to 100 K depending on the vibronic transition.[48] The coupling constants depend on the interaction of a particular vibronic state with the phonons and thus could well be different for different vibronic lines. However, the Debye temperature is a property of the host crystal and should be independent of the vibronic transition (see Section C). Thus it seems that although Eq. (8.1) provides a reasonable two-parameter description of the data, it does not provide a consistent interpretation of dephasing by acoustic phonons. Moreover, for *none* of the vibronic lines is the weak coupling condition $|W| \ll 1$ satisfied, and thus the use of Eq. (5.7) or Eq. (8.1) is not justified. This could provide a possible explanation for the nonconstancy of T_D.

To see if this is true, we have reanalyzed[40] some of the data[48] with the

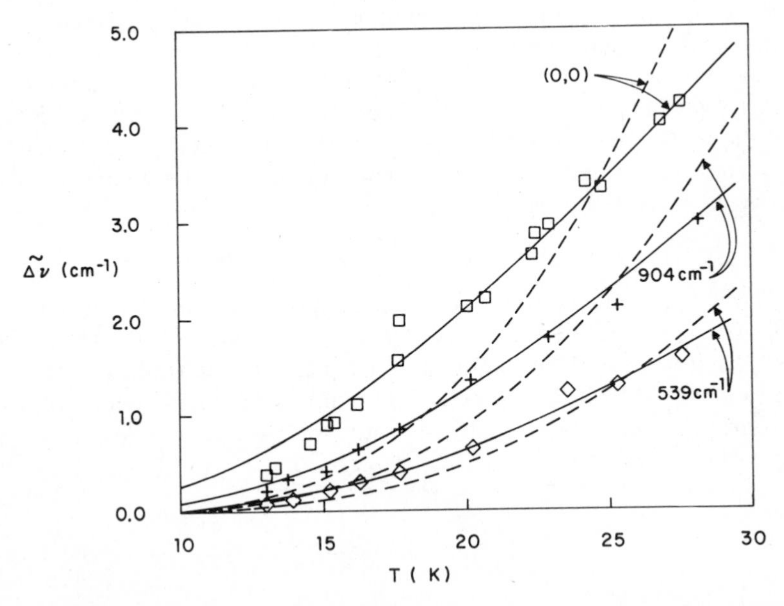

Fig. 9. Homogeneous widths (FWHM) of the (0, 0), 539 cm^{-1}, and 904 cm^{-1} vibronic lines of the $S_1 \leftarrow S_0$ transition of 1,3-diazaazulene in naphthalene. The experimental points (□, ◇, +) for the (0, 0), 539 cm^{-1}, and 904 cm^{-1} lines, respectively, were taken from Fig. 3 of ref. 48. The theoretical curves (—) were obtained from Eq. (5.4) with $T_D = 113$ K and $W = -0.785$, -0.489, and -0.639 for the three vibronic lines. Also shown are the perturbative results (---) from Eq. (5.7) with $T_D = 113$ K and $|W| = 0.967$, 0.572, and 0.769, respectively.

nonperturbative expression, Eq. (5.4). In order to reduce the number of adjustable parameters we have taken the value of $T_D = 113$ K for naphthalene from independent experimental[64] and theoretical[65] studies.[66] The data were then fitted with the single adjustable parameter W. The results[40] are shown in Figs. 9 and 10. In all cases, we see that this procedure gives a satisfactory fit of the data. (Our fits are about as good as the *two*-parameter fits of Burke and Small.[48]) Moreover, the values of W obtained are between -0.49 and -0.79 and are therefore consistent with the nonperturbative regime. For comparison we also show the best fit of the weak coupling expression, Eq. (5.7), when $T_D = 113$ K. It is seen that in all cases the fit is unsatisfactory. We claim that the nonperturbative theory provides a more satisfactory explanation of the data than does the weak coupling theory because: (1) all vibronic lines can be fit with the *same* Debye temperature, which is independently determined, and (2) the values of the parameter W obtained show that the system is not in the weak coupling limit.

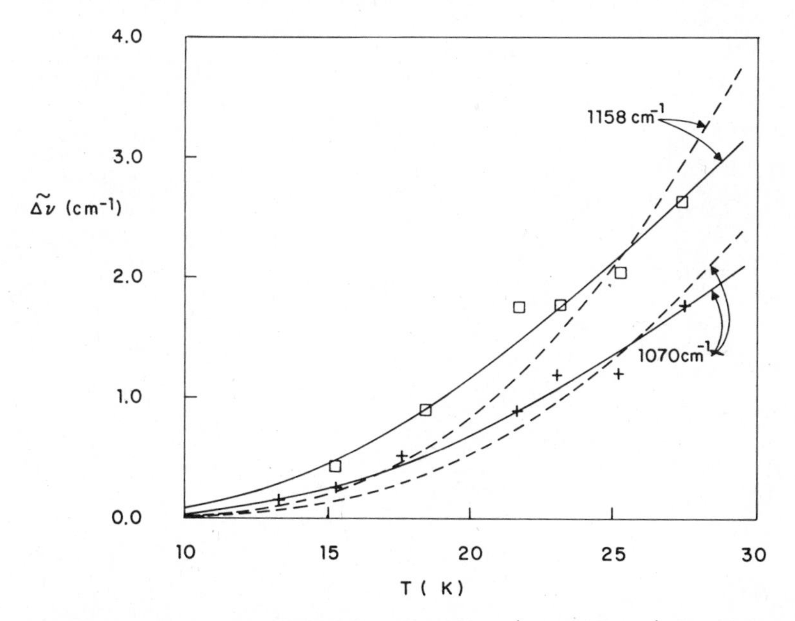

Fig. 10. Homogeneous widths (FWHM) of the 1070 cm^{-1} and 1158 cm^{-1} vibronic lines of the $S_1 \leftarrow S_0$ transition of 1,3-diazaazulene in naphthalene. The experiment points (+, □) for the 1070 cm^{-1} and 1158 cm^{-1} lines, respectively, were taken from Fig. 4 of ref. 48. The theoretical curves (—) were obtained from Eq. (5.4) with $T_D = 113$ K and $W = -0.50$ and -0.62, respectively. Also shown are the perturbative results (---) from Eq. (5.7) with $T_D = 113$ K and $|W| = 0.59$ and 0.74, respectively.

One might wonder whether vibronic transitions can be adequately modeled by a two-level electronic system. That is, one might think that vibrational relaxation could contribute to the vibronic linewidths. While we cannot rule this out, the fact that, for these low-lying vibronic levels, the spacing between nearby levels is greater than the Debye frequency means that one (acoustic) phonon processes are prohibited, and thus one would expect vibrational relaxation to be slow.

The quality of the data,[48] and hence the accuracy of our analysis, must be accepted with some caution for two reasons. First, especially at the lower temperatures, it is difficult to determine accurately the homogeneous linewidth from the observed line shape, which is a convolution of homogeneous and inhomogeneous lines. Second, while the experimental results for a single crystal were reproducible, different crystals seemed to give different results. The reason for this is not well understood.[48]

B. Cr^{3+} in Al_2O_3 (Ruby)

Here we discuss the classic experiments of McCumber and Sturge,[23] who measured the thermal widths of the *R* lines in dilute ruby from 77 to

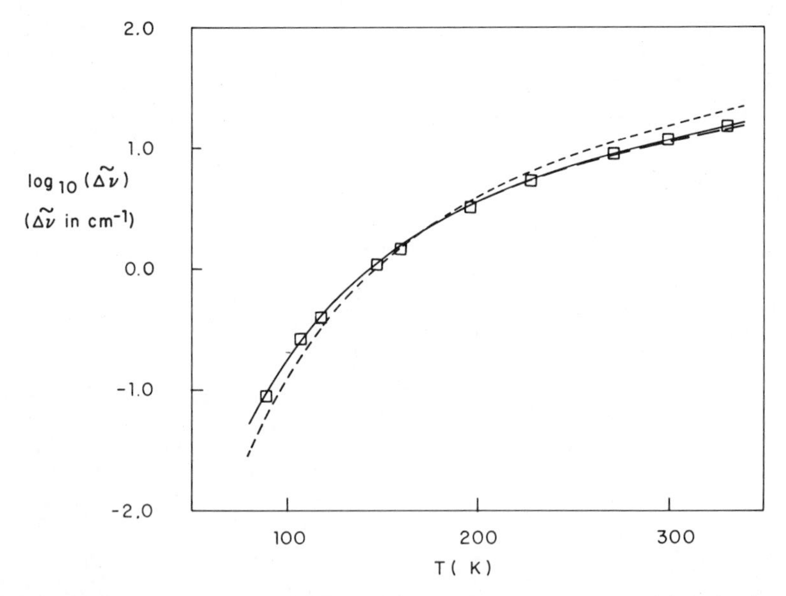

Fig. 11. Homogeneous width (FWHM) of the R_1 line of ruby. The experimental points (□) were obtained from Fig. 2 of ref. 23 by subtracting the low-temperature residual width. The theoretical curves (—) and (– –) were obtained from Eq. (5.4) with $T_D = 935$ K and $W = -0.312$ and 3.01, respectively. Also shown (---) is Eq. (5.7) with $T_D = 935$ K and $|W| = 0.373$.

350 K. Although there has been a substantial amount of later work[54–58] on line broadening in dilute ruby, all these experiments agree with McCumber and Sturge, and so we discuss their data here. This was the first work to be analyzed with the weak coupling theory, Eq. (8.1). They found a good fit to the data with $T_D = 760$ K and with $|W| = 0.27$ and 0.24, respectively, for the R_1 and R_2 lines. As in the organic crystal case, it is not obvious that these values of W are in the perturbative limit. This again led us to analyze the data with the nonperturbative theory, Eq. (5.4). As before, instead of treating T_D as an adjustable parameter, we used the Debye temperature determined from heat capacity measurements.[67] Taking $T_D = 935$ K we performed a one-parameter fit of the data omitting points below 90 K since they did not unambiguously result from the homogeneous linewidth.[23] The results[40] are shown in Figs. 11 and 12. For both lines a good fit could be obtained for both positive and negative values of W: for R_1, $W = -0.31$ or 3.01; for R_2, $W = -0.28$ or 1.39. We also show the weak coupling result of Eq. (5.7) with $T_D = 935$ K for comparison. It is clear that there are systemic deviations of the data from

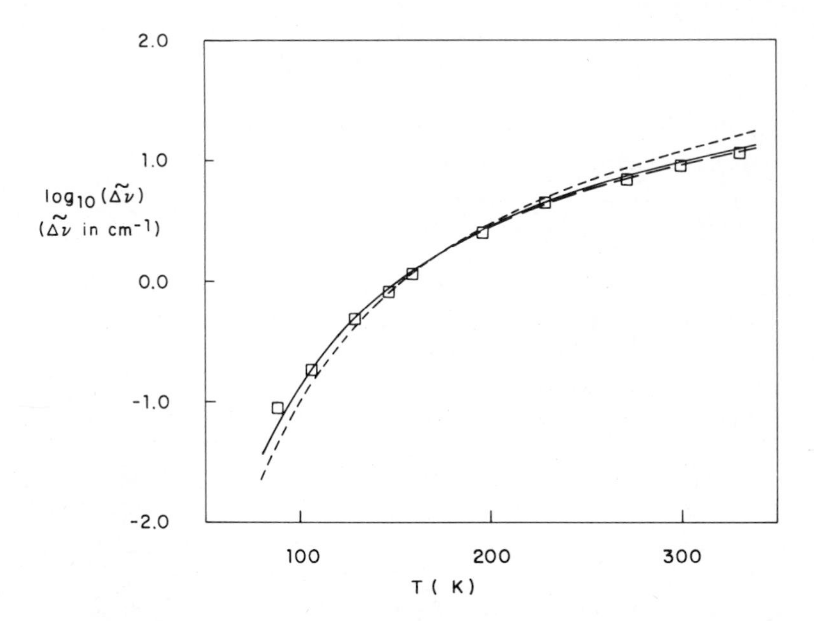

Fig. 12. Homogeneous width (FWHM) of the R_2 line of ruby. The experimental points (□) were obtained from Fig. 2 of ref. 23 by subtracting the low-temperature residual width. The theoretical curves (—) and (– –) were obtained from Eq. (5.4) with $T_D = 935$ K and $W = -0.282$ and 1.39, respectively. Also shown (- - -) is Eq. (5.7) with $T_D = 935$ K and $|W| = 0.330$.

the fit. Thus we feel that the nonperturbative theory provides a more convincing explanation of these experiments.

It should be noted, however, that in the case of ruby, the two-level approximation is not at all obviously valid since the splitting between the ^2E states responsible for the R lines is only 29 cm^{-1}. In fact at very low temperatures the "direct" process (one phonon emission or absorption) between these levels is the dominant contribution to the homogeneous linewidth.[57,58] However, at higher temperatures this process saturates, and it appears[23,57,58] that for $T > 90$ K its contribution is small compared to the pure dephasing that we have calculated. Other workers[54,56,68] have attributed the thermal broadening of the R lines to a combination of the Raman process and direct processes to the $^2T_1$ and $^4T_2$ levels. We cannot, of course, rule out this possibility.

C. Additional Remarks

We believe that the strong coupling theory provides a more consistent interpretation of the above experiments than does the perturbative theory, partly because the values of T_D used in the former approach are consistent with heat capacity measurements. Others[23,55,69] have argued that the value of T_D obtained from linewidth measurements does not necessarily correspond to T_D from the specific heat since "phonons of different frequencies may interact differently with the impurity ion."[55] A corollary would be that different vibronic lines could have different Debye temperatures. We feel that within the spirit of the Debye model this is not correct.

The Debye approximation replaces the three acoustic branches with the single dispersion relation $\omega_{\mathbf{q}s} = c|\mathbf{q}|$. This linearity of the dispersion relation is exact in the long-wavelength limit. The Debye model, simple as it is, provides a reasonably good description of the heat capacity of many crystals. In the case of optical dephasing , in the same spirit of the Debye picture, one makes the long-wavelength approximation[8,23,38] for the expansion coefficients of the strain field in the normal modes: $h_{\mathbf{q}s} \sim (\omega_{\mathbf{q}s})^{1/2}$. In Section V, we showed that within these approximations, even if some phonons do not couple to the optical excitation, the appropriate cutoff in the weighted density of states is determined by the usual Debye temperature. Thus we are not arguing that the Debye model is correct, only that the long-wavelength approximation is consistent with the Debye picture, and within that picture T_D is the same for the specific heat and optical dephasing experiments.

IX. COMPARISON WITH EXPERIMENT: PSEUDOLOCAL PHONONS

Hole burning, photon echo, and absorption experiments have been performed on a wide variety of mixed-crystal systems[70–89] at low temperatures. For most of these experiments the linewidth is found to have an Arrhenius temperature dependence (at least over a limited temperature range) with activation energies from 7 to 30 cm^{-1}. It seems likely that this temperature dependence results from coupling to low-frequency pseudolocal phonons. (Presumably at low temperatures the coupling to acoustic phonons is unimportant due to the small number of thermally populated low-frequency modes; indeed, for this process the linewidth vanishes as T^7.)

These low-temperature experiments have been analyzed with a variety of theoretical expressions: a simple Arrhenius form, the weak coupling expression of Eq. (7.10), exchange theory [Eq. (7.16)], the deBree–Wiersma uncorrelated scattering result [Eq. (7.19)], and the Osad'ko nonperturbative theory.[15,36] All these theoretical expressions require as input the frequencies and lifetimes of the pseudolocal mode in the ground and excited electronic states. Unfortunately, since for most of the systems these parameters are undetermined, the theoretical analysis is ambiguous.

In this section, in the context of the theoretical results presented in Section VII, we discuss two experiments where some of this information *is* known. We have also examined a third system[79] where some of the parameters are known, but we have not been able to satisfactorily understand[40] the data.

A. 3,4,6,7-Dibenzopyrene in *N*-Octane

Absorption and fluorescence spectra near the 3962 Å zero-phonon line (ZPL) were taken for the above mixed-crystal system by Korotaev and Kaliteevskii[70] for $T = 4.2$–40 K. They obtained the temperature dependence of the homogeneous linewidth by deconvolution of the observed ZPL shape. From the peaks in the emission and absorption sidebands they determined that the pseudolocal phonon frequencies are $\tilde{\nu}_0 = \nu_0/c = 13.1\ \mathrm{cm}^{-1}$ and $\tilde{\nu}_1 = \nu_1/c = 15.6\ \mathrm{cm}^{-1}$ or $\omega_0 = 2.47 \times 10^{12}\ \mathrm{s}^{-1}$ and $\omega_1 = 2.94 \times 10^{12}\ \mathrm{s}^{-1}$. By deconvoluting the emission sideband they found the homogeneous FWHM to be $4.6\ \mathrm{cm}^{-1}$, which gives $\tau_0 = 1.15$ ps. They do not report the FWHM of the absorption sideband, although, from the spectrum, it appears to be approximately the same as the emission sideband; thus $\tau_1 \simeq \tau_0$.

Korotaev and Kaliteevskii have analyzed the experiment with

Osad'ko's nonperturbative theory,[15,36] which is essentially equivalent to ours. Here, primarily for completeness, we repeat the analysis with only a few minor differences.

First, we note that the experiments are not in the low-temperature ($kT \lesssim \hbar\omega_0, \hbar\omega_1$) limit and so we cannot use our simple formula, Eq. (7.12), but must resort to numerical evaluation of the exact expression, Eq. (4.35). W can be obtained from Eq. (7.8), which is valid in the limit $1/2\omega_0\tau_0 \ll 1$ (which is barely satisfied in this case); we find $W = 0.418$. Second, we note that from Eq. (7.9) this gives $\tau_0/\tau_1 = 1.2$, which is consistent with the observed spectrum. Equation (4.35) was evaluated[40] with the above values for W, ω_0, and ω_1. For comparison we also evaluated Eq. (7.12), which is valid at low temperatures. The results, along with the experimental points are shown in Fig. 13. The agreement between the theory [Eq. (4.35)] and experiment is adequate for $T < 25$ K. (Our low-temperature expression is only valid for $T < 10$ K.)

It should be noted that the theory is valid only if the separation of time scales $1/T_2 \ll 1/2\tau_0$, $1/2\tau_1$ is satisfied.[20,39] This requires that $\Delta\tilde{\nu} \ll 4.6\ \text{cm}^{-1}$. Thus it is not surprising that for $T > 25$ K, when $\Delta\tilde{\nu} > 2\ \text{cm}^{-1}$, the theory

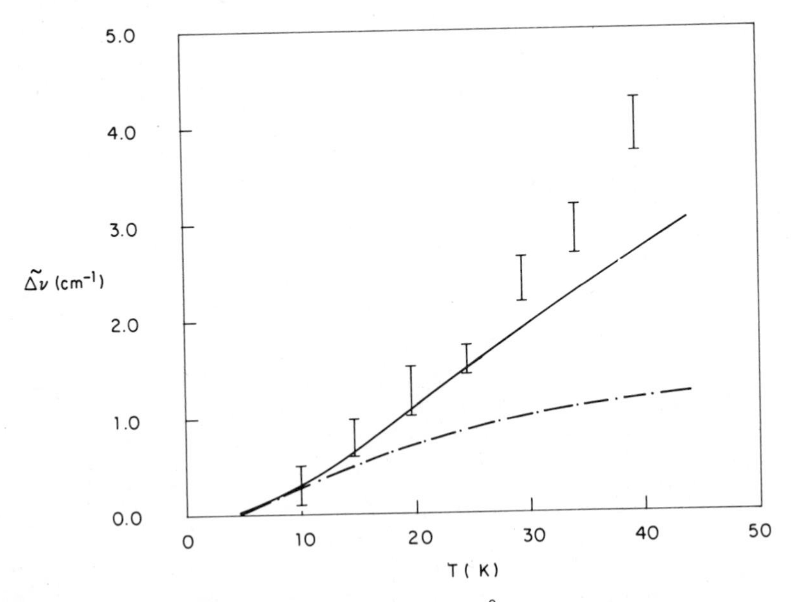

Fig. 13. Homogeneous width (FWHM) of the 3962 Å ZPL of 3,4,6,7-dibenzopyrene in *n*-octane. The experimental points (error bars) are from Fig. 2 of ref. 70. The theoretical curves are from Eq. (4.35) (—) and Eq. (7.12) (–·–) with $\omega_0 = 2.47 \times 10^{12}\ \text{s}^{-1}$, $\tau_0 = 1.15$ ps, and $W = 0.418$.

begins to break down. It should be emphasized that the theoretical results of Fig. 13 involve *no adjustable parameters*—all relevant parameters were determined from independent measurements of the sidebands. We also note that Korotaev and Kaliteevskii obtained slightly better agreement with the data, either because our normalization schemes are somewhat different or because they performed a more careful determination of the relevant parameters.

B. Pentacene in Benzoic Acid

Molenkamp and Wiersma[71] measured the ZPL width of pentacene (photosite I) in benzoic acid for $T = 1.5$–20 K by both photon echo and absorption techniques. From peaks in the sidebands they determined that $\tilde{\nu}_0 = 12\ \text{cm}^{-1}$ and $\tilde{\nu}_1 = 16.65\ \text{cm}^{-1}$. By both line shape deconvolution and photon echo experiments they determined that $\tau_1 = 2.5$ ps. They did not report a value for τ_0. Molenkamp and Wiersma interpreted their data with the uncorrelated phonon scattering result of deBree and Wiersma[30] [see Eq. (7.19)]. Since they did not know τ_0, they performed a one-parameter fit of the data to find $\tau_0 = 18$ ps.

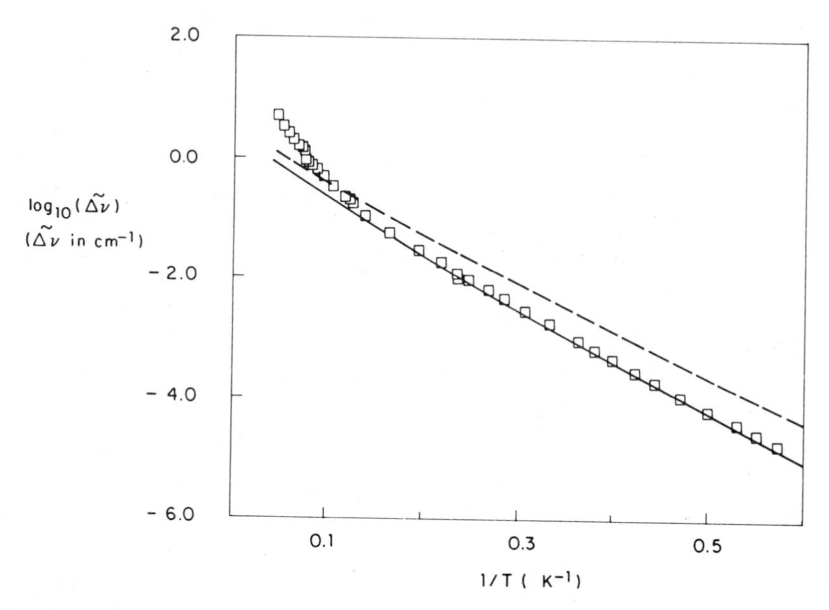

Fig. 14. Homogeneous width (FWHM) of the ZPL of photosite I of pentacene in benzoic acid. The experimental points (□) are from Fig. 6 of ref. 71. The theoretical curves are from Eq. (7.12) with $\omega_0 = 2.26 \times 10^{12}\ \text{s}^{-1}$, $\omega_1 = 3.14 \times 10^{12}\ \text{s}^{-1}$, $\tau_1 = 2.5$ ps, and $\tau_0 = 3.6$ ps (– –) and $\tau_0 = 18$ ps (—).

Equation (7.19) follows from our more general expression Eq. (7.12) in the limit $(\delta\omega\tau^*)^2 \gg 1$. Since this inequality is satisfied for the above parameters it follows that the analysis by Molenkamp and Wiersma is correct. It is interesting to note that if the harmonic theory is used to determine τ_0 from Eqs. (7.8) and (7.9) we find $\tau_0 = 3.6$ ps, which is not in good agreement with the value of 18 ps obtained by Molenkamp and Wiersma. The comparison of Eq. (7.12) (which we feel is valid even for anharmonic systems) for $\tau_0 = 3.6$ and 18 ps with the experimental data[71] is shown in Fig. 14. As seen, this expression with the harmonic value does not work well, while the anharmonic value of Molenkamp and Wiersma provides very good agreement, at least at low temperatures. We note that excellent agreement at all temperatures between Eq. (7.19) and the experiment has been obtained by including the measured temperature dependence of τ_1.[71] We also note that the observed temperature dependence of τ_1 supports the notion that anharmonicities are important since the harmonic theory predicts that τ_0 and τ_1 are temperature independent.[39]

C. Additional Remarks

It is clear that more work needs to be done in order to understand dephasing by pseudolocal phonons. First, to provide a rigorous test of the theory, one needs accurate measurements of ω_0, ω_1, τ_0, and τ_1. Of these parameters, τ_0 seems to be the most difficult to obtain, but perhaps with CARS experiments this will be possible.

Second, for systems where most of these parameters can be determined, it appears that our general low-temperature analytic expression, Eq. (7.12), will be useful for understanding dephasing. It should be valid for both harmonic and anharmonic systems (based on the theoretical work of deBree and Wiersma[30]) and thus should provide us with information about when anharmonicities are important. At higher temperatures one must analyze data numerically with Eq. (4.35). Unfortunately, this expression is only valid for harmonic systems.

X. CONCLUSION

It seems clear that our understanding of optical dephasing of impurities in crystals is at a fairly sophisticated level. As long as the dephasing rate constant is calculated nonperturbatively, the Debye model provides an adequate description of dephasing by acoustic phonons. The nonperturbative theory also provides a reasonably complete understanding of dephasing by pseudolocal phonons, although in this case, we must rely on the theoretical work of deBree and Wiersma to justify extending our results into the anharmonic regime.

To obtain a more complete understanding of optical dephasing in crystals, two experimental directions of approach seem important. First, it is clear that studies of zero-phonon linewidths are most informative when combined with studies of the sidebands. For example, if pseudolocal phonons dominate the dephasing, in order to make meaningful comparisons with theory, one must determine the frequencies and lifetimes of the local phonon for both the ground and excited electronic states of the impurity. These parameters can be obtained from time and/or frequency domain spectroscopy of the absorption and emission sidebands. Second, most of the experiments have been performed over a limited temperature range, primarily because of the limitations of any one experimental technique. In order to understand the crossover from dephasing by pseudolocal phonons to dephasing by acoustic phonons, it would be useful if experiments (photon echo or hole burning *and* absorption) on a single system were performed over a wide range of temperatures.

Theoretically, it is clear that one of the major unsolved problems is that of anharmonicities. For pseudolocal phonons, the deBree–Wiersma theory makes an important contribution to this problem. However, even this approach is limited to a truncated (four-level) system and hence is only valid at low temperatures. It will also be important to understand how to handle theoretically simultaneous dephasing by pseudolocal and acoustic phonons. Finally, it would be useful if we had at our disposal a nonperturbative theory of the phonon sideband to complement the nonperturbative theory of the zero-phonon line.

Acknowledgments

We thank Professor G. Small, Dr. M. Sturge, Dr. O. Korotaev, Dr. M. Kaliteevskii, and Professor D. Wiersma for allowing us to reproduce their published experimental results. We thank the National Science Foundation for support from Grant No. DMR 83-06429. J.L.S. is National Science Foundation Presidential Young Investigator, Camille and Henry Dreyfus Teacher-Scholar, and Alfred P. Sloan Fellow. Finally, J.L.S. thanks Professor Michael Fayer for introducing him to the exciting field of optical dephasing.

References

1. A. S. Davydov, *Theory of Molecular Excitons.* Plenum, New York, 1971.
2. L. Root and J. L. Skinner, *J. Chem. Phys.* **81**, 5310 (1984); *Phys. Rev. B* **32**, 4111 (1985).
3. R. Loring, H. C. Andersen, and M. D. Fayer, *J. Chem. Phys.* **81**, 5395 (1984).
4. W. S. Warren and A. H. Zewail, *J. Phys. Chem.* **85**, 2309 (1981); *J. Chem. Phys.* **78**, 2298 (1983).
5. D. E. Cooper, R. W. Olson, and M. D. Fayer, *J. Chem. Phys.* **72**, 2332 (1980); R. W. Olsen, H. W. H. Lee, F. G. Patterson, and M. D. Fayer, *J. Chem. Phys.* **76**, 31 (1982).

6. J. B. W. Morsink, B. Kruizinga, and D. A. Wiersma, *Chem. Phys. Lett.* **76**, 218 (1980).
7. T. Holstein, S. K. Lyo, and R. Orbach, in *Laser Spectroscopy of Solids*, W. M. Yen and P. M. Selzer (eds.), Springer–Verlag, Berlin, 1981.
8. B. DiBartolo, *Optical Interactions in Solids*. Wiley, New York, 1968.
9. R. M. Macfarlane, R. M. Shelby, and R. L. Shoemaker, *Phys. Rev. Lett.* **43**, 1726 (1979); S. C. Rand, A. Wokaun, R. G. DeVoe, and R. G. Brewer, *Phys. Rev. Lett.* **43**, 1868 (1979); R. M. Macfarlane, C. S. Yannoni, and R. M. Shelby, *Opt. Commun.* **32**, 101 (1980).
10. W. H. Hesselink and D. A. Wiersma, in *Spectroscopy and Excitation Dynamics of Condensed Molecular Systems*, V. M. Agranovich and R. M. Hochstrasser (eds.). North-Holland, Amsterdam, 1983.
11. M. J. Burns, W. K. Liu, and A. H. Zewail, in *Spectroscopy and Excitation Dynamics of Condensed Molecular Systems*, V. M. Agranovich and R. M. Hochstrasser (eds.) North-Holland, Amsterdam, 1983.
12. M. D. Fayer, in *Spectroscopy and Excitation Dynamics of Condensed Molecular Systems*, V. M. Agranovich and R. M. Hochstrasser (eds.). North-Holland, Amsterdam, 1983.
13. D. A. Wiersma, *Adv. Chem. Phys.* **47**. 2, 421 (1981).
14. A. A. Maradudin, *Solid State Phys.* **18**, 273 (1966).
15. I.S. Osad'ko, in *Spectroscopy and Excitation Dynamics of Condensed Molecular Systems*, V. M. Agranovich and R. M. Hochstrasser (eds.). North-Holland, Amsterdam, 1983; *Usp. Fiz. Nauk* **128**, 31 (1979) [*Sov. Phys.-Usp.* **22**, 311 (1979)].
16. K. K. Rebane, *Impurity Spectra of Solids*. Plenum, New York, 1970.
17. M. N. Sapozhnikov, *Phys. Status Solidi B* **56**, 391 (1973); **75**, 11 (1976).
18. R. Wertheimer and R. Silbey, *J. Chem. Phys.* **74**, 686 (1981).
19. L. Allen and J. H. Eberly, *Optical Resonance and Two-Level Atoms*. Wiley, New York, 1975.
20. J. L. Skinner, *J. Chem. Phys.* **77**, 3398 (1982).
21. H. deVries and D. A. Wiersma, *J. Chem. Phys.* **72**, 1851 (1980).
22. R. H. Silsbee, *Phys. Rev.* **128**, 1726 (1962).
23. D. E. McCumber and M. D. Sturge, *J. Appl. Phys.* **34**, 1682 (1963).
24. D. E. McCumber, *J. Math. Phys.* **5**, 222 (1964).
25. M. A. Krivoglaz, *Fiz. Tverd. Tela* **6**, 1707 (1964) [*Sov. Phys.-Solid State* **6**, 1340 (1964)]; *Zh. Eksp. Teor. Fiz.* **48**, 310 (1965) [*Sov. Phys.-JETP* **21**, 204 (1965)].
26. K. E. Jones and A. H. Zewail, in *Advances in Laser Chemistry*, A.H. Zewail (ed.). Springer, New York, 1978.
27. B. Halperin, *Chem. Phys.* **93**, 39 (1985).
28. G. J. Small, *Chem. Phys, Lett.* **57**, 501 (1978).
29. C. B. Harris, *J. Chem. Phys.* **67**, 5607 (1977); see also R. M. Shelby, C. B. Harris, and P. A. Cornelium, *J. Chem. Phys.* **70**, 34 (1979); S. Marks, P. A. Cornelius, and C. B. Harris, *J. Chem. Phys.* **73**, 3069 (1980); C. A. van't Hof and J. Schmidt, *Chem. Phys. Lett.* **36**, 460 (1975).
30. P. deBree and D. A. Wiersma, *J. Chem. Phys.* **70**, 790 (1979).
31. R. Kubo and K. Tomita, *J. Phys. Soc. Jpn.* **9**, 888 (1954).
32. P. W. Anderson, *J. Phys. Soc. Jpn.* **9**, 316 (1954).

33. A. G. Redfield, *Adv. Magn. Reson.* **1**, 1 (1965).
34. C. Cohen-Tannoudji, in *Frontiers in Laser Spectroscopy, Les Houches*, Vol. 1, R. Balian, S. Haroche, and S. Liberman (eds.). North-Holland, New York, 1977.
35. I. I. Abram, *Chem. Phys.* **25**, 87 (1977).
36. I. S. Osad'ko, *Fiz. Tverd. Tela* **13**, 1178 (1971); **14**, 2927 (1972); **17**, 3180 (1975) [*Sov. Phys.-Solid State* **13**, 974 (1971); **14**, 2522 (1973); **17**, 2098 (1976)]; *Zh. Eksp. Teor. Fiz.* **72**, 1575 (1977) [*Sov. Phys.-JETP* **45**, 827 (1977)]; I. S. Osad'ko and S. A. Zhdavov, *Fiz. Tverd. Tela* **18**, 766 (1976); **19**, 1683 (1977) [*Sov. Phys.-Solid State* **18**, 441 (1976); **19**, 982 (1977)].
37. D. Hsu and J. L. Skinner, *J. Chem. Phys.* **81**, 1604 (1984).
38. D. Hsu and J. L. Skinner, *J. Chem. Phys.* **81**, 5471 (1984).
39. D. Hsu and J. L. Skinner, *J. Chem. Phys.* **83**, 2097 (1985).
40. D. Hsu and J. L. Skinner, *J. Chem. Phys.* **83**, 2107 (1985).
41. D. Hsu and J. L. Skinner, to be published. See also ref. 15.
42. M. Lax, *J. Chem. Phys.* **20**, 1752 (1952).
43. R. G. Gordon, *Adv. Magn. Reson.* **3**, 1 (1968).
44. A. L. Fetter and J. D. Walecka, *Quantum Theory of Many-Particle Systems*, McGraw-Hill, New York, 1971.
45. R. Kubo, *J. Phys. Soc. Jpn.* **17**, 1100 (1962).
46. D. B. Fitchen, in *Physics of Color Centers*, W. B. Fowler (ed.). Academic, New York, 1968.
47. When $W \simeq -1$, one must use $\operatorname{arctg}(1/v) \simeq \pi/2 - v$ and Eq. (7.4) for $\Omega_0(\omega)$ is modified appropriately. It is then easily verified that for $W = -1$, $\omega_1 = \omega_0(v/\pi)^{1/2}$ [cf. Eq. (7.8)].
48. F. P. Burke and G. J. Small, *J. Chem. Phys.* **61**, 4588 (1974).
49. F. P. Burke and G. J. Small, *Chem. Phys.* **5**, 198 (1974).
50. V. A. Kizel' and M. N. Sapozhnikov, *Fiz. Tverd. Tela* **12**, 2083 (1970) [*Sov. Phys.-Solid State* **12**, 1655 (1971)].
51. I. S. Osad'ko, R. I. Personov, and E. V. Shpol'skii, *J. Lumin.* **6**, 369 (1973).
52. E. I. Al'shits, E. D. Godyaev, and R. I. Personov, *Fiz. Tverd. Tela* **14**, 1605 (1972) [*Sov. Phys.-Solid State* **14**, 1385 (1972)].
53. J. L. Richards and S. A. Rice, *J. Chem. Phys.* **54**, 2014 (1971).
54. J. A. Calviello, E. W. Fisher, and Z. H. Heller, *J. Appl. Phys.* **37**, 3156 (1966).
55. R. C. Powell, B. DiBartolo, B. Birang, and C. S. Naiman, *J. Appl. Phys.* **37**, 4973 (1966).
56. T. Kushida and M. Kikuchi, *J. Phys. Soc. Jpn.* **23**, 1333 (1967).
57. T. Muramoto, Y. Fukuda, and T. Hashi, *Phys. Lett. A* **48**, 181 (1974).
58. N. A. Kurnit, I. D. Abella, and S. R. Hartmann, in *Physics of Quantum Electronics*, P. L. Kelly, B. Lax, and P. O. tannewald (eds.). McGraw-Hill, New York, 1966.
59. Q. Kim, R. C. Powell, M. Mostoller, and T. M. Wilson, *Phys. Rev. B* **12**, 5627 (1975).
60. W. M. Yen, W. C. Scott, and A. L. Schawlow, *Phys. Rev.* **136**, A271 (1964).
61. J. P. Hessler, R. T. Brundage, J. Hegarty, and W. M. Yen, *Opt. Lett.* **5**, 348 (1980).
62. W. E. Moerner, A. R. Chraplyvy, and A. J. Sievers, *Phys. Rev. B* **29**, 6694 (1984).
63. G. F. Imbusch, W. M. Yen, A. L. Schawlow, D. E. McCumber, and M. D. Sturge, *Phys. Rev.* **133**, 1029 (1964).

64. D. W. J. Cruickshank, *Acta Crystallogr.* **9**, 1010 (1956).
65. G. S. Pawley, *Phys. Status Solidi* **20**, 347 (1967).
66. After the analysis of the experiment was performed with $T_D = 113$ K, we became aware of another estimate for naphthalene of $T_D = 131$ K. (See A. I. Kitaigorodskii, *Molecular Crystals and Molecules.* Academic, New York, 1973.) We then repeated the analysis for this value of T_D, finding results very similar to those displayed for $T_D = 113$ K.
67. J. T. Furukawa, T. B. Douglas, R. E. McCoskey, and D. C. Ginnings, *J. Res. Natl. Bur. Std.* **57**, 67 (1956).
68. B. Halperin, *J. Lumin.* **27**, 73 (1982).
69. B. Halperin, J. A. Koningstein, and D. Nicollin, *Chem. Phys. Lett.* **68**, 58 (1979).
70. O. N. Korotaev and M. Yu. Kaliteevskii, *Zh. Eksp. Teor. Fiz.* **79**, 439 (1980) [*Sov. Phys.-JETP* **52**, 220 (1980)].
71. L. W. Molenkamp and D. A. Wiersma, *J. Chem. Phys.* **80**, 3054 (1984).
72. W. H. Hesselink and D. A. Wiersma, *J. Chem. Phys.* **73**, 648 (1980); *Phys. Rev. Lett.* **43**, 1991 (1979); T. J. Aartsma and D. A. Wiersma, *Chem. Phys. Lett.* **42**, 520 (1976).
73. F. G. Patterson, W. L. Wilson, H. W. H. Lee, and M. D. Fayer, *Chem. Phys. Lett.* **110**, 7 (1984); H. W. H. Lee, F. G. Patterson, R. W. Olson, D. A. Wiersma, and M. D. Fayer, *Chem. Phys. Lett.* **90**, 172 (1982); R. W. Olson, H. W. H. Lee, F. G. Patterson, M. D. Fayer, R. M. Shelby, D. P. Burum, and R. M. Macfarlane, *J. Chem. Phys.* **77**, 2283 (1982).
74. R. M. Hochstrasser and C. A. Nyi, *J. Chem. Phys.* **70**, 1112 (1979).
75. K. Duppen, L. W. Molenkamp, J. B. W. Morsink, D. A. Wiersma, and H. P. Trommsdorff, *Chem. Phys. Lett.* **84**, 421 (1981).
76. G. Wäckerle, H. Zimmermann, and K. P. Dinse, *Chem. Phys. Lett.* **110**, 107 (1984).
77. A. A. Gorokhovski and L. A. Rebane, *Opt. Commun.* **20**, 144 (1977).
78. S. Völker, R. M. Macfarlane, and J. H. van der Waals, *Chem. Phys. Lett.* **53**, 8 (1978).
79. A. I. M. Dicker, J. Dobkowski, and S. Völker, *Chem. Phys. Lett.* **84**, 415 (1981).
80. A. Freiberg and L. A. Rebane, *Phys. Status Solidi B* **81**, 359 (1977).
81. L. A. Rebane, Estonian SSR Academy of Sciences, Proceedings of the International Symposium on Ultrafast Phenomena in Spectroscopy, Tallinn, USSR, Sept. 27–Oct. 1, 1978.
82. M. Glasbeek, D. D. Smith, J. W. Perry, Wm. R. Lambert, and A. H. Zewail, *J. Chem. Phys.* **79**, 2145 (1983).
83. S. Völker, R. M. Macfarlane, A. Z. Genack, H. P. Trommsdorff, and J. H. van der Waals, *J. Chem. Phys.* **67**, 1759 (1977).
84. A. I. M. Dicker, L. W. Johnson, S. Völker, and J. H. van der Waals, *Chem. Phys. Lett.* **100**, 8 (1983).
85. W. H. Hesselink and D. A. Wiersma, *Chem. Phys. Lett.* **50**, 51 (1977); **56**, 227 (1978).
86. T. J. Aartsma, J. Morsink, and D. A. Wiersma, *Chem. Phys. Lett.* **47**, 425 (1977).
87. T. J. Aartsma and D. A. Wiersma, *Chem. Phys. Lett.* **54**, 415 (1978).
88. L. A. Rebane, A. M. Freiberg, and Yu. Ya. Koni, *Fiz. Tverd. Tela* **15**, 3318 (1973) [*Sov. Phys.-Solid State* **15**, 2209 (1974)].
89. T. E. Orlowski and A. H. Zewail, *J. Chem. Phys.* **70**, 1390 (1979).

QUASI-TWO-DIMENSIONAL PHASE TRANSITIONS IN PARAFFINS

J. NAGHIZADEH†

Fachrichtung Theoretische Physik, Universität des Saarlandes, 6600 Saarbrücken, Federal Republic of Germany

CONTENTS

I. Introduction 46
II. Potential of Interaction 47
  A. Single Chain 48
  B. Salem Potential 51
  C. Orientational Energy 54
III. Motion of the Chain 55
  A. Infrared Spectra 56
  B. Equations of Motion of the Chain 57
    1. Generalized Coordinates 57
    2. Mass Tensor and Normal Modes 59
    3. Torsional Specific Heat 61
  C. Conformational Transitions 62
IV. Phase Transitions 67
  A. Melting Transition 69
  B. Rotator Transition 71
  C. Other Transitions 73
    1. Transition V → IV 75
    2. Transition IV → III 75
    3. Transition III → II 76
    4. Liquid–Liquid Transition 76
V. Theory of Phase Transitions 77
  A. Intralayer Melting Transition 79
    1. Ising Theory 80
    2. Dislocation Loops 83
    3. Chain Folding 86
  B. Theory of Rotator Transition 90
  C. Theory of Other Transitions 93
    1. Surface Melting Transition 93

† Present address: Department of Chemistry, University of Tennessee, Knoxville, TN 37996-1600.

2. V → IV Transition 93
3. IV → II Transition 94
D. Classification of Transitions 95
VI. Cycloparaffins and Phospholipids 97
A. Cycloparaffins 97
B. Phospholipids 100
Appendix A. Dislocations 103
Appendix B. Elastic Constants of Paraffins 108
References 110

I. INTRODUCTION

Normal as well as cyclic paraffins crystallize in lamellar form. They exhibit a number of phase transitions below the melting point. The lamellar structure of paraffin crystals and the large size of their molecules suggest that the phase transitions in these systems may be formulated within a structure incorporating liquid crystal systems. Actually, a theory of phase transitions in liquid crystals and paraffins of varying length and rigidity may be envisaged, bordered on one side by the classical point particle systems and on the other by the extended chain polymer and polymer liquid crystal systems. One should be able to develop a generalized phase diagram in which the molecular weight (molecule length) and rigidity appear as additional variables.

In recent years, the theory of two-dimensional melting has gained prominence through the works by Kosterlitz and Thouless[1] and Halperin and Nelson.[2] The *n*-alkanes or paraffins are ideal systems to which the two-dimensional melting theory may be applied. The paraffins exhibit a premelting transition known as the rotator transition where it is thought that the chains become free to perform hindered rotation around their long axes. This transition and the melting transition itself are special examples of the Kosterlitz–Thouless transition, which is discussed in detail later.

The study of *n*-alkanes or paraffins is also important because the system has resemblance to certain important biological systems, namely, bilayer membranes and monolayer fatty acids. A fatty acid consists of a saturated hydrocarbon chain attached at one end to an organic group or ionizable salt. Fatty acids form monolayers at a water surface and exhibit a number of phase transitions with increasing surface pressure and decreasing temperature. The synthetic or biological membranes are composed of a representative chemical with an ionizable head group and two attached paraffinic chains. These molecules, when dissolved in water in sufficient quantities, form vesicles that are the major structural blocks forming the biological cell. The biological character of the membrane is mainly determined by the motion and aggregation of the hydrocarbon chains in the vesicle. For these reasons, the study of paraffins is expected to lead to a better understanding of the membrane and fatty acid monolayer systems.

Whereas the transitions in membrane bilayer and monolayer systems are mainly of the smectic–smectic type, such liquid crystal type transitions are not observed in *n*-alkane aggregates. The absence of this phenomenon is also an intriguing problem which is briefly introduced.

In this review we treat the paraffin molecule mainly as a quasi-two-dimensional system in which the various phase transitions are driven by the instability of defect structures characteristic of the particular phase. Before proceeding to describe the phase transitions we briefly describe the characteristics of the single paraffin molecule. Thus, in Section II, the hindered rotation potential within a single chain as well as the pair potential between two parallel paraffin molecules are described. In Section III the motion of a single chain is considered. Within this context, the vibrational spectra are described briefly with emphasis on the localized spectra which has recently been used to identify gauche bonds. The equations of motion of a single chain are then treated in a general way, including all possible motions except the bond stretching motion. The emphasis in Section III is on the low-energy oscillations in the neighborhood of the internal rotation minima which have not been treated until recently. The torsional specific heat is calculated as an outcome of this theory. Finally, the conformational transitions of the chain are treated. In Section IV the experimental information on crystal structure and the structure's dependence on molecular weight and parity are described, followed by the description of melting transition, rotator transition, and other recently reported structural transitions. In Section V the melting transition is treated within the framework of the two-dimensional transition theory. Other transitions are treated briefly. In Section VI the membranes and cyclic paraffins are briefly described and their similarities with paraffins are stressed. In the appendixes a simple exposition of dislocation theory and a compilation of elastic constants of paraffins are given. The treatment is not exhaustive; rather it emphasizes the application of two-dimensional melting theory to paraffins.

II. POTENTIAL OF INTERACTION

The normal paraffins or normal alkanes have the chemical formula C_nH_{2n+2}. Their length may be characterized by the number of carbon–carbon links $n-1$. The simplest structural formula for this class is

$CH_3CH_2CH_2 \cdots CH_2CH_3$. The chain thus consists of two CH_3 end groups and a varying number of CH_2 middle groups. For the purpose of specifying the configuration of the chain, one may assume the carbon–carbon bond length and the C—C—C bond angles as constant. The configuration of the chain may thus be altered by internal rotation around each C—C bond.

A. Single Chain

The rotational potential may best be studied in the simplest paraffin, namely, ethane or CH_3CH_3. The rotational potential in this molecule around the C—C bond has a threefold symmetry as expected. The two extreme conformations in which hydrogen atoms have minimum distance—the eclipsed conformation—correspond to the state of the maximum energy, and the molecule usually finds itself in the stable minimum energy state—the staggered conformation. In the case of ethane, one notes that the distance between hydrogen atoms in the eclipsed conformation is 2.26 Å, which is slightly less than the sum of the van der Waals radii, 2.40 Å. Thus the barrier to rotation around the C—C bond may arise as a result of repulsion of overlapping hydrogen atoms. It has been established, however, that although the repulsive potential is important in determining the barrier height, the origin of the barrier energy appears to be quantum mechanical. It arises from the interaction between the sp^3 orbitals of the three C—H bonds on one carbon and the three C—H bonds on the neighboring carbon atom. In polar molecules such as CH_3OH, CH_3SH, CH_3SiH_3, and CH_3NH_2, the electrostatic interactions also make a significant contribution to the barrier height. Karplus and Parr[3] have calculated barrier heights for the above molecules based on the electrostatic model. A more general quantum mechanical treatment has been given by Sovers et al.[4] Thus the energy of a single ethane molecule as a function of the internal rotation angle may be described as a sum of three separate terms, namely,

$$U(\varphi) = U_{\text{steric}} + U_{\text{rot}} + U_{\text{val}} . \tag{2.1}$$

U_{steric} is the contribution of the repulsive forces, in this case of hydrogen atoms attached to the two carbon atoms. A Buckingham potential is usually used for this,

$$U_{\text{steric}} = \sum_{ij} \left[b_{ij} \exp\left(-\frac{R_{ij}}{\rho_{ij}}\right) - a_{ij} R_{ij}^{-6} \right], \tag{2.2}$$

where a_{ij}, ρ_{ij}, and b_{ij} are constants depending on the atoms i and j, and R_{ij}

is the distance between interacting atoms. Table I gives the values of the constants for the hydrogen–carbon system under consideration.

U_{rot} is the quantum mechanical part of the hindered rotation potential. Since the exact functional form of this is unknown, one may expand this part of the potential in a Fourier series, taking into account the threefold symmetry:

$$U_{rot}(\varphi) = E_1(1 - \cos 3\varphi) + E_2(1 - \cos 6\varphi) + \cdots . \tag{2.3}$$

The quantity E_1 is the barrier energy and is the most important contribution. E_2 is known experimentally to be small.

The third term, U_{val}, is only important when hydrogen atoms are replaced by larger substitutes; takes into account the valence angle deformation.

For most cases the quadratic approximation is adequate. The sum,

$$U_{val} = \sum \tfrac{1}{2} K_j (\Delta \theta_j)^2 , \tag{2.4}$$

is taken over all valence angles involved, and the force constants K_j are available from experimental information on vibrational spectra. For paraffins, Hägele and Pechhold[5] have performed detailed computations based on the parameters given in Table I. The parameters of Table I yield for ethane a barrier energy of 2.927 kcal/mole.

The rotational potential for a representative paraffin molecule may best be illustrated by examining the butane molecule $CH_3CH_2CH_2CH_3$. The potential of rotation around the central bond of butane may be calculated using the parameters of Table I. This is shown in Fig. 1. The most stable form of this molecule, the trans form, occurs with the CH_3

TABLE I
Potential Parameters for Paraffins

$a_{HH} = 3000$ kcal/mole	$a_{CC} = 53000$ kcal/mole
$b_{HH} = 33$ kcal/mole Å^6	$b_{CC} = 500$ kcal/mole Å^6
$\rho_{HH} = 2.267$ Å	$\rho_{CC} = 0.278$ Å
$a_{CH} = 14000$ kcal/mole	$K_{CCC} = 100$ kcal/mole
$b_{CH} = 128.5$ kcal/mole Å^6	$K_{CCH} = 52$ kcal/mole
$\rho_{CH} = 0.272$ Å	$K_{HC_H} = 75$ kcal/mole
$E_1 = 1.3$ kcal/mole	

Source: Reference 5, adapted from ref. 82.

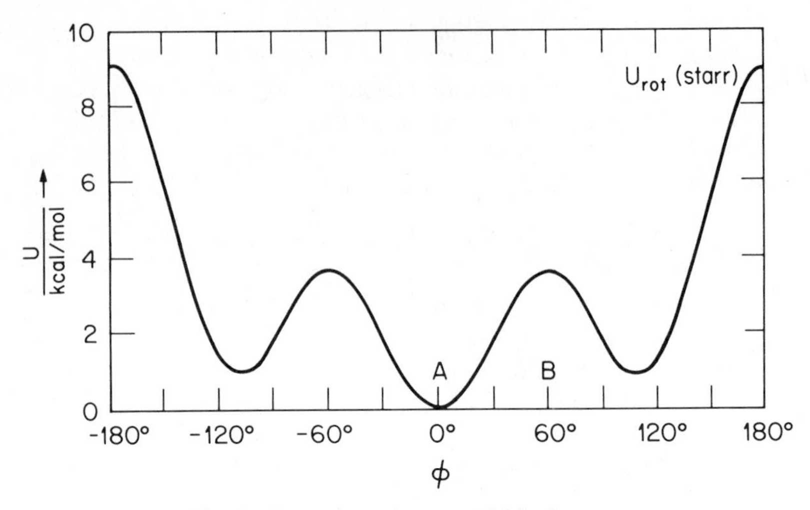

Fig. 1. Internal rotation potential for butane.

end group at maximum distance. This position is called the trans conformation. The position of a somewhat higher second minimum (gauche) is at 114.4° at an energy of 0.495 kcal/mole relative to the trans conformation. The energy barrier between the trans and gauche minima is 3.57 kcal/mole, and the highest energy state, that of eclipsed CH_3 terminal groups is 5.4 kcal/mole. The relative sharpness of the minima and the high barrier to rotation cause the molecule to remain most of the time in the vicinity of the minima, carrying out rotational oscillations.

Among the various conformations that a molecule may assume, one distinguishes several classes that have special meaning in the study of solid and liquid states. These are helical conformations, kinks, and jogs. The various possibilities are shown in Table II with the resultant net chain rotation and net chain shift relative to the all-trans conformation. An important conformation that apparently occurs in the solid state is the gtg^- kink which has a net shift of 4.5 Å, approximately equal to the lattice spacing.

The minimum energy state of the paraffin is the all-trans form, which occurs most prominently in the solid state. Figure 2 shows this form. It must be noted that the all-trans paraffin is symmetric with respect to a 180° rotation about the axis passing through the midpoint of the C—C bond plus a translational shift parallel to this axis by one C—C bond. The odd paraffins have an additional symmetry with respect to rotation about the axis perpendicular to the long axis passing through the midpoint of the molecule. This distinguishes odd and even paraffins because their

TABLE II
Some Conformations of the Paraffin Chain

Conformation	Symbolic arrangement	Relative shifts and notation of tails[a]			
		δ_x	δ_y	δ_z	ϕ
Planar	*tttttttt*				
2/1 Helix	*ggggggggg*				
3/1 Helix	*gtgtgtgtgt*				
Kink	*tttttgt$\bar{g}$tttt*	$b\sqrt{\frac{2}{3}}$	$b\frac{2}{\sqrt{3}}$	$b\sqrt{\frac{2}{3}}$	0
Jog	*ttttgtttg$\bar{g}$ttt*	$b\sqrt{\frac{8}{3}}$	$b\frac{4}{\sqrt{3}}$	$b\sqrt{\frac{8}{3}}$	0
Jog	*tttgt$\bar{g}$tgttt*	$b\frac{2\sqrt{2}}{\sqrt{3}}$	$b\frac{\sqrt{4}}{\sqrt{3}}$	$b\sqrt{\frac{8}{3}}$	0

[a] b = the bond length. Other conformations are given by S. Blasenbrey and W. Pechhold, *Ber. Bunsenges. Phys. Chem.* **74**, 784 (1970).

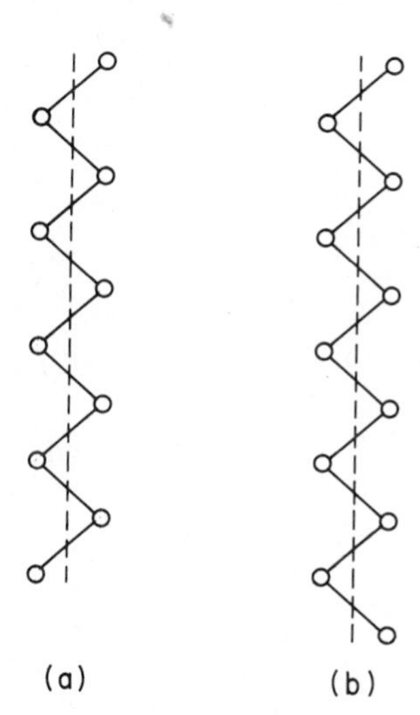

Fig. 2. The all-trans chain: (a) even number of carbons; (b) odd number of carbons.

properties are different, especially for lower molecular weight homologues.

B. Salem Potential

We now consider two parallel paraffin chains.[6] The dispersive energy is the sum of the dispersive energies of all CH_2 pairs units, assuming that each CH_2 acts as a single unit. Let the chains consist of N units, each of length λ and a total chain length $L = N\lambda$. Let the distance between the two chains be D. In general, $\lambda = 1.27$ Å and $D = 4.5$ Å for the paraffins. The London dispersion energy between a pair at a distance d is

$$w = \frac{A}{d^6}. \tag{2.5}$$

The distance between the nth unit in the first chain and the n'th unit in the second chain may easily be calculated:

$$d = [D^2 + \lambda^2(n - n')^2]^{1/2}. \tag{2.6}$$

Therefore the total attractive energy between the two chains, W, is given by

$$W = A \sum_{n=1}^{N} \sum_{n'=1}^{N} \frac{1}{[D^2 + (n - n')^2]^3}. \tag{2.7}$$

If we take $x = n - n'$, we may then write

$$W = A\left[\frac{N}{D^6} + 2 \sum_{x=1}^{N=1} \frac{N - x}{(D^2 + \lambda^2 x^2)^3}\right], \tag{2.8}$$

which may be transformed to an integral in the continuum limit:

$$W = \frac{2A}{\lambda^6 N^4} \int_0^1 \frac{(1 - y)\, dy}{(\rho^{-2} + y^2)^3}, \tag{2.9}$$

where $\rho = L/D$ and $y = x/N$. Carrrying out the integration, the attractive energy may be obtained from the simple expression

$$W = \frac{A}{4\lambda^2 D^4} \rho \left(3 \tan^{-1} \rho + \frac{\rho}{1 + \rho^2}\right). \tag{2.10}$$

The limiting expressions for $\rho = 0$ and $\rho = \infty$ are of interest. For small ρ, $\tan^{-1} \rho = \rho$ and

$$W_{\rho=0} \simeq \frac{A}{\lambda^2 D^4} \rho^2 = \frac{A}{D^6} N^2. \tag{2.11}$$

This is of the same form as the original London expression (2.5). For large ρ, $\tan^{-1} \rho = \pi/2$ and

$$W = A \frac{3\pi}{8\lambda^2} \frac{L}{D^5} = A \frac{3\pi}{8\lambda} \frac{N}{D^5} . \tag{2.12}$$

This would mean that, for long chains, the attractive energy is proportional to the inverse fifth power of the distance, whereas for short chains the attractive energy is proportional to the inverse sixth power of the distance. It is interesting to note that the Salem expression for short chains has an "end effect" represented by the expression in the bracket. The end effect in paraffins results from two sources: (1) the coupling constant for the CH_3 end groups is different from the coupling constant for two CH_2 groups and (2) the end groups interact only with internal groups of the chain and thus have higher potential energy. Only the latter effect has been taken into account in the Salem calculation. In order to assess the integration error, it was shown by Salem that for $N = 7$, the integrated energy is $W = 0.023(A/\lambda^6)$ compared to the exact calculation obtained by adding directly all the pair energies, $W = 0.021(A/\lambda^6)$. Thus the error due to integration is negligible for chains larger than $N = 7$.

A further result of this calculation is the application to circular or cyclic chains. Such chains occur in nature, and in the case of paraffins are known as cycloparaffins. The dispersion energy for two circular chains of N units each with circular circumference $L = N\lambda$ and a distance D between the centers of the two circles is given by

$$W = A \frac{N^2}{\pi D^6} \int_{-\pi/2}^{\pi/2} \frac{d\theta}{(1 + \rho'^2 \sin \theta)^3} , \qquad \rho' = L/\pi D . \tag{2.13}$$

Integration gives

$$W = A \frac{3\pi^2}{8\lambda^2 D^4} \frac{\rho'^2}{(\rho'^2 + 1)^{1/2}} \left[1 + \frac{2}{3} \frac{1}{\rho'^2 + 1} + \frac{1}{(\rho'^2 + 1)^2} \right] , \tag{2.14}$$

and the asymptotic expression for long chains is again

$$W = A \frac{3\pi^2}{8\lambda^2 D^4} \rho' = A \frac{3\pi}{8\lambda^2} \frac{L}{D^5} , \tag{2.15}$$

in agreement with asymptotic expression for noncircular chains. Salem has estimated the CH_2–CH_2 coupling constant as

TABLE III
Dispersion Energies of Various Fatty Acids in Monolayers at a Water Surface in kcal/mole[a]

Fatty acid	D (Å)	N	U_{disp}
$CH_3(CH_2)_{16}COOH$	4.8	~18	−8.4
$CH_3(CH_2)_{34}COOH$	4.8	~36	−16.8
CH_3—CH—$(CH_2)_{14}$—COOH	6.0	~18	−2.8

[a] According to Salem.[6]

$$w_{CH_2\text{-}CH_2} = \frac{A}{d^6} = \frac{1.34 \times 10^3}{d(\text{Å})^6}\ \text{kcal/mole}\ .$$

Based on this, the sublimation heat of paraffins, H_s, has been calculated to be

$$H_s = 1.71\ \text{kcal/mole/CH}_2\ ,$$

compared to the experimental value of 1.84 kcal/mole/CH_2. The attractive energy between the fatty acid monolayers has also been calculated using this scheme. The results are shown in Table III.

C. Orientational Energy

In the low-temperature phase, the structure in the plane perpendicular to the long axis is orthorhombic for odd numbered paraffins. The unit cell has the dimensions $a = 7.525$ Å and $b = 4.963$ Å, and the setting angle, the angle between the C—C zigzag plane and the a axis, is 42.5°. In this configuration the energy of a representative paraffin—the $C_{33}H_{66}$—has been calculated as a function of the orientation of the all-trans form with respect to the a axis, keeping all the neighbors in fixed orientation. The calculation, carried out by Strobl,[7] considers only the interaction between the hydrogen atoms attached to separate chains, in all-trans conformation. The different H–H interaction laws proposed have been examined, and it turns out that although the locations of potential minima are qualitatively the same for the various interaction laws, severe quantitative differences exist between absolute values of the energy depending on which interaction law is used. The various laws used were as follows:

1. The law proposed by Williams:[8]

$$w_{HH} = 2920\exp(-3.74R) - \frac{67.0}{R^6}\ .$$

TABLE IV
Energy Change ΔU for 180° Rotation of a Chain Within the Crystal According to Various Potential Laws (1–4)[a]

	1	2	3	4
ΔU(II)	11.75	5.30	2.848	1.179
ΔU(III)	11.25	5.016	2.680	1.674

[a] Because of slight differences in the crystal parameters in phases II and III, ΔU is slightly different for these two phases. ΔU is in kcal/mole. See ref. 7.

2. The law proposed by Warshel and Lifson:[9]

$$w_{\mathrm{HH}} = \frac{454}{R^9} - \frac{15.4}{R^6}.$$

3. The law proposed by Lifson and Warshel:[10]

$$w_{\mathrm{HH}} = \frac{1877}{R^{12}} - \frac{5.8}{R^6}.$$

4. The law of Admuir, Longmire, and Mason:[11]

$$w_{\mathrm{HH}} = \frac{33.2}{R^{6.18}}.$$

In all the above expressions, the interaction energy w is in kcal/mole and R, the mutual distance between the centers of the interacting hydrogen pairs, is in Å. The absolute minimum in almost all the curves is at an angle of 320°. There are three other minima, among them the minimum at 140°, 180° with respect to the absolute minimum. The energy difference between the two minima at 320° and 140° is shown in Table IV. Detailed calculations of potential energy surfaces for paraffins have been performed by McCullough,[12] who has also reviewed the literature on this subject.

III. MOTION OF THE CHAIN

The motion of paraffin molecules may first be examined by an analysis of an all-trans paraffin. In this molecule, in its isolated state, there exist a number of bond stretching and valence-angle deformation motions which give rise to the vibrational spectra of the molecule. Other types of motion are torsion around the skeletal C—C bonds, which are of much lower energy, and the jump over the torsional potential barrier, which causes conformational changes in the molecule. We treat these motions separately.

A. Infrared Spectra

The vibrational modes of normal paraffins in the all-trans conformation have been studied in detail by Snyder and co-workers.[13–15] Based on the observed infrared spectra, they have assigned force constants for the various motions and have assessed the variation of the force constants with the chain length. More recently, vibrational spectra for conformations other than the all-trans have been calculated and analyzed by the same authors.[15] The conformationally sensitive vibrational spectra for odd paraffins lie in the range 1400–700 cm^{-1}.

In discussing the spectra of *n*-alkanes, it is useful to classify vibrations into two types. First, there are the vibrations that are nonlocalized and that involve the entire chain. The nonlocalized modes give rise to band progressions which account for nearly all bands observed in this region of the IR spectrum. Second, there are vibrations that involve motion localized in certain parts of the chain. The methyl rocking at 890 cm^{-1} is an example of the localized vibration. The frequencies of these modes are dependent only on local conformations, and, in contrast to nonlocalized modes, the frequencies tend to be independent of both chain length and the conformation of the rest of the molecule. The IR spectra of isolated molecules have not been measured. The above features come out in the measurements of the IR spectra of low-temperature (all-trans) and high-temperature solid (kinked) and melt. Detailed study of the vibrational spectrum for C_{29} and lower odd paraffins have been made recently by Snyder and co-workers.[15] They are able to identify the localized frequencies which appear due to the introduction of gauche bond and kinks in the otherwise all-trans conformation. The most important frequencies are summarized in Table V. These frequencies and their intensities serve as a measure of the number of kinks and gauche bonds present in any aggregate and have been used as such by Snyder and co-workers to characterize the various solid or liquid phases in paraffins.

The theoretical treatment of the infrared spectra is somewhat similar to the treatment of torsional modes described in some detail in Section III B. The infrared spectra include the bond stretching modes, which are disregarded in the following treatment for simplicity (constant bond length). The discussion in the next section is more general than those presented earlier.[13–15]

TABLE V
Localized Frequencies[a]

	$\bar{\nu}_{obsd}$ (cm^{-1})	Description of mode
End gauche (gt_m)	1341	CH_2 wag
	1164	CH_2 rock, CH rock
	1078	CH_2 wag, C—C stretch
	955	CH_3 rock, C—C stretch
	873	CH_3 rock, C—C stretch
Kink ($t_m(gtg')t_m^*$	1366	CH_2 way
	1306	CH_2 wag
Double gauche ($t_m(gg)t_m^*$)	1353	CH_2 wag

[a] See ref. 15.

B. Equations of Motion of the Chain

1. *Generalized Coordinates*

The standard model of a molecule,[16,17] consisting of a chain of N masses (atoms) separated from each other by the bond length b, is used. The bond length is taken as constant throughout. The masses are numbered from 1 to N and their position vectors denoted by $\mathbf{r}_1, \mathbf{r}_2, \ldots, \mathbf{r}_N$. The *bond number* j is the line between the masses $j-1$ and j. The constancy of the bond length b reduces the number of degrees of freedom to $2N+1$. Therefore, $3N$ Cartesian coordinates, $\mathbf{r}_1, \ldots, \mathbf{r}_N$, are not the most appropriate coordinates to describe the dynamics of the system. Once $\mathbf{r}_1, \ldots, \mathbf{r}_{j-1}$ are given, $\mathbf{r}_j$ is totally defined by giving the *bond angle*, θ_j, between the bonds $j-1$ and j, and the *torsional angle*, ϕ_j, defined to be zero when the bond j is in the plane of the bonds $j-1$ and $j-2$. As shown by Flory,[18] two consecutive position vectors are related by

$$\mathbf{r}_i = \mathbf{r}_{i-1} + T_1 T_2 \cdots T_i \boldsymbol{\varepsilon}, \tag{3.1}$$

where

$$T_j = \begin{pmatrix} \cos\theta_j & \sin\theta_j & 0 \\ \sin\theta_j \cos\phi_j & -\cos\theta_j \cos\phi_j & \sin\phi_j \\ \sin\theta_j \sin\phi_j & -\cos\theta_j \sin\phi_j & -\cos\phi_j \end{pmatrix} \tag{3.2}$$

and

$$\boldsymbol{\varepsilon} = \begin{pmatrix} b \\ 0 \\ 0 \end{pmatrix}. \tag{3.3}$$

The bond angles θ_j are constrained to very small oscillations around a constant θ_0. The angles ϕ_j, on the other hand, may assume any value, but a potential well with three minima (see Fig. 1) ensures that ϕ_j oscillates most of the time, if the temperature is not too high, around 0 or $\pm 2\pi/3$ ($\phi_j \approx 0$ corresponds to the trans conformation and $\phi_j \approx \pm 2\pi/3$ to the gauche conformation). In symbols, a configuration will be described by the set of equilibrium angles $\alpha = (\alpha_1, \alpha_2, \ldots, \alpha_j, \ldots, \alpha_N)$, where $\alpha_j = 0$ or $\pm 2\pi/3$. If the energy (or temperature, in the equilibrium case) is not too high, the oscillations of both θ_j and ϕ_j have small amplitudes and the substitution of the real potentials by harmonic oscillators is an acceptable approximation.

For reasons of mathematical convenience, we artificially connect to one end of our chain three more masses, which are numbered 0, -1, and -2. These three masses will be kept in constant positions, that is, $\dot{\mathbf{r}}_{-2} = \dot{\mathbf{r}}_{-1} = \dot{\mathbf{r}} = 0$, where the dot represents the time derivative. The consequence of introducing these "artificial boundary conditions" are (1) the overall translations and rotations of the chain are inhibited and (2) additional potential energy is introduced, associated with the bonds -1 and 0. Now the expression for the potential energy of a bond in conformation α_j is

$$U_j = \tfrac{1}{2}k_\theta(\theta_j - \theta_0)^2 + \tfrac{1}{2}k_{\alpha j}(\phi_j - \alpha_j)^2 + C_{\alpha j}, \tag{3.4}$$

where $C_{\alpha j} = 0$ or Δ is valid for all bonds, $j = 1, 2, \ldots, N$. (Δ is the energy difference between the trans and gauche mimima.) It is believed that for very long chains these "boundary conditions" have a negligible effect on the internal vibrations associated with the thermal equilibrium of the polymer. This point is made clearer by analyzing the results.

To simplify notation a set of generalized coordinates q_j, appropriate for a given configuration α, are introduced

$$q_j = \phi_j - \alpha_j, \quad q_{N+j} = \theta_j - \theta_0, \qquad j = 1, \ldots, N. \tag{3.5}$$

The total potential energy will then be written

$$U = \sum_{j=1}^{2N} k_j q_j^2 + \sum_{j=1}^{N} C_{\alpha j}, \tag{3.6}$$

where

$$k_j = k_{\alpha j}, \quad \text{for } j = 1, \ldots, N,$$
$$k_j = k_\theta \quad \text{for } j = N+1, \ldots, 2N.$$

2. *Mass Tensor and Normal Modes*

To simplify notation the configuration dependent 3×3 matrix τ_j is defined

$$\tau_j = T_1 T_2 \cdots T_j. \tag{3.7}$$

The velocities are related to each other by the time derivative of Eq. (3.1):

$$\dot{\mathbf{r}}_j = \dot{\mathbf{r}}_{j-1} + \dot{\tau}_j \boldsymbol{\varepsilon} \tag{3.8}$$

with

$$\dot{\tau}_j = \sum_{i=1}^{2N} \tau_j^{(i)} \dot{q}_i \quad \text{and} \quad \tau_j^{(i)} = \frac{\partial \tau_j}{\partial q_i}. \tag{3.9}$$

Substituting Eq. (3.9) into (3.8) we obtain

$$\dot{\mathbf{r}}_n = \sum_{i=1}^{2N} \sum_{j=1}^{n} \tau_j^{(i)} \dot{q}_i \boldsymbol{\varepsilon}. \tag{3.10}$$

The kinetic energy is then given by the following expression:

$$\begin{aligned} K &= \frac{m}{2} \sum_{n=1}^{N} \dot{\mathbf{r}}_n^2 \\ &= \frac{m}{2} \sum_{i,l=1}^{2N} \dot{q}_i \dot{q}_l \sum_{n=1}^{N} \sum_{j,k=1}^{n} (\tau_j^{(i)} \boldsymbol{\varepsilon}) \cdot (\tau_k^{(l)} \boldsymbol{\varepsilon}). \end{aligned} \tag{3.11}$$

Defining the *mass tensor* as the $2N \times 2N$ symmetric matrix with elements

$$M_{il} = mb^2 \sum_{n=1}^{N} \sum_{j,k=1}^{n} \sum_{\mu=1}^{3} (\tau_j^{(i)})_{\mu 1} (\tau_k^{(l)})_{\mu 1}, \tag{3.12}$$

Eq. (3.11) becomes

$$K = \frac{1}{2} \sum_{i,\, l=1}^{2N} M_{il} \dot{q}_i \dot{q}_l \,. \tag{3.13}$$

Equation (3.12) may be written in a form more convenient for calculations:

$$M_{il} = mb^2 \sum_{j,\, k=1}^{N} \sum_{\mu=1}^{3} (\tau_j^{(i)})_{\mu 1} (\tau_k^{(l)})_{\mu 1} [N + 1 - \mathrm{Max}(j, k)] \,. \tag{3.14}$$

For a given configuration α, M_{il} is a function of all coordinates $q_1, \ldots, q_{2N}$. In a Taylor expansion of M_{il} in powers of the q's the zero-order term is calculated by taking all coordinates at the minima of the potential well. One expects that the effect of the linear terms on the observable quantities averages out and higher-order terms will be neglected because they are too small. With this plausibility argument one keeps only the zero-order terms, obtaining a mass tensor that is configuration dependent but time independent. This is in fact a textbook procedure[19] in the general treatment of small oscillations.

The Lagrangian for a given configuration is

$$L = K - U = \frac{1}{2} \sum_{i,\, l=1}^{2N} M_{il} \dot{q}_i \dot{q}_l - \frac{1}{2} \sum_{j=1}^{2N} k_j q_j^2 - \sum_{j=1}^{N} C\alpha_j \,. \tag{3.15}$$

Substituting this L into the Euler–Lagrange equations one obtains the following system of equations of motion:

$$\sum_{l=1}^{2N} M_{il} \ddot{q}_l + k_i q_i = 0 \,, \qquad l = 1, \ldots, 2N \,. \tag{3.16}$$

A general solution may be written in the form

$$q_l = \sum_{k=1}^{2N} D_{lk} \cos(\omega_k t + \beta_k) \,. \tag{3.17}$$

The normal coordinates are then given by

$$Q_k(t) = \sum_{1=l}^{2N} D_{kl}^{-1} q_l = \cos(\omega_k t + \beta_k) \,, \tag{3.18}$$

where D_{kl}^{-1} are the elements of the inverse matrix $\mathbf{D}^{-1}$ of $\mathbf{D}$, ω_k are the normal frequencies, and β_k are the initial phases. When Eq. (3.17) is substituted into Eq. (3.16) one obtains a system of linear homogeneous algebraic equations in the unknowns D_{lk}, for which nontrivial solutions are possible only if

$$\mathrm{Det}(W_{il} - \lambda_k \delta_{il}) = 0\,, \tag{3.19}$$

where

$$W_{il} = \frac{M_{il}}{k_i}\,, \qquad \lambda_k \equiv \frac{1}{\omega_k^2}\,, \tag{3.20}$$

that is, the normal frequencies ω_k are obtained from the eigenvalues of the (modified) mass tensor.

3. *Torsional Specific Heat*

The specific heat of a paraffin in the solid state has some contribution from intermolecular interactions which cannot be estimated from the present theory. However, the very important contribution from the *internal energy* of the molecule, $E = K + U$, is easily calculated. Although classical mechanics has exclusively been used to obtain the results, we follow now the usual procedure of considering each normal coordinate as an independent quantum oscillator. A state of the molecule is then characterized by its configuration and the excitation states n_k (number of phonons) of all oscillators, that is,

$$E = \sum_{k=1}^{2N} h\omega_k (n_k + \tfrac{1}{2}) + \sum_{j=1}^{N} C_{\alpha_j}\,. \tag{3.21}$$

The expectation value of E at thermal equilibrium at temperature T is obtained by first replacing n_k by its expectation value $\bar{n}_k$ for a given configuration,

$$\bar{n}_k = \frac{1}{e^{-\beta n\omega_k} - 1}\,, \qquad \beta = \frac{1}{k_B T} \tag{3.22}$$

and then averaging over all configurations, the probability P_α of each being given by the Boltzmann distribution,

$$P_\alpha \approx \exp\left(-\beta \sum_j C_{\alpha_j}\right). \tag{3.23}$$

Equation (3.23) contains the implicit assumption that the probability for the jth bond to be a gauche is independent of the other bonds. This is not exactly true, but it is a sufficiently good approximation especially at low temperatures, when the number of gauche bonds is small.

In what follows we call C_p the contribution of the phonons and C_c the contribution of the configuration energy to the specific heat:

$$C_p = \frac{\partial}{\partial T}\left\langle \sum_k h\omega_k(\bar{n}_k + \tfrac{1}{2}) \right\rangle_{\text{conf}} \tag{3.24}$$

and

$$C_c = \frac{\partial}{\partial T}\left\langle \sum_j C_{\alpha j} \right\rangle_{\text{conf}} , \tag{3.25}$$

where $\langle \cdots \rangle_{\text{conf}}$ means average over all configurations. For Eq. (3.24) this average may be performed, for example, by randomly selecting a number of configurations, with statistical distribution according to Eq. (3.23), and then taking the arithmetic average of the results obtained for each configuration. For Eq. (3.25) it may be done analytically as follows.

The probability P that the jth bond is a gauche, left or right, is

$$P = \frac{2\exp(-\beta\Delta)}{1 + 2\exp(-\beta\Delta)} \tag{3.26}$$

so that

$$C_c = \frac{\partial}{\partial T} N\,\Delta P = \frac{N\Delta^2}{k_B T^2} P(1-P) . \tag{3.27}$$

Computation of the mass tensor is particularly simple in the case of the all-trans chain. Numerical calculations have been performed for torsional modes of all-trans chains as well as chains with a fixed but small fraction of gauche bonds.[16,17]

C. Conformational Transitions

This type of transition has been treated in detail by Helfand and co-workers.[20–24] Consider a paraffin chain, which for the purpose of defining the conformational transition is denoted by P*gtt*Q, where *gtt* denote the bonds in the midportion of the chain whose detailed conformation is described in this case as a sequence of gauche–trans–trans. The P denotes the forward tail and Q the backward tail whose detailed conformation need not be described. An example of the conformational transition is the migration of a gauche bond denoted by

$$\mathrm{P}gtt\mathrm{Q} \longrightarrow \mathrm{P}'ttg\mathrm{Q}' . \tag{3.28}$$

In this transition a chain with the sequence gauche–trans–trans and tails P and Q is transformed into the sequence trans–trans–gauche while the tails P′ and Q′ have the same conformational structure as before, but differ in their orientation or distance relative to one another. A particular type of transformation which leaves the tails unchanged is the so-called crankshaft motion first proposed by Shatzki,[25] which in the symbolic form is represented by the equation

$$\mathrm{P}t(tg^+tg^+t)t\mathrm{Q} \rightarrow \mathrm{P}g^+(tg^+tg^+t)g^-\mathrm{Q} . \tag{3.29}$$

The conformational transitions are classified by Helfand to be of three types. The first type of transition is that which leaves the tails P and Q unchanged. The crankshaft transition defined above is an example of such a transition. The second type of transition results in a translation of Q relative to P. The gauche migration transition Eq. (3.28) is an example of such a transition. The third type of transition changes the orientation of Q relative to P. An example is

$$\mathrm{P}t\mathrm{Q} \rightarrow \mathrm{P}'g\mathrm{Q}' . \tag{3.30}$$

For the purpose of calculating the transition rate, the minimum energy path must be found. This includes, in addition to the rotational barrier, the stretching of the C—C bond and bending of the bond angle. For this purpose a simple potential, previously used in molecular dynamic calculations,[26] has been assumed.

The molecule is modeled as a chain of $N-1$ bonds that connect N identical vertices, labeled 0 to $N-1$. Each vertex represents a carbon atom and its substituents. The bond lengths b_i (distance from vertex $i-1$ to i) are maintained near a value b_0 by a potential V_b harmonic in bond length:

$$V_b(b_i) = \tfrac{1}{2}\gamma_b(b_i - b_0)^2 , \tag{3.31}$$

where γ_b is the bond stretching force constant. The bond angle θ_i is kept near a value θ_0 by a harmonic potential V_θ of the form:

$$V_\theta(\theta_i) = \tfrac{1}{2}\gamma_\theta(\cos\theta_i - \cos\theta_0)^2 . \tag{3.32}$$

The bond rotation is specified by a tortional angle ϕ_i, which is the angle between the plane of bonds $i-1$ and i and the plane of bonds i and $i+1$, taken as zero for the trans bond. Rotational motion is subject to the potential

$$V_\phi(\phi_i) = \gamma_\phi \sum_{n=0}^{5} a_n \cos^n \phi_i , \tag{3.33}$$

where $a_0 = 1$ and

$$\sum_{n=0}^{5} a_n = 0$$

so that the zero of energy is trans. The calculation of transition rates is made by assuming a harmonic approximation to the above rotational potential as follows:

$$V_\theta(\phi) = \tfrac{1}{2}\gamma_t \phi^2 , \tag{3.34}$$

$$V_\phi(\phi) = E^* - \tfrac{1}{2}\gamma^*(\phi \pm \phi^*)^2 , \tag{3.35}$$

$$V_\phi(\phi) = E_g + \tfrac{1}{2}\gamma_g(\phi \pm \phi_g)^2 , \tag{3.36}$$

near trans, near the trans–gauche barrier, and near gauche, respectively. The total potential is

$$V = \sum_{i=1}^{N-1} V_b(b_i) + \sum_{i=1}^{N-2} V_\theta(\theta_i) + \sum_{i=2}^{N-2} V_\phi(\phi_i) . \tag{3.37}$$

The same potential has been used in Brownian dynamic simulations carried out by Helfand and co-workers, the results of which were compared with the analytical rate calculations. The parameters of the potential function are those used by Rychaert and Bellemans[26] in their molecular dynamic calculations of butane, and these are presented in Table VI.

The rate of passage over the barrier is developed using the Kramers[27] picture of the reaction rate theory. One considers an ensemble of one-particle systems (one degree of freedom corresponding to the reaction coordinate x). In the neighborhood of the minimum (see Fig. 1) near x_A, let the potential be

$$V(x) = \tfrac{1}{2}\gamma_A(x - x_A)^2 , \tag{3.38}$$

TABLE VI
Ryckert–Bellemans Parameters[a]

$\beta = \xi/m$	1.00×10^5	ns^{-1}
M	0.014	kg/mole
B_0	0.153	nm
γ_b/m	2.5×10^9	J/kg nm^2 = ns^{-2}
θ_0	70.53°	
γ_θ/m	1.3×10^7	J/kg
γ_ϕ/m	6.634×10^5	J/kg
a_0	1	
a_1	1.3108	
a_2	−1.4135	
a_3	−0.3358	
a_4	2.8271	
a_5	−3.3885	
E^*	12.36	kJ/mole
	2.95	kcal/mole
E_g	2.933	kJ/mole
	0.70	kcal/mole
E_{cis}	44.833	KJ/mole
	10.7	kcal/mole
ϕ^*	±60°	
ϕ_g	±120°	
γ_t/m	5.412×10^6	J/kg
γ^*/m	1.903×10^6	J/kg
γ_g/m	7.530×10^6	J/kg

[a] Adapted from ref. 23. See ref. 26.

while near the barrier let it be

$$V(x) = E^* - \tfrac{1}{2}\gamma^*(x - x^*)^2 . \qquad (3.39)$$

Now the rate is obtained from the solution of the one-dimensional Fokker–Planck equation (FP) appropriate to the system. This is given by

$$\frac{\partial f}{\partial t} = -\frac{p}{m}\frac{\partial f}{\partial x} - F\left(\frac{\partial f}{\partial p}\right) + \frac{\xi}{m}\frac{\partial (pf)}{\partial p} + \xi k_B T\left(\frac{\partial^2 f}{\partial p^2}\right) . \qquad (3.40)$$

The solution is taken to be the product of the equilibrium solution and a correction factor g:

$$f = \frac{1}{d_A}(2\pi m k_B T)^{-1/2} \exp\left[-\frac{(p^2/2m) + V(x)}{k_B T}\right] g . \qquad (3.41)$$

In the above equations, f is the distribution function normalized to be unity near the well A, $N_A = 1$, t is the time, x the reaction coordinate, F the x component of the force, m the particle mass, p the particle momentum, ξ the friction constant, and d_A is given by

$$d_A = \int_{-\infty}^{+\infty} dx \exp\left[-\frac{1}{2}\frac{\gamma_A(x-x_A)^2}{k_B T}\right] = \left(\frac{2\pi k_B T}{\gamma_A}\right)^{1/2} . \tag{3.42}$$

The solution of the FP equation is obtained by seeking the steady-state solution for g, $dg/dt = 0$, with the boundary conditions

$$g = 1 \quad \text{as } x - x^* \rightarrow -\infty ,$$

$$g = 0 \quad \text{as } x - x^* \rightarrow +\infty .$$

In the highly viscous medium the FP equation for g becomes

$$\frac{\partial g}{\partial t} = 0 = -\frac{p}{m}\frac{\partial g}{\partial q} - \gamma^* q \frac{\partial g}{\partial p} - \frac{\xi}{m} p \frac{\partial g}{\partial p} + \xi k_B T \frac{\partial^2 g}{\partial p^2} , \tag{3.43}$$

TABLE VII
Rate of the Central $t \longrightarrow g^+$ Transition in an Eight Vertex Chain for $E^*/k_B T = 4$

Initial-state conformation	Rate (ns^{-1})
$(tt)t(tt)$	7.20
$(g^+t)t(tt)$	9.05
$(g^-t)t(tt)$	7.69
$(tg^+)t(tt)$	5.83
$(tg^-)t(tt)$	4.83
$(tg^+)t(g^+t)$	2.60
$(tg^-)t(g^*t)$	2.12
$(tg^-)t(g^-t)$	1.80
$(g^+t)t(tg^+)$	9.83
$(g^-t)t(tg^-)$	8.24
$(g^-g^-)t(g^-g^-)$	2.40
$(g^+g^+)t(g^+g^+)$	3.39
$(g^+g^+)t(g^-g^+)$	3.38
$(g^+g^-)t(g^+g^-)$	3.70

with $q = x - x^*$. The rate constant or the current over the barrier is calculated as

$$w = \int dp \frac{p}{m} f(q=0, p)$$
$$= \left[\frac{(\gamma_A \gamma^*)^{1/2}}{2\pi\xi}\right]\left[\frac{1}{2} + \left(\frac{1}{4} + \frac{m\gamma^*}{\xi^2}\right)^{1/2}\right]^{-1} \exp\left(\frac{-E^*}{k_B T}\right). \quad (3.44)$$

Actually, the reaction coordinate is determined by the saddle point of the potential (3.37). Helfand and co-workers have constructed the appropriate partition functions near the saddle point; they have also simulated the system as a Brownian bistable oscillator. The results of the calculation are a set of tables for the transition rate, a sample of which is presented in Table VII.

IV. PHASE TRANSITIONS

The first investigator of the structure of n-paraffins was A. Müller[28] who published his X-ray investigation of n–$C_{29}H_{60}$ in 1925. His basic findings have been confirmed by later researchers. Müller found that in the low-temperature phase the molecules have a zigzag plane structure (all-trans) and their long axes are parallel. Moreover, the parallel molecules form lamella with end groups of the two adjacent lamella forming a close-packed structure.

Thus every second layer is translationally identical to the previous one; that is, the structure is composed of two layers (Fig. 3). Actually, paraffin

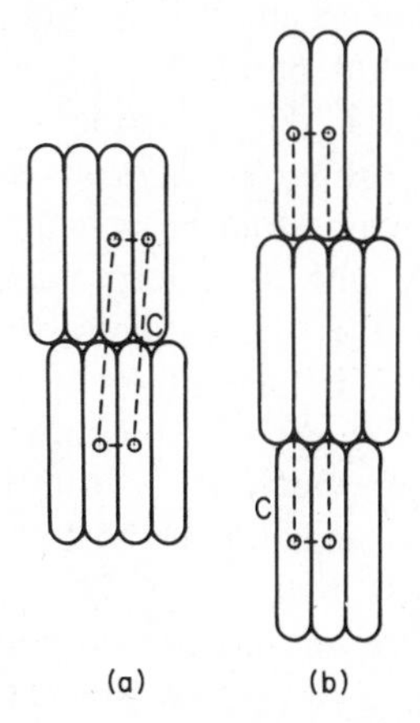

Fig. 3. Lamellar structure for (a) monoclinic and (b) orthorhombic modifications.

crystals undergo a variety of phase transitions before melting, and structural changes and defect mechanisms accompanying each transition are discussed later. It is instructive, however, to discuss first the crystal structure of low-temperature modifications. The crystal structure of the low-temperature phase depends on the parity of the chain (odd, even). For a representative member, $C_{29}H_{60}$, the low-temperature crystal structure is orthorhombic; the unit cell parameters of this crystal are

$$a = 7.45\ \text{Å}\,, \quad b = 4.97\ \text{Å}\,, \quad c = 7.72\ \text{Å}\,.$$

For even paraffins, the low-temperature crystal structure depends on chain length. For shorter chains up to $C_{28}H_{58}$ the structure is triclinic. A representative of this group, $C_{18}H_{38}$, was investigated by Müller and Lonsdale[28b] who found the parameters of the unit cell to be

$$a = 4.28\ \text{Å}\,, \quad b = 4.82\ \text{Å}\,, \quad c = 23.07\ \text{Å}$$

$$\alpha = 91°\,, \quad \beta = 92°\,, \quad \gamma = 107°\,.$$

For even paraffin C_{28} and higher the crystal structure is monoclinic. A representative of this group, n–$C_{36}H_{74}$, was investigated by Shaerer and Vand,[29] who found the parameters of the unit cell to be

$$a = 5.57\ \text{Å}\,, \quad b = 7.42\ \text{Å}\,, \quad c = 48.35\ \text{Å}$$

$$\beta = 119°\,.$$

The paraffin $C_{26}H_{54}$ is found in both monoclinic and triclinic structures. The triclinic structure of even numbered paraffins in contrast to the orthorhombic structure of odd paraffins and the change of structure of even numbered paraffins at $C_{26}H_{54}$ is explained in terms of symmetry arguments and packing energy by Kitaigorodskii.[30] He argues that molecules with symmetry $2m$ can retain their center of inversion $\bar{1}$ in the crystal, whereas molecules with symmetry mm only retain one minor plane (m). The symmetry $\bar{1}$ is more favorable than m as regards the compactness of the packing: it allows more orientational freedom, and thus the triclinic structure is the most compact one. Even numbered paraffins, the symmetry of which is $\bar{1}$, can be packed with triclinic symmetry. However, this packing is forbidden for odd numbered paraffins, the symmetry of which is $2m$. The theory can be verified by comparing the density of even numbered paraffins ($\approx$0.970 g/cm^3) against the density of odd numbered paraffins (0.930 g/cm^3). Also, the heat of fusion of even numbered paraffins is generally larger than the odd numbered ones. The origin of the co-existence of two crystalline forms for $C_{26}H_{54}$ has been explained by Mnyukh.[31] In general, the energy of the crystal is the sum of two terms. The first term arises from the interaction of parallel chains within a layer and is proportional to chain length. The second term is due to interlayer packing energy which is higher for a triclinic structure. The monoclinic packing, however, gives a higher contribution to the first term. Thus for short chains up to C_{24} the interlayer energy dominates and determines the structure, whereas for chains C_{28} and higher the structure is dominated by the packing within each single lamella and thus is monoclinic. For $C_{26}H_{54}$ both structures may co-exist since both monoclinic and triclinic structures happen to have the same free energy. The crystal structure of paraffins has been reviewed earlier by Boistelle.[32]

A. Melting Transition

Broadhurst[33] has reviewed the data on the melting transitions of paraffins. Measurements by Timmermann[34] show that the melting temperature of paraffins increases linearly for short chains, then flattens out and reaches an asymptotic value corresponding to the melting point of extended chain polyethylene. Attention was first focussed on this work by researchers seeking to establish the melting point of the extended chain polyethylene (Fig. 4).

Garner et al.,[35] from measurements of heat of crystallization H_f, concluded that beyond the first few members of the homologous series the heats and entropies of fusion increase linearly with molecular weight, excepting the well-known alteration for the even and odd numbers of

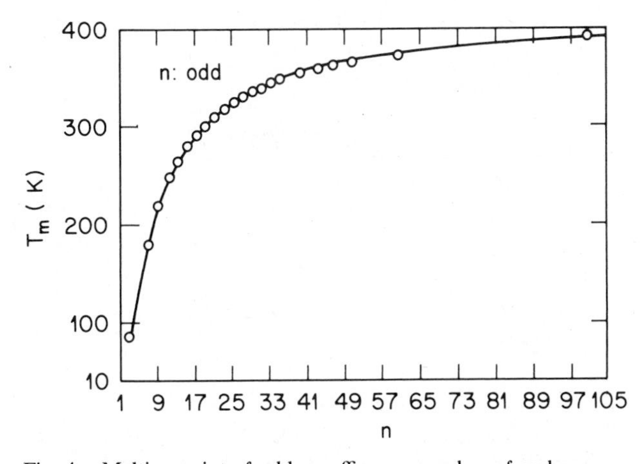

Fig. 4. Melting point of odd paraffins vs. number of carbons n.

chain atoms. Based on these data, Garner et al. proposed the following formula for normal paraffins and concluded a convergence temperature of 135°C for the melting point of the infinite molecular weight polyethylene (n = chain length):

$$H_f = 2.54n - 7.32 \text{ kJ/mole},$$

$$\frac{H_f}{T_M} = 0.006238n + 0.0169 \text{ kJ/K mole}.$$

From these equations one can derive a melting temperature (T_M) equation that gives reasonable agreement with paraffin melting points from 15 to 60 carbon atoms [standard deviation (SD) = ±0.7 K]. A detailed early discussion of the change of thermodynamic properties of long chain compounds was also given by Huggins.[36]

A final analysis based on the linearity of H_f and H_f/T_M was given by Broadhurst. His melting temperature equation,

$$T_M = 414.3 \frac{n - 1.5}{n + 5.0} \text{ K}, \qquad (4.1)$$

was derived for paraffins with orthorhombic crystal structure. This equation reproduces the melting points of $C_{44}H_{90}$ to $C_{100}H_{202}$ with a SD = ±0.3 K. Flory and Vrij[37] have applied polymer theory to obtain a corrected form of the empirical equation for melting point. They argue that the melting occurs as a two-step process. In the first step, melting of the chain occurs with ends still attached (smectic–smectic transition). This is then followed with melting of the ends. The molar free energy change for this two-step process, for paraffins of length n, can then be approximated by

$$G = ng_b + g_e - RT \ln n,$$

where g_b is the bulk free energy of fusion per unit length and g_e is the free energy change for unpairing the ends. Based on this analysis, the corrected empirical equation becomes

$$T_M = 419.6 \frac{n - 3.478}{n + \ln n - 0.425}. \qquad (4.2)$$

This equation has a SD = ±0.42 K. The fitting is poor for low molecular weight homologues. Several other fittings based on Eq. (4.2) have been presented.[34] Depending on the region of the melting temperature fitted, the asymptotic temperature varies and fittings remains poor in the low molecular weight region. More elaborate empirical equations have been proposed by Hoffman.[34] In general, the melting process is associated with a latent heat and the transition is first order.

B. Rotator Transition

Müller[28,29] was the first to observe that orthorhombic paraffin crystals tend toward, and in some cases reach, hexagonal symmetry with increasing temperature. This process is associated with a first-order phase transition below the melting point. Müller proposed that the molecules in the hexagonal phase rotate as rigid rods along their long axes. Thus he coined the name *rotator phase* for the hexagonal crystal just below the melting point. The rotator phase and rotator transition has since received considerable attention. The most recent experimental investigations of rotator transition are due to Doucet and co-workers and Unger. Unger[38] has investigated the odd paraffins $C_{11}H_{24}$ up to $C_{25}H_{52}$ by X-ray crystallography and thermal analysis. Denicolo and Doucet and co-workers[39–42] have investigated the odd paraffins $C_{12}H_{36}$ through $C_{25}H_{52}$ and the even paraffins $C_{18}H_{38}$ through $C_{26}H_{54}$. The transition behavior of odd and even paraffins are different, as expected. In general, the low-temperature phase, which for odd numbered paraffins is rhombohedral, shows an anisotropic thermal expansion in such a way that the unit cell parameters a and b perpendicular to the long axis approach a ratio $b/a \approx \sqrt{3}$, characteristic of the hexagonal lattice. The anisotropic thermal expansion in the case of $C_{33}H_{38}$ has been extensively studied by measurements of the Gruneisen constant and X-ray crystallography by Strobl.[43] In the case of the above mentioned homologue, X-ray methods were applied for direct measurements of the parameters a, b, c and appropriate ratios. The conclusions drawn are as follows: As the temperature increases, all three cell parameters expand slightly. At the transition point both a and b increase sharply toward a ratio of $\sqrt{3}$. In the rotator phase, a expands further while b decreases such that the ratio a/b approaches the hexagonal close-packed value of $\sqrt{3}$. In the shorter homologues below $C_{23}H_{48}$ the hexagonal close-packed structure is never reached. Starting from $C_{23}H_{48}$ a further weak transition is observed within the rotator phase on reaching the hexagonal close-packed structure. Accordingly, Doucet and co-workers claim that for odd numbered paraffins $C_{23}H_{48}$ and higher two rotator phases exist, the lower-temperature phase having a pseudo-hexagonal structure and the higher-temperature phase having the hexagonal structure with $a/b = \sqrt{3}$. The structure below the first rotator transition is described to be a herringbone orthorhombic structure with chains executing small oscillations about the long axis. At the rotator transition the herringbone structure is destroyed as disorder about the long axis sets in. Dynamic experiments, such as inelastic neutron scattering and proton spin relaxation, have been performed on $C_{19}H_{40}$ which establish a rotational jump relaxation time of the order of 3×10^{-12} s[44,45] in the rotator phase. The situation is different in even numbered paraffins. These paraffins, as stated above, have triclinic or monoclinic structure at low temperatures. On increasing the temperature, the compounds n-$C_{22}H_{46}$, n-$C_{24}H_{50}$, and n-$C_{26}H_{54}$ exhibit crystal rotator transitions. For n-$C_{20}H_{42}$ the rotator phase appears as a metastable phase and $C_{18}H_{38}$ and lower even numbered paraffins do not exhibit this phase transition. The behavior of even paraffins higher than $C_{26}H_{54}$ is presumably similar to that of $C_{26}H_{54}$. Whereas the odd numbered paraffins show two rotator phases, the even numbered ones exhibit a sharp transition from triclinic (monoclinic) structure to the hexagonal structure with $a/b = \sqrt{3}$. Moreover, the hydrocarbon chains in the rotator phase are rotationally disordered in the rotator phase. The transition to the hexagonal structure ($a/b = \sqrt{3}$) lies on a single straight line irrespective of parity. For odd numbered paraffins which exhibit a crystal pseudo–hexagonal rotator transition, the transition points define a separate line. The transition temperatures are depicted in Fig. 5.

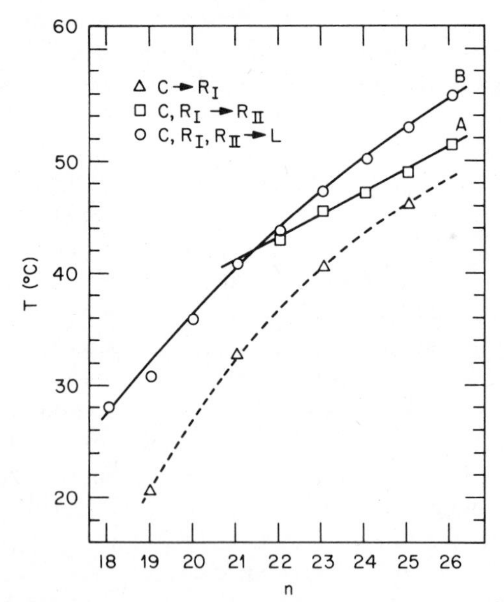

Fig. 5. Rotator transition temperatures. C, Crystal; R_I, pseudo-hexagonal rotator phase; R_{II}, hexagonal rotator phase. Line A defines the rotator transition temperatures for even paraffins as well as R_I–R_{II} transitions for odd paraffins. (Source: Ref. 41.)

C. Other Transitions

Since the discovery of the rotator transition by Müller,[28] a detailed study of the odd paraffin $C_{33}H_{68}$ was undertaken by Strobl.[46] In a series of articles, experiments on this compound using a variety of methods have been described. Accurate differential calorimetry identifies, in addition to the melting point at 71.8°C, three further transitions below the melting point. In order to classify the various phases the thermogram is reproduced in Fig. 6. Strobl, Fischer and co-workers,[46,47] in a series of small-angle X-ray, NMR and IR measurements, have characterized the structure of various phases and the defect mechanisms responsible for the transitions. Before going into the details of these findings we point out that the experiments of Strobl and co-workers were repeated by Oyma et al. and Synder et al.[15] on pure samples of $C_{25}H_{52}$ up to $C_{45}H_{92}$ (odd paraffins), and additional transitions below the melting point were also observed by these authors. In addition, for paraffins $C_{25}H_{52}$, $C_{27}H_{56}$, and $C_{29}H_{60}$ a further transition in the low-temperature region was observed. Figure 7 summarizes the existing data on the observed transitions with the nomenclature adopted from Synder et al.[15] The present information as to the structure of various phases derives from X-ray, NMR, IR, and detailed thermal analysis measurements.[15]

Conclusions drawn from the IR measurements were based mainly on the analysis of the bands assigned to the end gauche, kink defect, and double gauche. Table VIII gives defect concentration estimates based on the analysis of intensities of these bands as a function of temperature. In a descriptive way one may summarize the conclusions as follows:

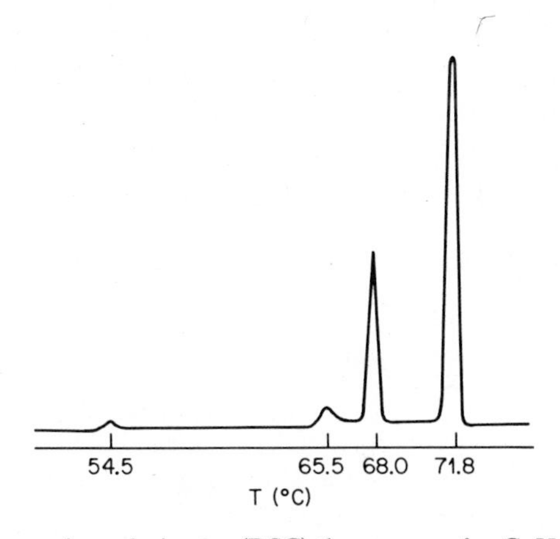

Fig. 6. Differential scanning calorimeter (DSC) thermogram for $C_{33}H_{68}$. (Source: Ref. 46.)

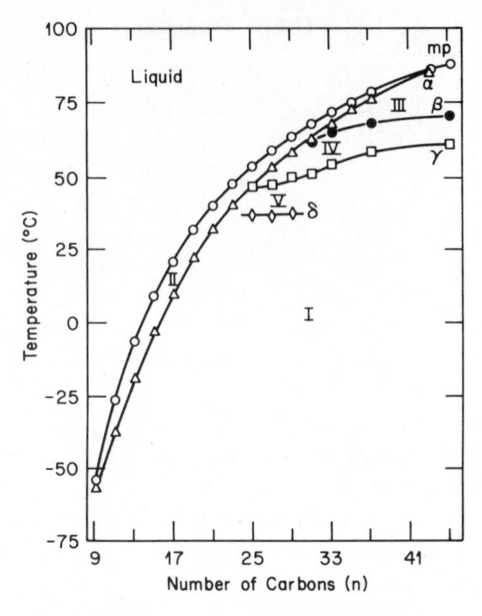

Fig. 7. Transition temperatures for odd n-paraffins as a function of number of carbons n. The phases are denoted by roman numerals and the co-existence curves by greek letters. (Source: Ref. 15.)

1. The end gauche concentration changes by a few percent at the δ transition (Fig. 7). The change in the concentration of this defect is much larger in going through the rotator transition. However, since other defects are generally absent in the low-temperature phase I, the δ transition may be a result of the melting of the interlayer surface which increases the end gauche freedom of rotation. This transition is weak and probably second order. That the δ transition must be associated with interlayer melting is also seen from its chain length independence. This is strictly a surface

TABLE VIII
Estimated Defect Concentrations in Rotator Phase (Solid) for n-$C_{27}H_{56}$[a]

Defect	$\bar{\nu}$ (cm^{-1})	$\bar{n}_{liquid}$	$\bar{n}_{solid}$
$gt\ldots$	1341	0.624	0.15–0.30
$\ldots(gtg)\ldots$	1306	2.55	0.38–0.64
$\ldots(gg)\ldots$	1353	2.34	≤0.002

[a] $\bar{n}$ = defect/chain. See ref. 15.

phenomenon and must be independent of the chain length. However, it is apparently difficult to detect this transition by differential scanning calorimetry (DSC) and the apparent transition has only been observed for $C_{25}H_{52}$, $C_{27}H_{56}$, and $C_{29}H_{60}$.

2. The kink concentration associated with the band at $1306\,cm^{-1}$ is very slight and remains constant in the low-temperature phase I. It remains constant in going through the surface melting transition. It changes slightly in passing through the γ transition ($V \rightarrow IV$) and it changes appreciably in going to phase II (the rotator transition). This picture is consistent with the observations of Strobl et al.[7] who based their conclusion on the analysis of X-ray measurements.

1. *Transition V→IV*

This transition denoted by the co-existence line γ in Fig. 7 has been more carefully studied by Strobl and co-workers[46–48] by X-ray, NMR and IR measurements on oriented samples. They conclude that phase V is a two-dimensional ordered array of zigzag all-trans chains as shown in Fig. 8a. Because of the asymmetry of the zigzag plane to 180° rotation, the projection of the chain on the two-dimensional plane must be identified by an arrow. Thus the low-temperature phase V has a herringbone structure in the plane perpendicular to the long axis. According to these authors, the defect mechanism driving the transition $V \rightarrow VI$ is the 180° rotation of the chains as shown in Fig. 8b. The latter mechanism will lead to surface roughening and the X-ray experiment apparently showed no surface roughening.

2. *Transition IV→III*

The structural change in this transition is unambiguously established through X-ray analysis. These experiments show that the system experiences a tilt during this transition.[7] The all-trans hydrocarbon chains that are essentially perpendicular to the lamellar plane in phase IV go into a tilted form relative to the lamellar plane in phase III, the chains

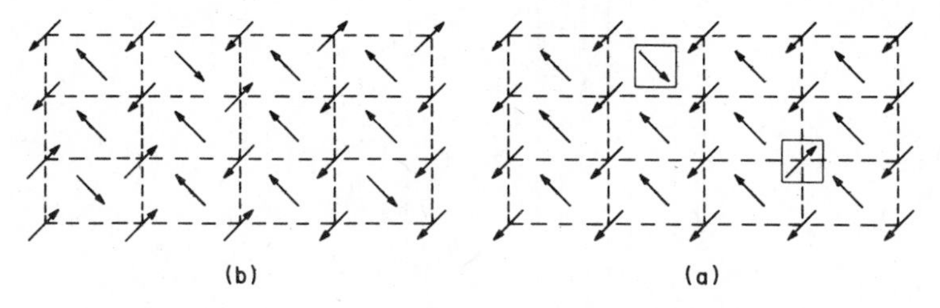

Fig. 8. Order–disorder transition (γ) for $C_{33}H_{68}$. (After Strobl, ref. 7.)

retaining their all-trans structure (Fig. 9). The lattice constant in the direction parallel to the chains expands with $\Delta C/C_0 \approx 10^{-2}$. It is estimated that a C—C bond at each head becomes free to rotate, leading to a roughening of the interlamellar surface. This roughening is also concluded from analysis of the X-ray experiments.

3. *Transition III → II*

During this transition a significant lateral expansion of the lattice takes place. The change in the area per hydrocarbon chain ΔA increases as $A/A_0 \approx 4\cdot 7\times 10^{-2}$. The free volume thus introduced permits the occurrence of kinks. The average number of kinks per chain may be estimated by noting that the lattice constant parallel to the chain decreases by approximately 0.97 Å.[7] A kink of *gtg* type would cause an effective shortening of the chain by 1.27 Å. Thus the effective number of kinks per chain for $C_{33}H_{68}$ is estimated to be $0.97/1.27 \approx 0.7$. Existence of kinks and their average number may also be estimated by IR spectroscopy.[15] This transition is the rotator transition discussed earlier.

4. *Liquid–Liquid Transition*

This transition has been reported by Krüger et al.[49] who used Brillouin spectroscopy to measure the sound velocity in the liquid range. For $C_{24}H_{50}$ they observed a discontinuity in the slope of sound velocity against temperature at 110°C, well above the melting point. It is postulated that this apparent transition is due to cooperative trans bond enrichment in the melt, presumably the trans–gauche ratio falls off discontinuously above this temperature. Other authors have attempted to directly measure the trans–gauche ratio to confirm this result. Fischer et al.[50] use Rayleigh depolarization experiments and were unable to detect a significant change of the depolarization (a measure of cooperative trans

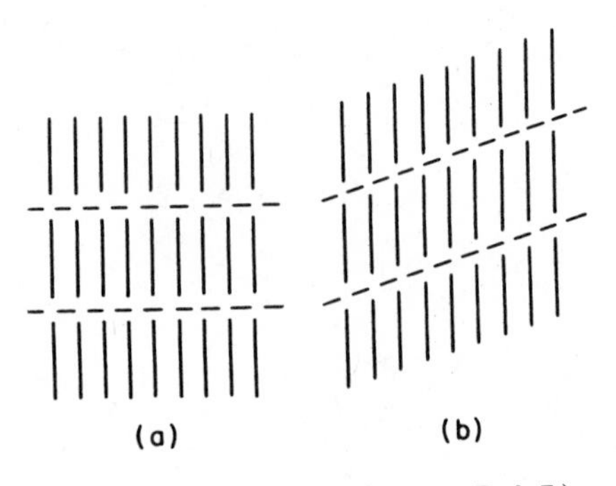

Fig. 9. Tilt transition. (Source: Ref. 7.)

enrichment). Synder[51] used Raman spectroscopy of the deuterated paraffin in solution and in melt. Through Raman spectroscopy the trans–gauche ratio may be inferred, and no significance difference in trans–gauche ratio of the paraffin in solution and in melt was found. Thus the molecular mechanism for the transition is still unclear.

V. THEORY OF PHASE TRANSITIONS

Analysis of the phase transitions in paraffins is partly based on the liquid crystal theory. The lamellar structure of paraffins justifies this. However, the elastic energies of a paraffin layer and a smectic A liquid crystal differ and are given as follows.[52]

$$E_p = \frac{1}{2}\int d^3r \left[\kappa_\perp \left(\frac{\partial u_z}{\partial z}\right)^2 + \kappa_\parallel \left\{\left(\frac{\partial u_z}{\partial x}\right)^2 + \left(\frac{\partial u_z}{\partial y}\right)^2\right\}\right] \tag{5.1}$$

and

$$E_{SA} = \frac{1}{2}\int d^3r \left[\kappa'_\perp \left(\frac{\partial u_z}{\partial z}\right)^2 + \kappa'_\parallel \left(\frac{\partial^2 u_z}{\partial x^2} + \frac{\partial^2 u_z}{\partial y^2}\right)^2\right], \tag{5.2}$$

where u_z is the elastic displacement normal to the layer plane assumed to point in the z direction, and $\kappa_\perp$, $\kappa_\parallel$, $\kappa'_\perp$, $\kappa'_\parallel$ are the elastic constants. The special form of the second term in Eq. (5.2) is a consequence of the fact that, for short molecules, gradient terms of u_z with respect to x and y only contribute anharmonic terms to the elastic energy because neighboring molecules are able to rotate into a parallel position. For long chains this leads to steric constraints in regions of finite curvature of the lamella. In classical point particle systems, the material in regions of high curvature yields by producing arrays of dislocations that move material from high- to low-pressure regions. In layers consisting of long chain molecules, this process could be replaced by the motion of dislocations with the Burgers vector parallel to the molecular axis. Therefore, to distort a paraffin lamella into a configuration similar to a smectic A layer under the same boundary conditions, inelastic processes are required, such as the creation and motion of vertically oriented dislocations. This implies that with increasing molecular length a transition from Eq. (5.2) to Eq. (5.1) must occur via anharmonic terms; that is, Eq. (5.2) applies only for infinitesimal distortions. The vertical dislocations in paraffin degenerate into the vertical dislocations in liquid crystals (with a Burgers vector of one layer thickness) and these, according to Hubermann et al.,[53] Helfrich,[54], and Nelson and Toner,[55] drive the smectic A–nematic transition.

Thus the intimate relation between the two problems is obvious. Because the elastic interaction between dislocations depends on the laws described by Eqs. (5.1) and (5.2), cross over behavior between the two systems may lead to interesting phase transition properties. For example, the anisotropic scaling law postulated by Helfrich[54] and calculated by Nelson and Toner,[55] so far not in agreement with experiment, may be explained as a crossover phenomenon of that type. Furthermore, it has been observed by McMillan[56] that with increasing molecular length the width of the nematic phase decreases and at some point the smectic *A*–nematic phase transition becomes first order; and the smectic *A*–nematic and nematic–isotropic transitions coalesce. We classify the various possible defects in paraffins as follows:

1. Vortex loops.
2. Horizontal dislocation loops, vertical dislocation loops, kinks, vacancies.
3. Defects in equally spaced flat lamellar structure.
4. Disclination loops.

Defects of type 1 are associated with the anisotropic shape of the paraffin molecule and drive the so-called rotator transition as discussed later. Defects of type 2 are those which are able to impart liquid state properties to the sample. Horizontal dislocations can occur in the lamellar plane with a Burgers vector $b_{\|} = m\, d_{xy}$ with d_{xy} a lateral lattice unit and m an integer. Disordering of the vertical positional order can occur within the layer by loops of dislocations with a Burgers vector $b_{\perp} \cong nd_z$ with d_z of the order of C—C bond length and n an integer. Disordering of the lamellar structure can occur by means of loops of dislocations with a Burgers vector $b_{\perp} = mL$ with L the lamellar thickness. The latter process is the driving mechanism for the smectic–nematic transition. Because horizontal dislocation loops do not produce surface step lines, it can be described as a pair of dislocation lines of edge type. The intralayer melting is essentially driven by such dislocations. These dislocations interact weakly over the lamellar interface; thus the melting process is quasi-two-dimensional.

In contrast to the horizontal dislocation loops, the vertical dislocation loops produce surface step lines and lead to interlayer coupling. The vertical dislocation will lead to liquid state properties longitudinal to the molecular axis. The Burgers vector can have a modulous $mb_{\perp}$ with m taking a maximum value of L, the length of the molecule. $Lb_{\perp}$ will shift stacks of lamellar planes relative to one another leading to the smectic *A*–nematic transition.[53–55] In addition to dislocation of pure vertical or horizontal, there will be dislocations of mixed character with a Burgers vector $[(nb_{\perp})^2 + (mb_{\|})^2]^{1/2}$ with m and n a pair of integers. Defects of type

3 are common in liquid crystal smectic phases and are known as *focal defects* (described, e.g., in de Gennes' book[52]). They depend on an elastic law as given by Eq. (5.2) and can be realized in paraffins only if additional defects are present, such as vertical dislocation loops or folded polymer molecules, playing the role of edge type dislocation arrays in classical point particle systems.[56] Defects of disclination type, as listed under 4 above, will drive the nematic–isotropic transition.

A. Intralayer Melting Transition

This is the same as the main melting transition. The melting point of paraffins lies on two smooth curves depending on parity (even–odd effect). We consider the melting of odd numbered paraffins (Fig. 4). The melting point increases linearly with chain length for short chains then flattens out and reaches a plateau value (convex shape). For short chains, kink excitation is prohibited due to geometrical constraints and low melting point (much below the kink excitation energy). For these chains the melting process is similar to the smectic–smectic transition in liquid crystals. Hubermann, Lublin, and Doniach[53] have applied the dislocation theory to the smectic–smectic transition in liquid crystals. Considering the smectic A (S_A) to smectic B (S_B) transition and neglecting the weak interlayer coupling, the free energy F_B per layer in terms of two-dimensional dislocation density n_B (per atom) may be written

$$F_B = E_c n_B - A \ln n_B + k_B T[n_B \ln n_B + (1 - n_B)\ln(1 - n_B)]\,, \qquad (5.3)$$

where the first term denotes the dislocation core energy and the second term the mean strain energy of the dislocation. The strain energy of the dislocation is calculated by assuming that dislocations are uniformly distributed and by cutting off the strain field at the mean interdislocation distance. The bracket contains the configurational entropy, in an approximation where dislocations are treated as independent. The constant A is given in terms of the shear modulus and poisson ration, or it may be expressed in terms of the elastic constants as follows:

$$A = \frac{C_{66}(C_{11} - C_{66})b^2 L}{4\pi C_{11}}\,, \qquad (5.4)$$

where L is the thickness of the layer. Minimizing Eq. (5.3) and assuming $E_c < A$ one obtains

$$n_B = 1 - \exp\left(-1 + \frac{E_c}{A}\right) \qquad (5.5)$$

and the transition temperature is given by

$$T_{AB} = \frac{A}{k_B} . \tag{5.6}$$

If the lattice is hexagonal (with the Burgers vector b equal to the lattice spacing), one may express A in terms of molecular weight and density (b^2L being the volume occupied by one molecule). Then

$$T_{AB} = \frac{C_{66}(C_{11} - C_{66})M}{2\pi\sqrt{3}\, k_B \rho C_{11}} , \tag{5.7}$$

where M is the mass of one molecule and ρ the density.

The transition is associated with an entropy change,

$$\Delta S = -k_B[n_B \ln n_B + (1 - n_B)\ln(1 - n_B)] , \tag{5.8}$$

which may be calculated explicitly using the formula for n_B. Reasonable agreement was found between the calculated and experimental values for the liquid crystal EEBAC. For paraffins, using the experimental value of $C_{66} = 8 \times 10^8$ dynes/cm^2 measured by Pechhold and co-workers,[57] $C_{11} = 15 \times 10^{10}$ dynes/cm^2 based on low-temperature measurements of Krüger et al.,[58] and the experimental value of $\rho = 0.7$ g/cm^3, one obtains for the initial slope of the melting curve a value of 21 K/bond.[59] This compares well with the experimental value of 18.7 K/bond quite reasonably. The applicability of the Kosterlitz–Thouless melting criterion is thus reasonably established and further work must focus on the role of chain flexibility (kinks) in modifying the melting criterion. This is explained in the following sections.

1. Ising Theory

The role of kinks on dislocations has been investigated by Peredecki and Statton[60] and by Keith and Passaglia.[61] Kink bands tend to redistribute the dislocations in such a manner that the edge dislocation core no longer runs through the entire layer thickness L. This is demonstrated in Fig. 10, where it is shown that the edge dislocation in Fig. 10a which runs through the entire depth L has been broken into two parts in Fig. 10b. This process will sometimes lead to the annihilation of a part of the dislocation of opposite charge. In general, the factor L appearing in Eq. (5.4) must be modified and replaced by a mean length, $\langle l \rangle$. The value of $\langle l \rangle$ may be calculated by analogy to a similar problem.[62] In terms of $\langle l \rangle$ the

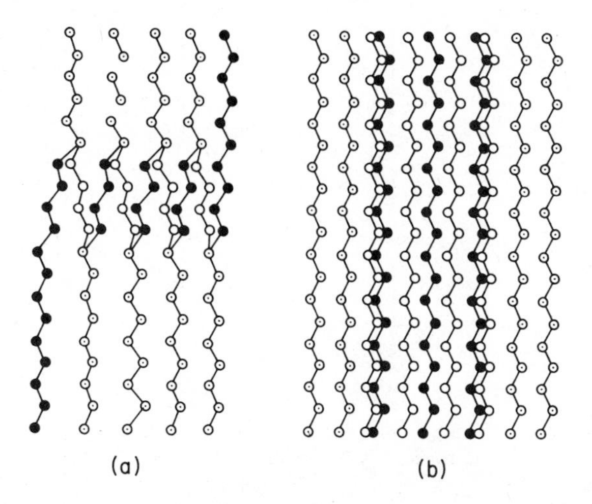

Fig. 10. Possible dislocation in paraffins with the Burgers vector perpendicular to chain: (a) Screw dislocation; (b) edge dislocation. Two other possible dislocations with the Burgers vector parallel to the chain are not shown. Chains with an open circle and dot represent a plane of parallel chains in a plane perpendicular to the paper.

melting criterion may be written

$$T_M = \frac{K^e}{k_B} \langle l \rangle , \tag{5.9}$$

where K^e is the coefficient of L in Eq. (5.4).

We now investigate the behavior of T_M/L for short chains:

$$\lim_{L\to 0} \left(\frac{T_M}{L} \right) = \frac{K^e}{k_B} . \tag{5.10}$$

Introducing the dimensionless parameter $\chi = T_M k_B / K^e$, Eq. (5.9) may be written

$$f(\chi) = \frac{\langle l \rangle}{L} , \tag{5.11}$$

where $f(\chi)$ goes to unity as $L \to 0$.

The redistribution of the dislocations occur by appearance of kink pairs as shown in Fig. 10. In Fig. 11a and b, the two configurations have an energy difference U equal to the energy of a kink pair, which may be regarded as the nearest-neighbor interaction. The system may be for-

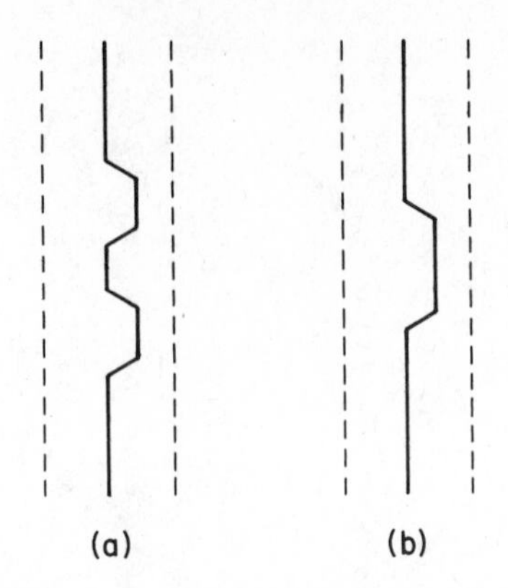

Fig. 11. Comparison of energy edge dislocations decorated by kinks: (a) Two neighboring kink decorated edge dislocations each of unit length. (b) One such dislocation with twice the unit length.

mulated as a one-dimensional Ising problem with a set of parameters t_i (for a unit length), where $t_i = 0$ for the condensed sections and $t_i = 1$ elsewhere.

The energy is thus given as

$$E\{t\} = -U \sum_{i=1}^{n} t_i t_{i+1} - J \sum_{i=1}^{n} t_i , \tag{5.12}$$

where J is the elastic energy per unit length. The grand partition function Z_G is given by

$$Z_G(Z, L, T) = \sum_{\{t\}} Z^{\Sigma_i t_i} \exp\left(-\beta U \sum_{\{t\}} t_i t_{i+1}\right), \tag{5.13}$$

where Z is the internal partition function for a unit length (in which the energy J is included). We now identify $f(\chi)$ as

$$f(\chi) = Z \frac{\partial \ln Z_G}{\partial Z} = \frac{\langle l \rangle}{L} . \tag{5.14}$$

Carrying out the calculation

$$f(\chi) = \frac{1}{2} - \frac{\sinh(\beta H)}{2[e^{-\beta U} + \sinh^2(\beta H)]^{1/2}} , \tag{5.15}$$

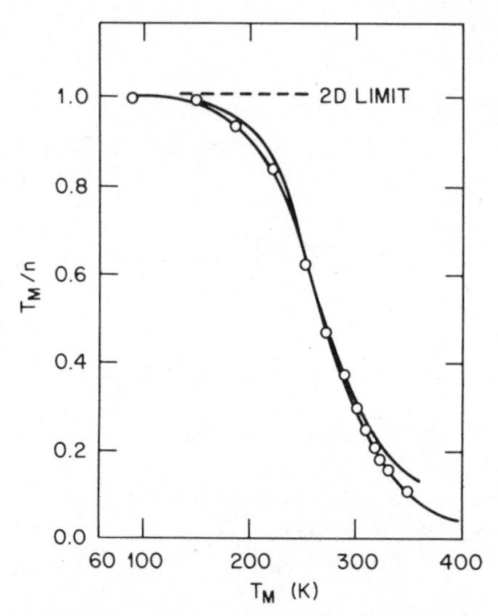

Fig. 12. Comparison of experimental and theoretical curves for a plot of T_M/n vs. T_M normalized to unity for short chains. The theoretical curve is the solid line without circles.

βH may be expanded as in similar problems;[62] thus

$$\beta H = \frac{T - T_0}{T} \tag{5.16}$$

with T_0 an adjustable parameter. Equations (5.15) and (5.16) are fitted to the experimental data of Fig. 12. The best fit is obtained for $U/k_B = 800$ K and $T_0 = 250$ K. The parameter U/k_B is unambiguously the energy for a kink pair. This agrees surprisingly well with the experimentally determined energy for a pair of kinks.[18] The role of screw dislocations has been neglected in the above treatment.

2. *Dislocation Loops*

The above analysis suffers from the disadvantage that the role of screw dislocations associated with kinks has been ignored. Actually, the redistribution of dislocations generate dislocation loops which are usually of three general types as depicted in Fig. 13. The bridging loops start at one surface and end in the other. The open loops begin at one surface and end in the same surface and closed loops appear in the bulk. The loops may be considered as random walks in the slip planes, with steps parallel to the chain axis being of edge dislocation character and steps parallel to the surface of screw character. In calculating the free energy of dis-

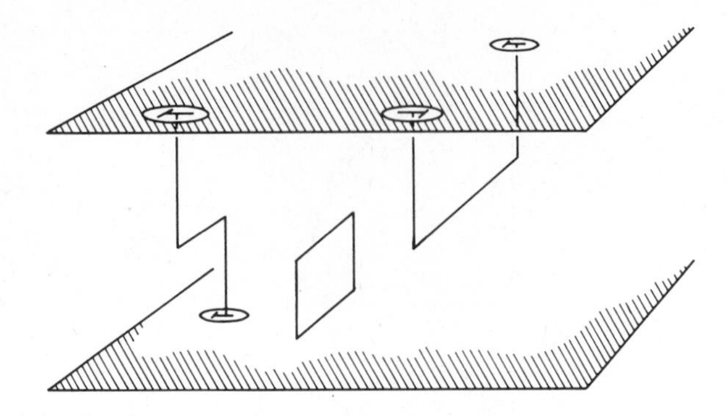

Fig. 13. Various possible dislocation loops in paraffins. The hatched surfaces are the upper and lower lamellar surfaces. The bridging dislocation loop begins at one surface and ends at another. The open loop begins and ends at the same surface. The closed loop begins and ends within the lamella. The vertical lines in the above are edge dislocations and horizontal sections screw dislocations as displayed in Fig. 10. (Source: Refs. 59 and 63.)

locations, the configurations of loops and their energy must be taken into account. Vigren[63] has analyzed this problem. He ignores the hexagonal structure of the lattice and assumes a square lattice parallel to the dislocation slip system. He further ignores the role of closed loops and considers the dislocation loops of bridge (B) and open (O) types.

The square lattice has N_0 column positions corresponding to the positions of chains at the lamellar surface and $K = 1, 2, \ldots, M$ row positions corresponding to M kink positions along the chain. The ith dislocation loop is represented by a random walk of N_i steps. Neglecting the excluded volume effects, the first step may be chosen in W_s ways,

$$W_s = N_0![n_B!n_0!(N_0 - n_B - n_0)!]^{-1}, \tag{5.17}$$

with n_B and n_0 the number of bridging and open loops, respectively. Using the Sterling formula, the free energy of the dislocation loop ensemble in the same approximation as Eq. (5.3) is given by

$$F = \beta^{-1}\left[(1 - \rho_0 - \rho_B)\ln(1 - \rho_0 - \rho_B) + \sum_{i=0,\,B} \rho_i \ln\left(\frac{\rho_i}{Z_i}\right)\right], \tag{5.18}$$

where $\rho_i = n_i/N_0$ and $[Z_i\colon i = 0, B]$ are single-loop partition functions. Fortunately, the Z_i have been calculated in the analogous problem of random walk between two parallel surfaces by DiMarzio and Rubin.[64,65] We define P_h, P_u, and P_d as the horizontal, upward, and downward unit step probabilities. Furthermore, we choose

$$P_h = \xi^{-1} e^{-\beta\varepsilon_h} ,$$
$$P_v \equiv P_d \equiv P_u = (2\xi)^{-1} e^{-\beta\varepsilon_v} , \tag{5.19}$$

where P_v denotes vertical step probability, $\xi = e^{-\beta\varepsilon_h} + e^{-\beta\varepsilon_v}$, and ε_h and ε_v are energies associated with creation of a unit step horizontal or vertical (screw or edge) dislocation.

Now the partition functions Z_0, Z_B are given as[64]

$$Z_0 = e^{-\beta\varepsilon_v^0} \sinh[(M-2)\phi_0]\{\sinh[(M-1)\phi_0]\}^{-1} , \tag{5.20}$$

$$Z_B = e^{-\beta\varepsilon_v^\beta} \sinh \phi_B \{\sinh[(M-1)\phi_B]\}^{-1} , \tag{5.21}$$

and

$$\cosh \phi_i = \tfrac{1}{2} e^{\beta\varepsilon_v^i}(1 - 2e^{-\beta\varepsilon_h^i}) . \tag{5.22}$$

The step energies in the open and bridging loops are necessarily different because of different geometries. One may use the approximate step energies given by Nabarro[66] as

$$\varepsilon^B = \varepsilon_c - A_B \ln(\rho_B + \rho_O) \tag{5.23}$$

and

$$\varepsilon^O = \varepsilon_c - A_0 \ln(\rho_B + \rho_O) . \tag{5.24}$$

Furthermore assuming

$$\rho_B = e^{-\gamma M} , \quad \rho_O = \rho(1 - e^{\gamma M}) , \quad \text{and} \quad \rho = \rho_O + \rho_B \tag{5.25}$$

with γ an adjustable parameter indicating the transition from bridging to open loop dominated melting, one obtains the simplified free energy expression

$$F = A\rho + B\rho \ln \rho + k_B T(1-\rho) \ln(1-\rho) \tag{5.26}$$

with A and B functions of M and T. Minimizing the free energy with respect to ρ one obtains the melting criterion:

$$k_B T_M = 2A_O + [(M-1)A_B - 2A_O] e^{-\gamma M} . \tag{5.27}$$

Vigren has chosen the parameters $A_O/k_B = 194$ K, $A_B/k_B = 10$ K, and $\gamma = 0.066$ and has reproduced the melting curve with these values. It must be pointed out that although this theory considers the dislocation configuration, it suffers from too many adjustable parameters.

3. *Chain Folding*

The above formulations are approximate in the sense that interactions between pairs have been neglected. In general, a pair of dislocations at a distance r interact with other pairs in the intervening or the surrounding area. Consideration of these interactions leads to a reduction of the energy constant. Thus the transition temperature obtained from the above approximate theory may be regarded as an upper limit and is usually larger than the experimentally observed transition temperature as in the case of the liquid crystal[53] mentioned above. Improvement of the theory has been attempted by Kosterlitz and Thouless[1] and by Holz et al.[67] Kosterlitz and Thouless consider the dislocation pairs located in the intervening distance between the test pair which screen the elastic interaction. Their net effect may be summarized by introducing a dielastic constant $\varepsilon(r)$. Thus the screened interaction leads to an effective potential U_{eff}:

$$U_{\text{eff}} = \frac{b^2 K^e \ln(r/r_0)}{4\pi\varepsilon(r)} . \tag{5.28}$$

The consequences of this potential have been investigated and it leads in the mean field approximation to a modified transition temperature given by

$$\frac{b^2 K^e}{4\pi k_B T_M} = 2\left[1 + 1.3\pi \exp\left(\frac{\mu_c}{k_B T_M}\right)\right], \tag{5.29}$$

where K^e is the elastic constant and μ_c is the chemical potential for thermally produced dislocations. Holz and Medeiros[67] using arguments borrowed from electrostatics, have explicitly calculated the dielastic constant as

$$\varepsilon = 1 - \frac{K^e(C_1 + C_2)(1 + \frac{1}{4}K^e C_1)}{1 + \frac{1}{2}K^e(C_1 + C_2)}, \tag{5.30}$$

$$K^e = \frac{\mu}{1-\nu}, \quad C_1 = -2\beta b^2 n_d \langle r^2 \rangle_0 ,$$

and

$$C_2 = \beta b^2 n_d \langle r^2 \cos(2\theta) \rangle ,$$

where μ is the shear modulus, ν the Poisson ratio, n_d the density of dislocations per unit area, $\langle r^2 \rangle_0$ the mean separation of a pair, and θ the angle between the Burgers vector and vector **r**.

In the following we investigate the melting of paraffins,[68] taking into account the interaction between dislocation pairs which had been neglected in previously described treatments. The analysis indicates that folding phenomena may be accounted for in this formulation. We begin by writing the expressions for the effective interaction energy between a pair of dislocation loops in paraffin:

$$U^{\parallel}_{\text{eff}}(r) = \frac{b_{\parallel}^2 K_{\parallel}^r L}{4\pi} \left[\ln\left(\frac{r}{r_0}\right) - \frac{1}{2}\cos(2\theta) \right] + 2\mu_{\parallel}^c L + 2F_{\parallel}^k(T, L) \tag{5.31}$$

and

$$U^{\perp}_{\text{eff}}(r) = \frac{b_{\perp}^2 K_{\perp}^r L}{4\pi} \ln\left(\frac{r}{r_0}\right) + 2\mu_{\perp}^2 L + 2F_{\perp}^k(T, L) , \tag{5.32}$$

where r_0 is the dislocation cutoff and μ_σ^c represents the core energy of a dislocation of type σ ($\sigma \equiv \parallel$ or $\perp$). The first expression represents the interaction energy between a pair of edge dislocations a distance r apart. $K_{\parallel}^r$ is the renormalized elastic coupling constant, θ is the angle between the Burgers vector **b** and **r**. $F_{\parallel}^k(T, L)$ is the free energy of the kinked dislocation core. Similar definitions apply to Eq. (5.32) for screw dislocations. The melting criterion may be stated in terms of the renormalized coupling constant:

$$k_B T_M^\sigma = \frac{b_\sigma^2 K_\sigma^r L}{16\pi} . \tag{5.33}$$

Using Eq. (5.30), the renormalized coupling constant K_σ^r may be written

$$K_\sigma^r = \frac{K_\sigma^e}{1 + [(b^2 K_\sigma^e \alpha_\sigma L / 4\pi k_B T) n_\sigma \langle r_\sigma^2 \rangle_0]} , \tag{5.34}$$

where K^e is the unrenormalized elastic constant (in the absence of screening) and α_σ is a numerical factor of order unity.

By defining the reduced parameter

$$\xi_\sigma = \left(\frac{16\pi}{b_\sigma^2 K_\sigma^e}\right) k_B T_M^\sigma \quad \text{and} \quad Z_\sigma = \frac{\xi_\sigma}{L}, \tag{5.35}$$

Eq. (5.33) becomes

$$Z_\sigma = \left(1 + \frac{4\alpha_\sigma}{Z_\sigma} n_\sigma \langle r_\sigma^2 \rangle_0\right)^{-1}. \tag{5.36}$$

The product $n_\sigma \langle r_\sigma^2 \rangle_0$ may be calculated in the noninteracting approximation[1]

$$n_\sigma \langle r_\sigma^2 \rangle_0 = \frac{\pi}{2} \frac{Z_\sigma}{1 - Z_\sigma} \exp\left\{-\frac{2[\gamma_\sigma^c + f_\sigma(\xi_\sigma, L)]}{Z_\sigma}\right\}, \tag{5.37}$$

where γ_σ^c is the reduced core energy per unit length given by

$$\gamma_\sigma^c = \frac{16\pi}{b_\sigma^2} \frac{\mu_\sigma^c}{K_\sigma^e} \tag{5.38}$$

and $f_\sigma(\xi, L)$ is the reduced configurational free energy of loop per unit length given by

$$f_\sigma(\xi_\sigma, L) = \frac{\gamma_\sigma}{\mu_\sigma^c L} F_\sigma^k(T, L). \tag{5.39}$$

Combining Eqs. (5.33)–(5.39) one obtains the reduced melting criterion:

$$[f_\sigma(\xi_\sigma, L) + \gamma_\sigma] = -\frac{Z_\sigma}{2} \ln \frac{(1 - Z_\sigma)^2}{\alpha_\sigma' Z_\sigma} \equiv \varepsilon(Z_\sigma), \tag{5.39a}$$

$$\alpha_\sigma' = (\pi/2)\alpha_\sigma. \tag{5.39b}$$

For short chains of length L_0 or less, where kinks cannot occur, the chains may be regarded as stiff. In this case only straight bridging edge dislocations are possible and f vanishes. If we neglect the screening ($\alpha_\sigma' = 0$), it follows that $Z_\sigma = 1$, and we recover the linear dependence of the melting point on chain length discussed earlier. For $\alpha_\sigma' > 0$ and $L < L_0$ (screened rigid chain limit), Eq. (5.39) gives two solutions—one with z greater than 1 and one with Z less than 1. Since the screening must reduce the melting temperature, only the solution $Z < 1$ is physically admissible. This still gives a linear dependence of the melting point with chain length. For $L > L_0$ kinks are possible and the term $f_\sigma(\xi_\sigma, L) < 0$. For $L \gtrsim L_0$ the contribution of f is small and it may be assumed $|f_\sigma(\xi_\sigma, L)| \ll \gamma_\sigma$. In this limit Eq. (5.39a) takes the simplified form

$$\frac{Z_\sigma}{Z_\sigma^0} \cong 1 + 2\left(\frac{1 - Z_\sigma^0}{Z_\sigma^0}\right) f_\sigma(\xi_\sigma^0, L), \tag{5.40}$$

where Z_σ^0 and ξ_σ^0 are the values of these parameters for $L = L_0$. Since f_σ decreases with increasing L (more kink possibilities become available), Eq. (5.40) gives the correct length dependence of the melting point qualitatively.

We now consider the behavior of the function $\varepsilon(Z_\sigma)$. The qualitative behavior is shown in Fig. 14. This function is negative in the interval $[0, Z_\sigma^*]$ and positive in the interval $[Z_\sigma^*, 1]$ with a minimum in the interval $[0, Z_\sigma^*]$, where Z_σ^* is given by

$$Z_\sigma^* = \left(1 + \frac{\alpha'_\sigma}{2}\right)\left\{1 - \left[1 - \left(1 + \frac{\alpha'_\sigma}{2}\right)^{-2}\right]^{1/2}\right\}. \tag{5.41}$$

The behavior of ξ as a function of L may be qualitatively analyzed by considering the total derivative of Eq. (5.39):

$$\begin{aligned} \frac{\partial f_\sigma}{\partial \xi_\sigma}\xi'_\sigma + \frac{\partial f_\sigma}{\partial L} &= \frac{1}{2}(\xi'_\sigma L^{-1} - \xi_\sigma L^{-2})g(Z_\sigma), \\ g(Z_\sigma) &\equiv -\ln\left[\frac{(1 - Z_\sigma)^2}{\alpha'_\sigma Z_\sigma}\right] + \frac{1 + Z_\sigma}{1 - Z_\sigma}, \end{aligned} \tag{5.42}$$

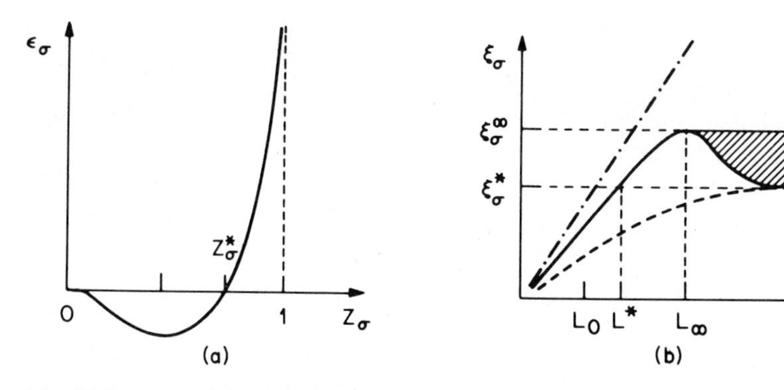

Fig. 14. (a) Function $\varepsilon(Z)$. (b) Reduced temperature ξ vs. chain length (lamellar length, L). L_0 is the rigidity limit. L_∞ is the folding length.

where $\xi'_\sigma = \partial \xi'_\sigma / \partial L$. Since $\partial f_\sigma / \partial \xi_\sigma < 0$ because of the entropic nature of f_σ, it may be shown that ξ'_σ changes sign at some $L \approx L_\infty$.

One may expand Eq. (5.39) about

$$\xi_\sigma^\infty = L_\infty Z_\sigma^\infty \,. \tag{5.43}$$

Thus one obtains in the neighborhood of the maximum (Fig. 14b)

$$\xi_\sigma - \xi_\sigma^\infty = -\frac{1}{2}\left|\frac{\partial f_\sigma}{\partial \xi_\sigma}\right|_{\xi_\sigma^\infty}\left[\frac{Z_\sigma^\infty(2 - Z_\sigma^\infty)}{(1 - Z_\sigma^\infty)^2}\right]\left[\frac{L - L_\infty}{L_\infty}\right]^2 . \tag{5.44}$$

The existence of maximum is interpreted as the onset of folding which is observed in long chain polymers. The data for the melting point of paraffins do not go far enough to show the folding phenomenon although folding is usually observed in polyethylene. Better data on the length at which folding occurs exist for polyethylene oxide.[69] One needs an explicit function $f_\sigma(\xi_\sigma, L)$ in order to make a comparison with experiment. This has not yet been carried out.

B. Theory of Rotator Transition

There have been numerous attempts to formulate a theory of rotator transition in paraffins. A review of the work up to 1968 is given by McClure.[70] The model proposed by Hoffman[71] is illustrative of most of the theories. In this model the rotational potential is assumed to have a deep well and a degenerate number Ω of shallow wells with energy difference V. Moreover, the neighboring rotators are assumed to be interacting. If the population of chains in the ground state and rotator state are assumed to be n_2 and n_1, respectively, one may write

$$\frac{n_2}{n_1} = \Omega \exp\left(\frac{-V}{k_B T}\right) . \tag{5.45}$$

The nearest-neighbor interactions is treated in the mean field approximation by defining

$$x = \frac{n_1}{n_1 + n_2} \quad \text{and} \quad V = V_0 x \,, \tag{5.46}$$

where V_0 is a constant. Thus one obtains

$$x = \left[1 + \Omega \exp\left(-\frac{x V_0}{k_B T}\right)\right]^{-1} \tag{5.47}$$

and for the free energy

$$F = \frac{V_0}{2}(1 - x^2) - k_B T[(1-x)\ln\Omega - x\ln x - (1-x)\ln(1-x)]\,. \tag{5.48}$$

Minimization of the free energy gives for the rotator transition temperature T_R

$$T_R = \frac{V_0}{2k_B \ln\Omega}\,. \tag{5.49}$$

Hoffman, using a value of $\Omega \cong 12$ and $V_0/k_B \cong 300$ K, calculated the transition temperatures and dielectric behavior for a number of paraffins and paraffin bromides. More recent work by Kobayashi[72] considers a chain potential

$$U(\theta, \ell) \cong \exp\left[-\frac{V(\theta, \ell)}{k_B T}\right] \tag{5.50}$$

in which chain translation parameter ℓ is also included. Kobayashi has calculated the free energy of the system based on the above potential using experimental potential parameters and lattice constants. Furthermore, he has included the vibrational free energy and has nearly reproduced the transition points and their length dependence in fair agreement with experimental data. A different study by Ishinabe[73] has considered the chain as having translational motion parallel to the long axis plus two orientational displacements of the chain axis for the rotational motion about the long axis.

In this work the rotational transition temperature T_R is found as a function of the molecular weight M:

$$T_R = T_{R,\infty}\left(1 - \frac{a}{M} + \frac{b}{M^2}\right) \tag{5.51}$$

where a, b, and $T_{R,\infty}$ are parameters. Satisfactory fit to experimental data is found by choosing $a = 6.5$, $b = 19.4$, and $T_{R,\infty} = 415$ K. Experimentally, the odd paraffins show rotational transitions starting at C_9H_{20} up to $C_{45}H_{92}$. Below C_9 and above C_{45}, the rotational transition temperature presumably lies above the melting point and is not observed. The latter theory attempts to explain this phenomenon.

The behavior of the rotational transition points for odd paraffins as a function of chain length is surprisingly similar to that of melting point.

The initial portion of T_R versus L rises linearly with L, then flattens out, but saturation is not observed above C_{45} because the rotator transition curve crosses the melting curve. One may conjecture that the mechanism for rotator transition may be explained in similar fashion as the melting transition. This is equivalent to assuming that the system is a paired vortex–antivortex lattice which goes into a state of vortex plasma at the transition point.

Unfortunately, the coupling constant for vortex–antivortex pairs is not available in terms of the elastic constants. The system of the classical and quantum two-dimensional rotators has been treated by Brezinski[74] and by Holz.[75] Also thin films of superfluid helium and superconductors presumably behave as two-dimensional rotator systems whose transition temperatures have been estimated using the Kostelitz–Thouless theory.[1] The

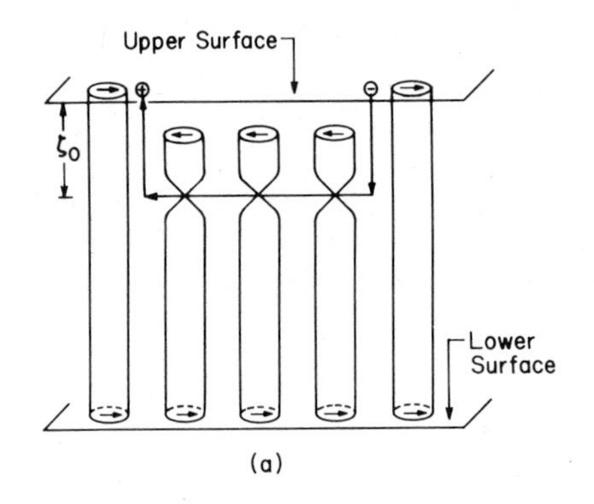

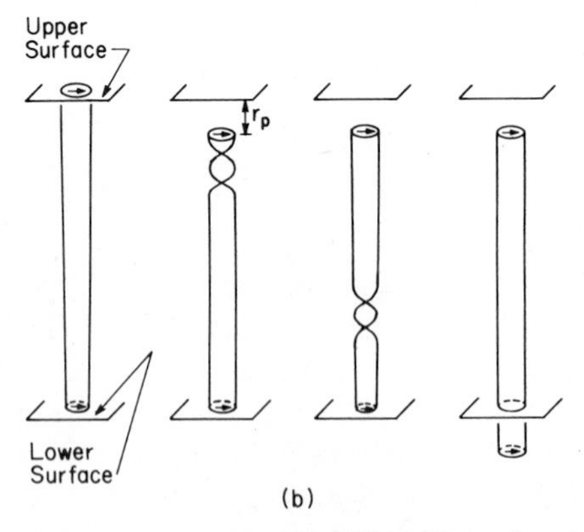

Fig. 15. (a) Vortex–antivortex pair in paraffins. (b) Drilling of vortex along the chain and consequent roughening.

second rotator transition observed by Doucet and co-workers has never been explained theoretically.

A further peculiarity of the rotator transition is that it is accompanied by large surface roughening. This aspect of the transition has been analyzed earlier. It was shown that in order for long chains to perform hindered rotation about their long axis a drilling mechanism is necessary in view of the large energy barrier for the rotation of the entire rod. The hindered rotation process thus begins at one surface with creation of a kink and intrusion of the chain end into the surface. The rotation process can then travel along the chain, and upon the completion of this drilling process a portion of the chain will extrude at the other surface. This leads to the roughening observed experimentally (Fig. 15).

C. Theory of Other Transitions

1. *Surface Melting Transition*

This transition observed by Snyder et al.[15] has been shown by spectroscopic methods to involve the rotation of the end group such that the all-trans chain at the low-temperature region goes in to the *gtt*. . . chain at the high-temperature side. It must be associated with the melting of the interlamellar surface and is strictly molecular weight independent as expected. This transition is denoted by δ (see Fig. 7).

2. *V→IV Transition*

This transition (see Fig. 7) is observed for paraffins C_{23} and higher. Assuming a two-dimensional order–disorder transition as postulated by Strobl (Fig. 8), its molecular weight dependence may be calculated as follows. The interaction between a pair of all-trans paraffin chains may be estimated from Eq. (2.10). The notable feature of Eq. (2.10) is that the attractive energy is proportional to chain length, L, except for end effects. Now if the order–disorder mechanism depicted in Fig. 8 is assumed for this transition, the transition temperature is given by the two-dimensional Ising theory which is

$$T_\gamma \approx \frac{\Delta U}{k_B}, \tag{5.52}$$

where ΔU is the energy difference between two chains with one of them executing a 180° rotation (Table IV). The absolute value of the energy for each configuration and the energy difference depend on the potential function assumed. However, regardless of which potential function one assumes, the linearity of ΔU with chain length will remain as is evident from Selem's calculation. Thus, using ΔU proportional to L, one obtains

$$T_\gamma \approx \frac{L}{D^5}\left(3\tan^{-1}\rho + \frac{\rho}{1+\rho^2}\right). \tag{5.53}$$

This is in contrast to the experimental findings of Fig. 7 (γ curve). The experimental curve shows a slight increase of the transition temperature for short chains which may be ascribed to end effects. The transition temperature for longer chains rapidly flattens out and becomes independent of chain length for $C_{45}H_{92}$. One may argue that the use of the Ising theory to calculate this transition temperature is erroneous because of the slight volume change associated with this transition. Equation (5.52) of the Ising theory is valid for the incompressible lattice. Attempts have been made to modify the Ising theory for compressible systems. Larkin and Pikin[79] have treated the compressible Ising system and have explicitly calculated the contribution of finite compressibility and acoustic phonons to the heat capacity and transition temperature. They give a simple formula for the change of transition temperature, namely,

$$T_c = T_c^0 + cP, \tag{5.54}$$

where T_c^0 is transition temperature of the incompressible system, T_c that of the compressible system, P the pressure, and c a parameter related to volume change. The precise nature of c is not of interest here, but Eq. (5.54) shows that consideration of volume change will not alter the linearity of the transition temperature with chain length derived in Eq. (5.53). One may now wonder if the two different mechanisms proposed for this transition, namely, the flip–flop 180° chain rotation and the translational chain shift, produce essentially the same effect and result in the same energy difference ΔU proportional to the chain length L. Thus both mechanisms give a linear dependence of the transition temperature with chain length—a result in disagreement with experimental findings.

3. *IV→II Transition*

This transition (designated by β in Fig. 7) is associated with an 18° tilt of the chains with respect to the axis perpendicular to the lamellar plane (for $C_{33}H_{68}$ Fig. 9). As a result, a C_2H_4 bond at one end becomes free to rotate and the interlamellar surface roughens. X-ray data on the electron density in the interlamellar surface are available for $C_{33}H_{68}$ and on this basis Strobl[7] has computed the energy and entropy of transition. The transition therefore involves an increase of Salem energy which is compensated by the disordering of the end C_2H_4 groups due to their rotational freedom. A rough estimate of the Salem energy difference as a function of the tilt angle $\Delta U(\theta)$ is easy to obtain:[78]

$$\Delta U(\theta) = \frac{B \tan \theta}{D_B^5} \left(3 \tan^{-1} \rho + \frac{\rho}{1+\rho^2}\right), \tag{5.55}$$

where B is a coupling constant. This is independent of L except for end effects represented by the terms in parentheses. A quantitative estimate of the transition entropy for the disordering of the ends has been made for the case of $C_{33}H_{68}$. We assume that the tilt angle is independent of the molecular weight, thus the transition entropy will also be assumed independent of chain length. On this basis, the molecular weight dependence of the transition temperature is fully included in the end effect term in the parentheses. Expanding the parentheses terms in inverse powers of ρ (for large ρ), one obtains by simple algebraic manipulation

$$T_\beta = T_\beta^\infty \left(1 - \frac{4}{3\pi\rho}\right), \tag{5.56}$$

where T_β^∞ is the value of T_β for $\rho \to \infty$. Using the value of $T_\beta = 338.5$ K for $C_{33}H_{68}$, one obtains $T_\beta^\infty = 356.3$ K, and the values of T_β for various homologues calculated from Eq. (5.56) are in satisfactory agreement with experiment.

D. Classification of Transitions

Based on defect structures that may occur in paraffins, one may conceive of the following transitions:

1. Intralamellar melting transition.
2. Rotator transition.
3. Surface melting transition.
4. Lamellar surface roughening transition.
5. Lamellar smectic *A* transition.
6. Smectic *A*–nematic transition.
7. Nematic–isotropic transition.

The intralamellar melting transition and the rotator transition were discussed in Sections V.A and V.B. Among the transitions discussed in Section V.C, which are experimentally observed, the surface melting transition may be identified as the δ co-existence curve of Fig. 7. Other transitions cannot be unambiguously assigned to any of the above classes. Holz[80] has analyzed this problem. He begins by writing an expression for the elastic energy of dislocation loops as follows:

$$U(r, L, \{s\}) = U^{\perp}_{\text{eff}}(r, L, \{s\}) + U_{vh}(r, L, \{s\}) + V_e(r, L, \{s\}), \quad (5.57)$$

where

$$U^{\perp}_{\text{eff}}(r, L, \{s\}) = \frac{b_{\perp}^2 K_{\perp}^b(T, r)L}{4\pi} \ln\left(\frac{r}{r_\rho^b}\right) + 2\mu_{\perp}^b L + 2F_{\perp}^b(T, L) + 2\mu_{\perp}^s + 2G_{\perp}^s(T, L, \{s\}). \quad (5.58)$$

In Eq. (5.57) the first term represents the effective interaction energy of the loop which is located in the lamella considered. The second and third terms represent the interaction energy of the vertical and horizontal dislocation loops and the elastic energy located in the neighboring lamellas, respectively. The first term of Eq. (5.58) represents the elastic interaction energy between the vertical segments of the dislocation loop, the second term the core energy of these segments, and the third term the free energy of the kinked dislocation core running in the bulk of the layer as in Eq. (5.32). The fourth and fifth term of Eq. (5.58) are additional surface terms. The fourth term represents the core energy of the two surface segments of the dislocation loop of length s and the fifth term their elastic interaction energy which will depend on their shape described by $\{s\}$. All possible shapes $\{s\}$ for given r, therefore represent the additional degrees of freedom of the vertical loop with respect to the horizontal loop. Based on this general energy expression, Holz concludes the following:

1. Surface roughening transition is possible for a multilamellar system, the transition temperature given approximately by[81]

$$k_B T_r(L) = \frac{2\mu_{\perp}^s(L)}{\ln z} \quad (5.59)$$

with $\mu_{\perp}^s(L)$ being the core energy of the surface dislocation loops of length s, and $z \approx 3$.

2. The smectic A phase may occur as a state located between the intralayer molten state and a nematic or isotropic state. It can only be stable for chain lengths belonging to the straight part of the melting curve $b_{\perp} < L_0$ where kinks are prohibited on geometrical or energy grounds (rigid rods). As soon as a high density of kinks is present, dislocation loops with $b_{\perp}$ and $b_{\parallel}$ get strongly intercoupled and the smectic A type may only survive as a short-time memory effect. This may be an

explanation of the effect observed in polyethylene where the melt crystallizes into the original solid morphology when kept liquid for only a short time.[82]

3. The smectic–nematic transition is also conceivable for short (rigid) paraffins because the occurrence of kinks introduces complicated geometrical factors which make it difficult to define a director. The same problem arises in discussion of nematic–isotropic transitions. Discussion of the latter transitions leads into the problem of polymer liquid crystals which have been amply treated in recent years.[83] The liquid–liquid transition observed by Krüger and disputed by a number of other experimentalists cannot be a nematic–isotropic transition. Denny and Boyer[84] have conducted high sensitivity DSC measurements on a number of paraffins as well as other polymers and consistently find a liquid–liquid transition. The overwhelming evidence for the existence of this transition has been reviewed by Boyer.[85] The precise molecular mechanism for this transition is not clear at this point. It does not appear from the experimental evidence[51] that this transition is nematic–isotropic.

VI. CYCLOPARAFFINS AND PHOSPHOLIPIDS

A variety of systems exhibit properties similar to paraffins. Among these systems, cycloparaffins and phospholipids are important because they are structurally similar to paraffins yet are significantly different.

A. Cycloparaffins

Cycloparaffins have attracted attention for some time because they are the simplest folded structures, and it is thought that their study might shed light on the folding of chains in polymer crystals. Newman and Kay[86,87] have carried out an X-ray structure analysis of $(CH_2)_{33}$ and $(CH_2)_{34}$. Similar X-ray investigations of the structure for $(CH_2)_n$ ($n = 36$, 48, 60, 72) have been performed by Trzebiatowski[88] and Groth.[89] From the X-ray investigations, one concludes that, except for short homologues, the crystal structure of cycloparaffins are lamellar and similar to normal molecules with a uniform stem orientation and the surface is occupied by chain folds with conformation *tggtggt.* Stems have slightly twisted all-trans conformation. The deformation energy of cycloalkane rings have been calculated by Dale.[90]

Thermal and spectroscopic investigations have been reported on cycloparaffins. Höcker and Riebel[91] have measured the melting point and enthalpies of the fusion of cycloparaffins. The enthalpy and entropy of fusion behave similar to normal paraffins and are linear with chain length. The corresponding empirical melting equation for cycloparaffins is given by

$$T_M = \frac{\Delta H}{\Delta S} = T^* \frac{n + a}{n + b} \tag{6.1}$$

with $T^* = 414.3$, $a = -11.9$, and $b = -7.38$ as in normal paraffins. Höcker and Riebel maintain, however, that Eq. (6.1) poorly fits the experimental data for low molecular weight homologues. Grossmann[92] has made careful DSC measurements of cycloparaffins and his data together with those of others[93] are shown in Fig. 16. Analysis of this plot reveals a number of peculiarities as follows:

1. A plot of inverse melting point versus inverse chain length gives an approximate straight line for normal paraffins.
2. For cycloparaffins three different classes of molecules are distinguishable with the high molecular weight group behaving as normal

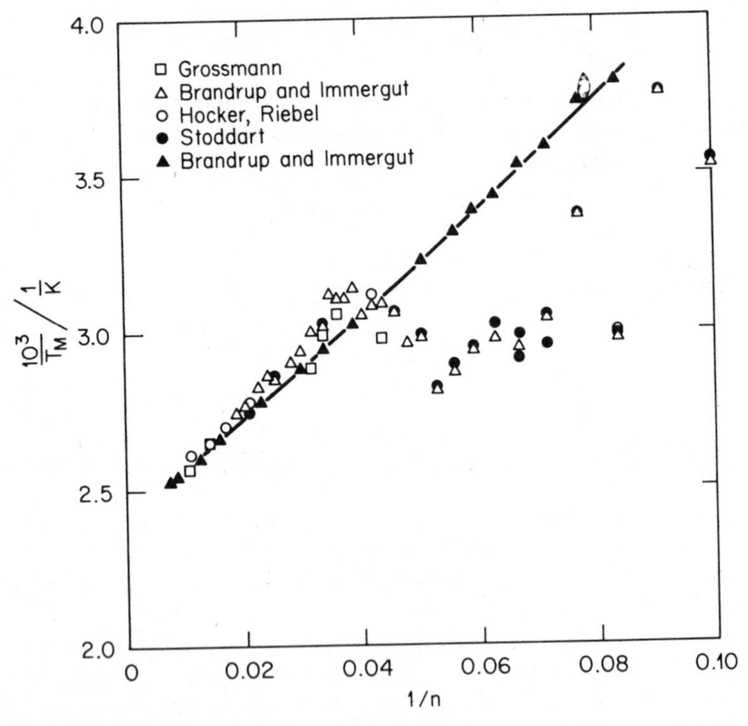

Fig. 16. Melting point of normal paraffins (straight line) in comparison to that of cycloparaffins.

paraffins ($n > 40$) and an intermediate molecular weight group ($18 < n < 24$) deviating from linear behavior but appearing to show behavior characteristic of their own. The low molecular weight group ($n < 18$) shows chaotic behavior of melting points due to geometrical constraints for collapsed ring formation which is a prerequisite for lamellar structure. The absolute value of the melting point for n-$C_{96}H_{194}$ appears to be smaller than the melting point of its cyclic counterpart $C_{96}H_{192}$ according to Grossmann.[92] This is interpreted to be due to higher rigidity of the cyclic paraffin which is caused as a result of folding. However, in terms of dislocation driven two-dimensional melting theory it is not clear how one would reconcile this difference. In fact, the study of the defect mechanism of the melting transition in cycloparaffins will reveal much information towards the quasi-two-dimensional melting theory. No attempt in this direction has been taken according to information available to the author. Further peculiarities of the cyclic paraffins appear in the study of premelting transition. For $(CH_2)_{22}$ a premelting transition is found at 298 K about 25 degrees below the melting transition.[92] A similar transition is observed for $(CH_2)_{24}$. The entropy of transition estimated from DSC data is four times higher for the premelting transition than for the melting transition itself. However, the situation is different for higher molecular weight homologues such as $(CH_2)_{96}$ where the premelting transition has a small entropy associated with it compared to melting entropy.[94] The molecular mechanism accompanying the premelting transition appears to be significantly different for shorter and longer chains. This difference comes out in the investigation of IR and Raman spectra.[95] As in normal paraffins one is able to estimate the trans–gauche ratio as well as some particular conformations as kinks and double gauche sequences. Investigation of IR spectra in the range 1200–1300 cm^{-1} shows a rapid increase of gauche conformations in passing through the premelting transition point for short chains whereas the trans–gauche ratio does not change appreciably for the melting transition in these short molecules. It appears that for short chains $18 < n < 24$, the so-called premelting transition may be identified as a lamellar smectic transition whereas the higher-temperature transition may be a smectic–nematic or smectic–isotropic transition. The possibility of the existence of the smectic phase in cycloparaffins is greater because of added rigidity of molecules brought about by the collapsed ring structure. For long chains, the system behaves essentially as for normal paraffins with the premelting transition having a small entropy change compared to the melting entropy. Furthermore, the crystal structure goes from monoclinic to hexagonal with the appearance of kinks.[96,97] Rotational symmetry around the long axis prevails in the hexagonal phase indicating a rotator phase. The mechanism of achieving this symmetry can only be the stepwise rotation (as in Fig. 15b) because ring constraint would prohibit other mechanisms. NMR studies confirm these views.[98]

Another related homologous system, the *n*-perfluoroalkanes,[94,99] have been studied, but information on the phase changes are still meager for these systems.

B. Phospholipids

Another quasi-two-dimensional system of great experimental and theoretical interest is that of biological membranes. As stated in the introduction, this system resembles in many ways the paraffin system because it is composed of paraffinic chains attached at one end to a polar group (Fig. 17). A member of this class, namely, DPPC (dipalmytol phosphatidylcholin), has been extensively studied both as monolayers at a water surface[100] and as vesicles (membrane).[101] We sketch here the main experimental and theoretical information relating to phase transition in these systems for the purpose of indicating similarities and differences with the paraffin systems. Recent reviews on the subject are available.[101,102] At low surface density the monolayer forms a two-dimensional gas (with surface pressure Π inversely proportional to surface area). At

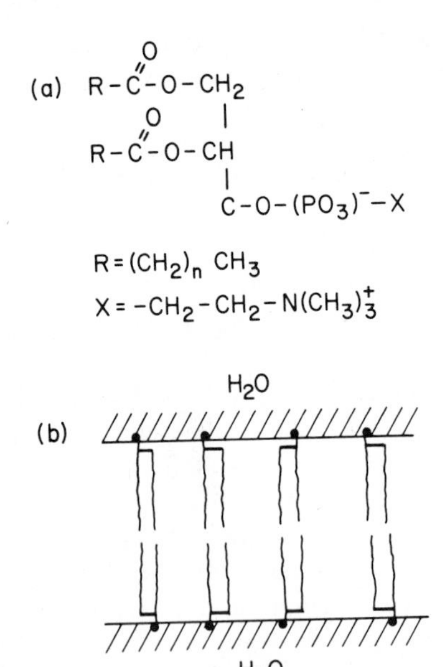

Fig. 17. (a) Chemical composition of phospholipid molecule. (b) Bilayer membrane with charged group in water. Half the bilayer would represent a monolayer at water surface.

higher surface densities, there is evidence of a phase transition to a two-dimensional liquid. Investigations have focused on the possibility of liquid crystalline phases in this system. Albrecht et al.[100] have found the following:

1. The gas–liquid co-existence curve terminates at a critical point.
2. At higher surface densities a nematic liquid crystal phase appears.
3. The system goes into a smectic phase at higher surface densities, the transition being of first order. With increasing temperature a tricritical point is encountered where the heat of transition goes to zero and the order of transition goes to second order. This phase is also encountered in bilayer membranes.
4. At still higher surface densities the system goes into a tilted liquid crystalline phase.
5. The low-temperature high density phase is one of nontilted close-packed crystalline structure.

The bilayers are formed at high concentrations of the lipid molecule in a water–salt solution in the form of vesicles. They are model systems for biological membranes. Lipid bilayers exists essentially in two different states, a disordered fluid phase at high temperatures and an ordered phase at low temperatures. The ordered state has positional order of the end groups as well as hydrocarbon chains which form a two-dimensional densely packed lattice. Furthermore, chains in this state are in the all-trans state, parallel to one another as in paraffins. In the fluid state, the positional order is destroyed and the chain conformational order is also destroyed through the appearance of gauche bonds. The two disordering events apparently occur simultaneously at the transition temperature T_M. It is assumed that, at least for membranes consisting of long hydrocarbon chains, the main contribution to transition entropy derives from chain disordering, and most of the theoretical work addresses this aspect of melting. However, in some membranes a weak "pretransition" is observed, the origin of which still remains controversial. The major theoretical contribution in this area was formulated by Marcelja.[103] In his model, the potential energy of the system is given by

$$E_{\text{total}} = E_{\text{intra}} + E_{\text{disp}} + \Pi A \,, \tag{6.2}$$

where E_{intra} is the contribution of single-chain conformational potential (gauche–trans states), E_{disp} the contribution of the London potential between neighboring chains, and ΠA represents the net contribution of other forces such as the repulsive forces and the forces due to the salt–water interaction with end groups. The latter term is cast into a product of lateral pressure Π and area per head A. The dispersive energy is

$$E_{\text{disp}} = \phi \frac{n_{\text{trans}}}{n} \sum_i \left(\tfrac{3}{2}\cos^2 \theta_i - \tfrac{1}{2}\right) \tag{6.3}$$

and the parameter ϕ is defined as

$$\phi = V_0 \sum_i \frac{n_{\text{trans}}}{n} \left\langle \tfrac{3}{2}\cos^2 \theta_i - \tfrac{1}{2} \right\rangle , \tag{6.4}$$

V_0 being equal to 680 cal/mole derived from the freezing energy of polyethylene. The statistical problem is formulated using the single-chain partition function

$$Z = \sum_{\substack{\text{all} \\ \text{conformations}}} \exp\left[\frac{-E_{\text{tot}}(\phi, \Pi)}{k_B T}\right], \tag{6.5}$$

where ϕ is given by the self-consistent equation

$$\phi = \frac{1}{Z} \sum \left\{ \left[\frac{n_{\text{trans}}}{n} \sum \left(\tfrac{3}{2}\cos \theta_i^2 - \tfrac{1}{2}\right)\right] \exp\left[\frac{-E_{\text{tot}}(\phi, \Pi)}{k_B T}\right] \right\} . \tag{6.6}$$

The system was calculated by generating all conformations of a chain of N groups in a computer. Marcelja further assumed that the system has constant density (density changes slightly at the transition point) and used the relation

$$AL = A_0 L_0 , \tag{6.7}$$

where L is the thickness of the bilayer. From the above relationships isotherms of the system were constructed which agree well with experimental information. Furthermore, the variation of order along the chain as well the average order in the two aforementioned states were calculated in fair agreement with experiment. A simpler modification of this model was formulated by Jahnig[104] and Massih and Naghizadeh.[105] Other theoretical attempts have been made by Doniach[106] who used a two-state Ising model, by Pink et al.[107] who used a multi-state Ising model, and Nagle[108] who used a lattice gas model.

In principle, the theoretical methods described above are applicable to monolayers with the lateral pressure Π and the area per head A determined externally. These variables are fixed in the bilayer system by the salt–head interaction mainly. Jahnig[104] has observed that in certain membranes with large surface charge density a further tilted structure exists known as the tilt transition. The various phases observed in monolayers and bilayers are reminiscent of the phases in paraffins, however, a detailed comparison has never been attempted.

An attempt to treat the membrane and monolayer problems in terms of dislocation melting theory has recently been made by Holz et al.[109] These authors consider the melting of the head groups only and formulate the two-dimensional dislocation driven melting theory for dimers which constitute the head groups of phospholipids. The dimer network imposes constraints on the motion of dislocations, thus dissociation of a dislocation pair in such a network cannot proceed by a single slip plane. It must traverse a random walk course due to constraints imposed on slip at certain configurations of dimer. The melting point T_M^{**} for such a dimer solid is calculated to be

$$T_M^{**} = \frac{2\gamma\mu b^2}{\ln(Z_0 Z^*)},$$

where γ is the core energy parameter $0.01 < \gamma < 0.1$, μ the shear constant, b the Burgers vector, Z_0 the coordination number of random walk, and Z^* a configurational degeneracy. This theory ignores the role of paraffin chains which are thought to be paramount in accounting for the entropy of transition.

APPENDIX A. DISLOCATIONS

The concept of linear lattice imperfections called dislocations arose primarily from the study of plastic deformations in crystals. The two important types of dislocations are edge and screw dislocations. Figure 18a illustrates a model that approximates a simple cubic crystal. Atoms are held together by a binding mechanism which is represented by flexible springs. The application of shear stress to this crystal will produce an elastic distortion. If all the atomic bonds intersected by plane ABCD were broken, the lattice would separate along this plane if pulled in tension. The arrangement of atoms around an edge dislocation can be found by inserting in the slot so formed an extra plane of atoms as shown in Fig. 18b. The line DC is a positive edge dislocation represented symbolically by $\perp$. A negative edge dislocation would be obtained by inserting an extra plane of atoms below plane ABCD and is represented by $\top$. A screw dislocation is produced by displacing the crystal on one side of the ABCD relative to the other side in the direction AB as in

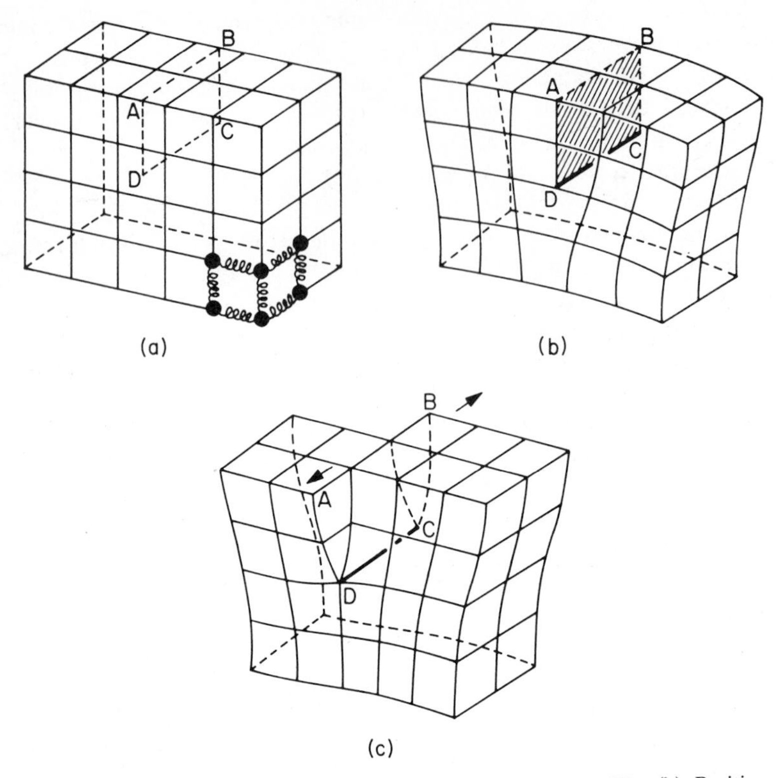

Fig. 18. (a) Model of simple chains lattice with a cut at ABCD. (b) Positive edge dislocation DC. (c) Left-handed screw dislocation DC.

Fig. 18c. Consider the line AD fixed along DC and rotate it in a counter-clockwise direction in the (100) plane of the crystal. After a rotation of 360°, it has moved down one lattice spacing on an unbroken plane and it can continue rotating while moving through the crystal on a helical surface. DC is a screw dislocation. Looking down the dislocation line, if the helix advances one plane when a clockwise circuit is made round it, it is referred to as a right-handed screw dislocation, and if the reverse is true, it is a left-handed screw dislocation.

In general, the definition of a dislocation does not require the assumption of an atomistic nature of matter. Such a definition was given by Volterra who showed that any state of internal stress of a classical (homogeneous, continuous, isotropic) elastic solid could be described as due to a continuous distribution of dislocation lines each with an infinitesimal strength. Dislocation lines in a continuous elastic medium could thus be constructed by the Volterra process which includes the following set of operations:

1. Cut the solid along the surfaces ABCD bondered on line CD.
2. Displace the lips (AB)′ and (AB)″ of the cut by an infinitesimal amount with respect to each other without changing their form.
3. Fill the empty space thus created, if any, with unstressed matter, or remove the matter created in excess.
4. Stick along the lips (AB)′ and (AB)″ and remove the applied stress that was used for displacement.

In what follows we still consider the system as atomistic and confine ourselves only to the simplest examples encountered in the theory of dislocations. The reader is referred to classical works of Nabarro,[66] Friedel,[110] and Hirth and Lothe[111] for further details on the dislocation theory.

The most useful concept in the dislocation theory is the concept of the Burgers circuit. A Burgers circuit is any atom to atom closed path taken around the dislocations. Such a path, MNOPQ, is illustrated in Fig. 19. If the same atom to atom sequence is traversed in the dislocation free crystal, the vector required to complete the circuit is called the Burgers vector. In an edge dislocation as shown in Fig. 19, the Burgers vector is

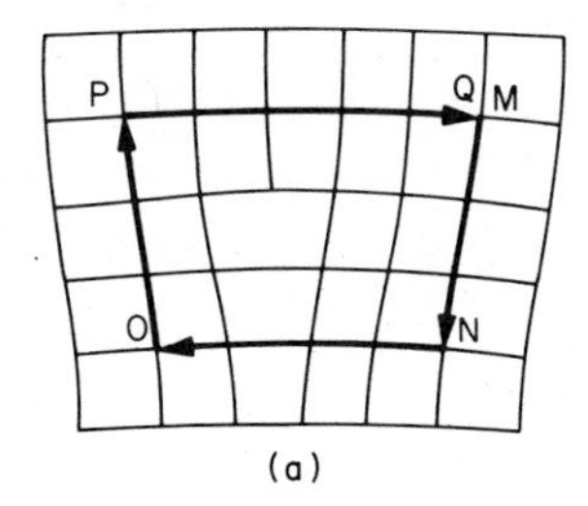

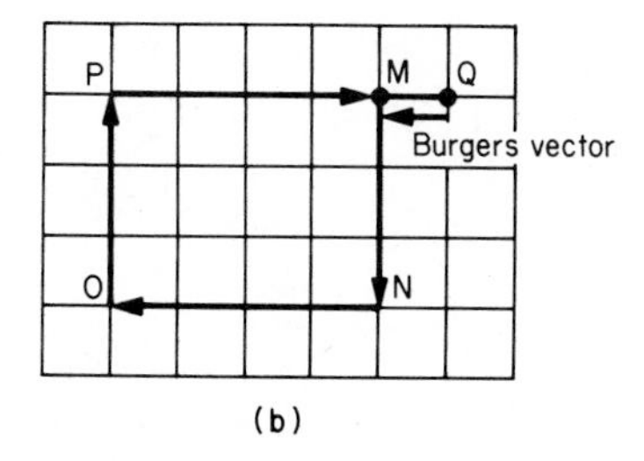

Fig. 19. (a) Burgers circuit around a dislocation. (b) Same circuit in perfect crystal with closure failure indicating the Burgers vector.

perpendicular to the dislocation line. If the same operation is made around a screw dislocation it is found that the Burgers vector is parallel to the dislocation line. In general, the dislocation line lies at an arbitrary angle to the Burgers vector and thus the dislocation line has a mixed edge and screw character. However, the Burgers vector of a dislocation is always the same and is independent of the position of the dislocation or the circuit chosen. The symbol **b** is used for the Burgers vector in which the directions of the vector relative to dislocation line defines the nature of dislocation and its absolute magnitude defines the strength of the dislocation. In crystals the magnitude of the dislocation is usually of the order of lattice distance. It may be shown that at a node, where several dislocation lines meet,

$$\sum_{i=1}^{n} \mathbf{b}_i = 0 . \tag{A.1}$$

Dislocation lines can end at the surface of a crystal and at grain boundaries, but never inside the crystal. Thus a dislocation must either form a closed loop or branch into another dislocation. Dislocations may be experimentally studied by electron microscopy or X-ray techniques. We now consider the elastic energy of a dislocation. In general, the energy of a dislocation consists of two parts:

$$U_{\text{total}} = U_{\text{core}} + U_{\text{elastic}} . \tag{A.2}$$

The elastic energy per unit length of a screw dislocation in an isotropic crystal, for the region enclosed between cylinders of radius r_0 and r, is given by

$$U_{\text{el(s)}} = \frac{\mu b^2}{4\pi} \ln\left(\frac{r}{r_0}\right) . \tag{A.3}$$

Thus the total elastic energy depends on radius r or the extent of the crystal. In the above formula b is the magnitude of the Burgers vector and μ is the shear modulus. The radius r_0 is the limit very close to the dislocation line beyond which distortions are large and Hooke's law is no longer valid. In this region the stored energy is called the core energy and it has been calculated for a screw dislocation (per unit length) to be of the order of

$$U_{\text{cor(s)}} = \frac{\mu b^2}{4\pi} . \tag{A.4}$$

The corresponding formulas for an edge dislocation are

$$U_{\mathrm{el(e)}} = \frac{\mu b^2}{4\pi(1-\nu)} \ln\left(\frac{r}{r_0}\right) \tag{A.5}$$

or

$$U_{\mathrm{cor(e)}} = \frac{\mu b^2}{4\pi(1-\nu)}, \tag{A.6}$$

where ν is the Poisson ratio. The above equations are valid for an isotropic crystal. For a mixed dislocation of the Burgers vector **b**, the system may be resolved into an edge dislocation of the Burgers vector $\mathbf{b}_1$ and a screw dislocation $\mathbf{b}_2$ with $b_1 = b \sin\theta$ and $b_2 = b\cos\theta$, θ being the angle between **b** and the perpendicular to the tangent to the dislocation line. The elastic energy of such a dislocation is given by

$$U_{\mathrm{el(mixed)}} = \frac{\mu b^2}{4\pi(1-\nu)} \ln\left(\frac{r}{r_0}\right)(1-\nu\cos^2\theta). \tag{A.7}$$

Another notion of interest is the force acting between two dislocations. The force between two edge dislocations of opposite sign is attractive, and its components in polar coordinate are given by

$$F_r = \frac{\mu b^2}{2\pi(1-\nu)} \frac{1}{r} \tag{A.8}$$

and

$$F_\theta = \frac{\mu b^2}{2\pi(1-\nu)} \frac{\sin(2\theta)}{r}, \tag{A.9}$$

where θ is the angle between **r** and **b**. It may be seen from the above force relationships that dislocations act very similar to electrical charges and have similar force laws. The forces between dislocations have, however, angular dependence arising from the tensor character of the elastic constants. The attractive energy between a pair of edge dislocations at a distance r from one another is

$$U_{\mathrm{pair}} = \frac{\mu b}{4\pi(1-\nu)}\left[\ln\left(\frac{r}{r_0}\right) - \frac{1}{2}\cos(2\theta)\right]. \tag{A.10}$$

There are two types of basic dislocation movements—glide or conservative motion in which the dislocation moves in the surface defined by its line and the Burgers vector, and climb or nonconservative motion in which the dislocation moves out of the glide surface. The problem may be simplified by introducing the concept of slip. Plastic deformation in a crystal occurs by the movement or sliding of one plane of atoms over another on "slip" planes. Discrete blocks of crystals between two slip planes remain undistorted. The slip plane is normally the plane with highest density of atoms and the direction of slip is the direction in the slip plane in which the atoms are most closely packed. Thus in the close-packed hexagonal crystal, slip usually occurs on the (0001) basal plane in directions such as $(\bar{1}2\bar{1}0)$. It may be shown that theoretical shear stress for slip is many times greater than the experimentally observed stress. The observed low value may be accounted for by dislocations. Thus slip may be associated with formation or movement of dislocations.

Dislocations in polymers and paraffins have been extensively studied using electron microscopic techniques.[112] The general deformation of a long chain molecular crystal has been described by Bonart.[113] The mechanism of deformation, as in nonpolymeric solids, must be based on molecular slip. The covalent bonds along the molecular chain restrict slip mechanism. Slip is only permitted in two independent directions parallel to the chain axis. In the direction perpendicular to the chain axis slip must be associated with the formation of kinks. Peredecki and Statton[60] and Keith and Passaglie[61] have investigated possible dislocations in linear chain molecular crystals, taking into account the constraints imposed by chain connectivity. They conclude that four basic dislocations are possible in such crystals as illustrated in Fig. 10.

APPENDIX B. ELASTIC CONSTANTS OF PARAFFINS

The elastic tensor of orthorhombic polyethylene crystal at $T = 0$ K has been calculated by Wobser and Hägele[114] by lattice dynamic methods and the force constants derived from Table I. They find the following values for the components of the elastic tensor c_{ij}:

$$c_{ij} = \begin{pmatrix} 13.7 & 7.3 & 2.46 & 0 & 0 & 0 \\ & 12.5 & 3.96 & 0 & 0 & 0 \\ & & 3.25 & 0 & 0 & 0 \\ & & 0 & 3.19 & 0 & 0 \\ & & & & 1.98 & 0 \\ & & & & & 6.2 \end{pmatrix}.$$

Experimentally, the elastic constants have been measured by Müller using X-ray methods for $C_{23}H_{48}$ and $C_{29}H_{60}$.[28] Weir and Hoffman[115a] have measured the compressibility of a number of paraffins and Sakurada et al.[115b] have measured the elastic moduli by X-ray methods. The above theoretical treatment compares as follows with experimental data:

Moduli	Theoretical	Experimental
Shear moduli	3.66×10^{10}	3×10^{10} dynes/cm^2
Compressibility	1.0×10^{-10}	1.6×10^{-11} cm^2/dyne
Poisson ratio	0.56	0.54

Strobl[43] has summarized experimental moduli in his treatment of the anisotropic expansion of paraffins. More recently, the complete elastic tensor of the monoclinic $C_{36}H_{74}$ has been measured by Krüger and co-workers[58] by Brillouin spectroscopy. They give the following values for the components of elastic tensor:

$$c_{ij} = \begin{pmatrix} 15.6 & -2 & X & 0 & 4 & 0 \\ -2 & 7.5 & X & 0 & 0.9 & 0 \\ & & & 0 & 1.9 & 0 \\ & & & 1.6 & 0 & 0.6 \\ & & & & 6.4 & 0 \\ & & & & & 3 \\ & \multicolumn{4}{c}{\text{(X indicates unknown)}} \end{pmatrix} \times 10^{10}\ \text{dynes/cm}^2 .$$

The above elastic constants may not be used for the evaluation of the melting point because they represent the low-temperature values. It is known that at the rotator transition, the shear modulus drops drastically. Measurements of shear moduli in the rotator phase have been made by Pechhold and co-workers[57] for a number of paraffins. The shear moduli of odd paraffins from $C_{11}H_{24}$ through $C_{31}H_{64}$ have been measured. The modulus that is approximately 10^{10} dynes/cm^2 in orthorhombic low-temperature modification goes through a sharp drop at the rotator transition temperature and falls to about 5×10^8 dynes/cm^2.

Acknowledgments

The author is grateful to Professor Holz for making some unpublished material available for inclusion in this review. Stimulating discussions with Professors A. Holz, E. W. Fischer, J. Kovac, R. G. Strobl, and Dr. J. Krüger are acknowledged. Many thanks are due to Professors J. Petersson, H. Gleiter, H. Müser and H. G. Unruh who supported this project and Joann Hickson, who typed the final manuscript.

Financial support for this work was provided by Deutsche Forschungsgemeinshaft under SFB 130; partial support was provided by the Department of Energy, Division of Materials Sciences, Office of Basic Energy Sciences.

References

1. J. M. Kosterlitz and D. J. Thouless, *J. Phys. C* **6**, 1181 (1972).
2. B. I. Halperin and D. R. Nelson, *Phys. Rev. Lett.* **41**, 121 (1978).
3. M. Karplus and R. G. Parr, *J. Chem. Phys.* **38**, 1547 (1963).
4. O. J. Sovers, C. W. Kern, R. M. Pitzer, and M. Karplus, *J. Chem. Phys.* **49**, 2592 (1968).
5. P. C. Hägele and W. Pechhold, *Kolloid Z. Z. Polym.* **241**, 977 (1970).
6. S. Salem, *J. Chem. Phys.* **37**, 2100 (1962).
7. R. G. Strobl, *Colloid Polym. Sci.* **256**, 427 (1978).
8. D. E. Williams, *J. Chem. Phys.* **47**, 4680 (1967).
9. A. Warshel and S. Lifson, *J. Chem. Phys.* **53**, 582 (1970).
10. S. Lifson and A. Warshel, *J. Chem. Phys.* **49**, 5116 (1968).
11. T. Admuir, M. S. Longmire, and E. A. Mason, *J. Chem. Phys.* **35**, 895 (1961).
12. R. L. McCullough, *J. Macromol. Sci. Phys. B* **9**, 97 (1974).
13. R. G. Synder and J. H. Schachtschneider, *Spectrochim. Acta* **19**, 85, 117 (1963).
14. R. G. Synder, *J. Chem. Phys.* **47**, 1316 (1967); J. D. Barnes and B. M. Franconi, *J. Chem. Phys.* **56**, 5190 (1972).
15. R. G. Snyder, M. Maroncelli, S. P. Qi, and H. Strauss, *Science* **214**, 188 (1981); M. Maroncelli, S. P. Qi, H. L. Strauss, and R. G. Synder, *J. Am. Chem. Soc.* **104**, 6237 (1982).
16. C. Scherer and J. Naghizadeh, *J. Chem. Phys.* **75**, 5522 (1981).
17. J. A. Borges da Costa, C. Scherer, A. Holz, and J. Naghizadeh, *Z. Naturforsch.* **38a**, 1285 (1983).
18. P. Flory, *Statistical Mechanics of Chain Molecules.* Innterscience, London, 1969.
19. L. D. Landau and E. M. Lifshiftz, *Mechanics.* Pergamon, London, 1960.
20. E. Helfand, *J. Chem. Phys.* **54**, 4651 (1971).
21. E. Helfand, *J. Chem. Phys.* **69**, 1010 (1978).
22. E. Helfand, Z. R. Wasserman, and T. A. Weber, *Macromolecules* **13**, 526 (1980).
23. J. Skolnick and E. Helfand, *J. Chem. Phys.* **72**, 5489 (1980).
24. C. K. Hall and E. Helfand, *J. Chem. Phys.* **77**, 3275 (1982).
25. T. F. Shatzki, *J. Polym. Sci.* **57**, 496 (1962).
26. J. P. Ryckaert and A. Bellemans, *Chem. Phys. Lett.* **30**, 123 (1975).
27. H. A. Kramers, *Physica* **7**, 284 (1940).
28. (a) A. Müller, *Proc. R. Soc. A* **138**, 514 (1932); (b) A. Müller and K. Lonsdale, *Acta Crystallogr.* **1**, 129 (1948).
29. H. M. Shearer and V. Vand, *Acta Crystallogr.* **9**, 379 (1956).
30. A. J. Kitaigorodskii, *Molekül Kristalle.* Academie–Verlag, Berlin, 1979.
31. Y. V. Mnyukh, *J. Phys. Chem. Solid* **24**, 631 (1963).
32. R. Boistelle, in *Current Toffics in Material Science*, Vol. 4, El Kaldis (Ed.). North-Holland, Amsterdam, 1980.
33. M. G. Broadhurst, *J. Res. Natl. Bur. Stand.* **66A**, No. 3 (1962).
34. For a recent compilation see B. Wunderlich, *Polymer Physics*, Vol. 3. Academic Press, New York, 1980.

35. W. E. Garner, K. Van Bibber, and A. M. King, *J. Chem. Soc.* **1931**, 1533 (1931).
36. M. L. Huggins, *J. Phys. Chem.* **43**, 1083 (1939).
37. P. J. Flory and A. Vrij, *J. Am. Chem. Soc.* **85**, 3548 (1963).
38. G. Unger, *J. Phys. Chem.* **87**, 689 (1983).
39. J. Doucet, I. Denicolo, and A. F. Craievich, *J. Chem. Phys.* **75**, 1523 (1981).
40. J. Doucet, I. Denicolo, A. F. Craievich, and A. Collet, *J. Chem. Phys.* **75**, 5125 (1981).
41. I. Denicolo, J. Doucet, and A. F. Craievich, *J. Chem. Phys.* **78**, 1465 (1983).
42. I. Denicolo, H. F. Craievich, and I. Doucet, *Phys. Rev.* (in press).
43. G. Strobl, *Colloid Polym. Sci.* **254**, 170 (1976).
44. M. Stohrer and F. Noack, *J. Chem. Phys.* **67**, 3729 (1977).
45. J. D. Barnes, *J. Chem. Phys.* **58**, 5193 (1973).
46. G. R. Strobl, B. Ewen, E. W. Fischer, and W. Piesczek, *J. Chem. Phys.* **61**, 5257 (1974).
47. B. Ewen, E. W. Fischer, W. Piesczek, and G. R. Strobl, *J. Chem. Phys.* **61**, 5265 (1975).
48. G. R. Strobl, Habilitationsschrift. Mainz, 1975.
49. J. K. Krüger, L. Peetz, W. Wildner, and M. Pietralla, *Polymer* **21**, 620 (1980); J. K. Krüger, G. W. Hohne, and M. Pietralla, *Polym. Bull.* **9**, 252 (1983).
50. E. W. Fischer, R. G. Strobl, M. Dettenmaier, M. Stamm, and N. Steidle, *Faraday Discuss. Chem. Soc.* **68**, 26 (1979).
51. R. G. Synder, *J. Chem. Phys.* **76**, 3342 (1982).
52. P. G. deGennes, *The Physics of Liquid Crystals.* Oxford University Press, London, 1974.
53. B. A. Hubermann, D. M. Lublin, and S. Doniach, *Solid State Commun.* **17**, 485 (1975).
54. W. Helfrich, *J. Phys.* (*Paris*) **39**, 1199 (1978).
55. D. R. Nelson and J. Toner, *Phys. Rev. B* **24**, 363 (1981).
56. W. L. McMillan, *Phys. Rev. A* **7**, 1419 (1973).
57. W. Pechhold, W. Dollhopf, and A. Engel, *Acustica* **17**, 61 (1966).
58. J. K. Krüger, H. Bastian, G. C. Asbach, and M. Pietralla, *Polym. Bull.* **3**, 633 (1980).
59. J. Naghizadeh and A. Holz, *Solid State Commun.* **43**, 573 (1982); J. Naghizadeh, *Int. J. Quantum Chem.* **16**, 205 (1982).
60. P. Peredecki and W. O. Statton, *J. Appl. Phys.* **38**, 4140 (1967).
61. H. D. Keith and E. Passaglia, *J. Res. Natl. Bur. Stand. Sect. A* **68**, 513 (1964).
62. E. W. Montroll and N. S. Goel, *Biopolymers* **4**, 844 (1966).
63. D. T. Vigren, *Phys. Rev. B* **27**, 2932 (1983).
64. E. A. DiMarzio and R. J. Rubin, *J. Chem. Phys.* **55**, 3418 (1971).
65. R. J. Rubin, *J. Chem. Phys.* **43**, 2392 (1965).
66. F. R. N. Nabarro, *Theory of Crystal Dislocations.* Oxford University Press, London. 1967.
67. A. Holz and J. F. N. Medeiros, *Phys. Rev. B* **17**, 1161 (1978).
68. A. Holz, J. Naghizadeh, and D. T. Vigren, *Phys. Rev. B* **27**, 512 (1983).
69. A. Kovacs and J. F. N. Straupe, *Discuss. Faraday Soc.* **68**, 225 (1980).

70. D. W. McClure, *J. Chem. Phys.* **49**, 1830 (1968).
71. J. D. Hoffman, *J. Chem. Phys.* **20**, 541 (1952).
72. M. Kobayashi, *J. Chem. Phys.* **68**, 145 (1978).
73. T. Ishinabe, *J. Chem. Phys.* **72**, 353 (1980).
74. W. L. Brezinskii, *Sov. Phys.-JETP* **34**, 610 (1971).
75. A. Holz, *Physica* (*Utrecht*) **97A**, 75 (1979).
76. B. A. Huberman and S. Doniach, *Phys. Rev. Lett.* **43**, 950 (1979).
77. S. Doniach and B. A. Huberman, *Phys. Rev. Lett.* **42**, 1169 (1979).
78. J. Naghizadeh, unpublished work.
79. A. I. Larkin and S. A. Pikin, *Sov. Phys.-JETP* **29**, 891 (1969).
80. A. Holz, *Phys. Lett. A* **96**, 475 (1983).
81. A. Holz, unpublished work.
82. B. Wunderlich, *Macromolecular Physics* 1. Academic Press, New York, 1973.
83. E. T. Samulski, *Phys. Today*, May (1982).
84. L. R. Denny and R. Boyer, *Polym. Bull.* **4**, 527 (1981).
85. R. Boyer, *Polymer Yearbook*. Gordon & Breach, New York (in press).
86. B. A. Newman and H. F. Kay, *J. Appl. Phys.* **38**, 4105 (1967).
87. B. A. Newman and H. F. Kay, *Acta Crystallogr. B* **24**, 615 (1968).
88. T. Trzebiatowski, Dissertation, Mainz (1980).
89. P. Groth, *Acta Chem. Scand. A* **33**, 199 (1979).
90. J. Dale, *Acta Chem. Scand.* **27**, 1115, 1130 (1973).
91. H. Höcker and K. Riebel, *Makromol. Chem.* **179**, 1967 (1978).
92. H. P. Grossmann, *Polym. Bull.* **5**, 137 (1981).
93. J. Brandrup and E. H. Immergut, *Polymer Handbook*. Wiley, New York, 1975; J. R. Stoddart, *Comparative Organic Chemistry*. Pergamon Press, New York, 1979.
94. G. R. Strobl, H. Schwickert, and Trzebiatowski, *Ber. Bunsenges. Phys. Chem.* **87**, 274 (1983).
95. G. Zerbi, R. Magni, M. Gussoni, K. Holland-Moritz, A. Bigotto, and S. Dirlikov, *J. Chem. Phys.* **75**, 3175 (1981).
96. W. Picsczek, G. Strobl, and K. Malzahn, *Acta Crystallogr. B* **30** 1278 (1974).
97. H. P. Grossmann, R. Arnold, and K. R. Bürkle, *Polym. Bull.* **3**, 135 (1980).
98. M. Müller, H. J. Cantow, J. K. Krüger, and H. Höcker, *Polym. Bull.* **5**, 125 (1981).
99. C. W. Bunn and E. R. Howells, *Nature* **174**, 549 (1954).
100. O. Albrecht, H. Gruler, and E. Sackmann, *J. Phys.* (*Paris*) **39**, 301 (1978).
101. A. Caille, D. Pink, F. Deverteuil, and M. J. Zuckermann, *Can. J. Phys.* **58**, 581 (1980).
102. J. F. Nagle, *Annu. Rev. Phys. Chem.* **31**, 157 (1980).
103. S. Marcelja, *Biochim. Biophys. Acta* **367**, 237 (1974).
104. F. Jahnig, *J. Chem. Phys.* **70**, 3279 (1979).
105. A. R. Massih and J. Naghizadeh, *Mol. Cryst. Liq. Cryst.* **90**, 145 (1982).
106. S. Doniach, *J. Chem. Phys.* **68**, 4912 (1978).
107. D. Pink, T. J. Green, and D. Chapman, *Biochemistry* **19**, 345 (1980).
108. J. F. Nagle, *J. Chem. Phys.* **63**, 1255 (1975).

109. A. Holz, D. T. Vigren, and M. J. Zuckermann, *Phys. Rev.* **B 31**, 420 (1985).
110. J. Friedel, *Dislocations.* Pergamon, New York, 1964.
111. J. P. Hirth and J. Lothe, *Theory of Dislocations.* McGraw-Hill, New York, 1968.
112. V. F. Holland and P. H. Lindenmeyer, *J. Appl. Phys.* **36**, 3049 (1965); P. H. Lindenmeyer, *J. Polym. Sci. C* **15**, 109 (1966).
113. R. Bonart, *Kalloid Z. Z. Polym.* **231**, 438 (1969).
114. G. Wobser and P. C. Hägele, *Ber. Bunsenges. Phys. Chem.* **74**, 896 (1970).
115. (a) C. E. Weir and J. D. Hoffman, *J. Res. Natl. Bur. Stand.* **55**, 307 (1955); (b) J. Sckurda, T. Ito, and K. Nakama, *J. Polym. Sci. C* **15**, 75 (1966).

CORRELATION EFFECTS IN THE IONIZATION OF MOLECULES: BREAKDOWN OF THE MOLECULAR ORBITAL PICTURE

L. S. CEDERBAUM, W. DOMCKE, J. SCHIRMER, AND W. VON NIESSEN‡

Lehrstuhl für Theoretische Chemie
Institut für Physikalische Chemie
Universität Heidelberg
D-6900 Heidelberg, West Germany

CONTENTS

I. Introduction 115
II. Outline of Theory 119
  A. Ionization Spectrum 119
  B. Computational Aspects and Results 126
III. Correlation Effects in the Ionization of Molecules 129
  A. Representative Examples 129
  B. Discussion 143
IV. Other Effects and Outlook 148
References 154

I. INTRODUCTION

Photoelectron (PE) spectroscopy has long since been a versatile tool to investigate the electronic structure of atoms, molecules, solids, and surfaces.[1–5] More recently the classic techniques, namely, X-ray induced PE spectroscopy (XPS) and ultraviolet PE spectroscopy (UPS), have been complemented by dipole ($e, 2e$) spectroscopy[6] and PE spectroscopy using synchrotron radiation from storage rings,[7,8] which allow continuous tuning

‡ Permanent address: Institut für Physikalische und Theoretische Chemie, Technische Universität Braunschweig, D-3300 Braunschweig, West Germany.

of the photon energy over a wide range. These instrumental developments have produced a wealth of data on the magnitude and energy dependence of photoionization cross sections as well as on the angular distribution of photoelectrons. An alternative technique, which has contributed considerably to our understanding of the electronic structure of atoms and molecules, is the so-called binary $(e, 2e)$ spectroscopy.[9,10]

An extremely useful concept to rationalize the huge amount of data provided by PE and $(e, 2e)$ spectroscopies is the single-particle picture or shell model. In the simplest case of closed-shell atoms, for example, each line in the PE spectrum can be associated with an electronic orbital to a first approximation. In molecules, rotational and vibrational excitation, Jahn–Teller splittings, and so on lead to additional complications, but still the resulting somewhat broader "bands" in the PE spectrum can be associated with individual molecular orbitals (MOs). In this sense, PE spectroscopy is considered a "spectroscopy of orbitals" and helps us to understand the electronic structure and reactivity of compounds of chemical interest.[1–5] The fine structure of the bands in molecular PE spectra gives, moreover, direct information on the bonding properties of the MOs.[3–5]

Corrections to this simple picture arise because of electron correlation, which is not considered in the self-consistent field (SCF) model of independent electrons. As has been known since the early days of PE spectroscopy, these residual correlation and relaxation effects lead to the appearance of additional weak bands in the PE spectrum. These so-called shake-up or satellite bands correspond to excitation processes accompanying the ionization.[11] Associated with the appearance of satellite bands is a shift and a reduction of the intensity (compared to the independent-particle model) of the main band representing the MO. This simple picture of main lines and accompanying shake-up lines has been confirmed by numerous studies of valence and core ionizations in atoms and molecules (e.g., see refs. 1–6, 9, 10).

The first indications that the concept of main ionizations and accompanying shake-up transitions may lose its validity for ionization out of deeper valence orbitals of molecules came from theoretical studies on H_2O,[12] CO,[13,14] and N_2.[14] These calculations, which were based on the Green's function method[12,14] or the configuration-interaction (CI) technique,[13] predicted satellite lines on *both* sides of the main line representing the inner-valence orbital, in contradiction to the familiar shake-up picture, which can explain only satellite lines at the higher binding-energy side of the main line.[11] The origin of the phenomenon is the quasi-degeneracy of the inner-valence single-hole configuration with certain configurations representing two holes in the outer-valence shell and one particle in a virtual orbital. The interaction of these quasi-degenerate configurations via the residual electron–electron interaction leads to a redistribution of the intensity associated with the inner-valence orbital. If the interaction is sufficiently strong, the intensity may be distributed over numerous lines and it is no longer possible to discern between main lines and satellite lines. This effect has been termed the *breakdown of the molecular orbital picture of ionization.*[14,15]

The calculations on N_2, CO, CO_2, CS_2, and other small molecules[14–17] explained *a posteriori* the observation (by XPS) of very broad bands with a half-width of several electron volts for the more tightly bound electrons.[2,3,18] The limited experimental resolution and the existence of additional broadening mechanisms in molecules, such as vibrational excitation and dissociation, usually prevent the direct observation of the theoretically predicted line splittings. Recent high-resolution studies employing synchrotron radiation have revealed, however, substructures of the broad inner-valence bands in N_2 and CO which are in qualitative agreement with the theoretical predictions.[19,20] The generality of the breakdown phenomenon in the inner-valence region has also been convincingly established by numerous $(e, 2e)$ spectroscopic studies (e.g., see refs. 16, 21–27).

In the last few years Green's function calculations based on the two-particle-hole Tamm–Dancoff approximation (2ph–TDA)[28,29] have been performed to study the complete valence-shell ionization spectra of numerous small- and medium-size molecules (see refs. 30–32 and references therein). Although the calculations are not of really quantitative accuracy owing to limitations of the method and the finite basis sets employed,[30,31] the breakdown of the MO picture of ionization as a general phenomenon in the inner-valence region is now well established. Depending on the size, symmetry, and chemical properties of the molecule, the outer-valence orbitals may also be affected, that is, the intensity of several main lines may be strongly reduced compared to the SCF model and low-lying and intense satellite lines may appear.

The breakdown of the MO picture as a qualitative phenomenon has also been confirmed by a number of subsequent theoretical studies using different computational techniques. Calculations based on Green's function or propagator methods have been performed, for example, by Herman et al.[33] (N_2), Mishra and Öhrn[34] (N_2, H_2O) Cacelli et al.[35] (HF, H_2O, NH_3), and Baker[36] (C_2H_4). Calculations based on CI techniques generally yield very similar results, as is shown, for example, by Honjou et al.[37] (N_2, CO, O_2, NO), Langhoff et al.[38] (N_2, CO), Müller and co-workers[39–41] (CS, C_2H_2, H_2O), and Nakatsuji and Yonezawa[42] (H_2O, CO_2, N_2O, CS_2, COS). Calculations using semiempirical model Hamiltonians were also quite successful.[43–47] It is generally found that strong correlation effects in the inner-valence region are accompanied by low-lying and intense satellite structure in the outer-valence region.[47–51]

A closely related phenomenon is the apparent missing of certain inner-shell ionization lines in the PE spectra of heavy atoms such as Xe.[2,52] Lundqvist and Wendin[53] and Wendin and Ohno[54] have shown that this effect is caused by an unusually strong coupling of the single-hole configuration with an underlying two-hole–one-particle continuum. This phenomenon has subsequently been discovered in a whole series of elements but seems to be confined to deep core levels of relatively heavy atoms. For a comprehensive review, the reader is referred to the recent article by Wendin.[55]

The breakdown of the MO picture in molecules is a fundamental effect and is of relevance far beyond the field of photoionization experiments. High-resolution X-ray emission, for example, is an alternative technique used to investigate valence hole states in molecules,[56] and the breakdown of the MO picture has been identified in the X-ray emission spectra of a variety of molecules.[57,58] Autoionization of core-excited states may also populate these levels.[59] Related correlation phenomena are expected to occur in more complicated excitation processes such as the Auger effect. The severe limitations of the single-particle picture in interpreting molecular Auger spectra have recently been pointed out by Ågren and Siegbahn,[60,61] Kvalheim,[62] Jennison and co-workers,[63] Aksela and co-workers,[64] and Liegener.[65] The breakdown of the MO picture also has important consequences for the understanding of nuclear dynamics in ionization processes, in particular, photodissociation[66] and photon-stimulated ion desorption.[67,68]

The breakdown of the MO picture may be classified as a final-state correlation effect, that is, as configuration interaction taking place in the cation. There are additional correlation phenomena that have to be taken into account in a quantitative description of photoionization experiments, for example, initial-state correlation and correlation in the continuum. These effects are separable only in an approximate sense. It is worthwhile, therefore, to begin this article with a general, though necessarily brief, introduction to correlation effects in photoionization. In particular, a general theory of line intensities in PE spectra is required for a quantitative comparison between theory and experiment in the inner-valence region. The many-body Green's function or CI methods which are used in the actual calculations are only briefly sketched, since comprehensive reviews can be found elsewhere.[28,32,69–71] The qualitative and quantitative aspects of the breakdown of the MO picture are then discussed for a few selected examples. Attempts are made to exhibit the general trends and to establish simple rules which possibly allow qualitative predictions, without the necessity of extensive calculations. Additional interesting phenomena associated with the breakdown of the MO picture are discussed in a concluding section.

II. OUTLINE OF THEORY

A. Ionization Spectrum

The ionization of a molecule by a photon of energy ω_0 is analyzed by recording the ionization spectrum. As a function of the kinetic energy ε of the ejected electron or, equivalently, as a function of binding energy $\omega_0 - \varepsilon$, this spectrum consists of a series of lines corresponding to the states $|\Psi_n^{N-1}\rangle$ of the residual ion. A line in the spectrum is fully characterized by its position and weight or, more precisely, by the ionization energy I_n and partial-channel ionization cross section σ_n. In contrast to the energy I_n, which is independent of the experimental conditions, the cross section σ_n depends on a number of external parameters like the kinetic energy ε, the scattering angle, and the polarization of the incident light. It is convenient to introduce a set of quantities alternative to σ_n, which depend solely on the internal properties of the molecular and ionic states. The sudden ejection of an electron out of the molecular orbital ϕ_q in the exact initial molecular state $|\Psi_0^N\rangle$ results in a pseudo-state $a_q|\Psi_0^N\rangle$, where a_q is the corresponding annihilation operator. The probability of finding the final ionic state $|\Psi_n^{N-1}\rangle$ in this pseudo-state is given by the absolute square of the *transition amplitude*:

$$x_q^{(n)} = \langle \Psi_n^{N-1} | a_q | \Psi_0^N \rangle . \tag{2.1}$$

This probability $|x_q^{(n)}|^2$ is often called *pole strength* or *spectroscopic factor.* Intuitively, it is clear that the spectroscopic factors are closely related to the weight of a line in the ionization spectrum. The relation between the spectroscopic factors and the partial-channel ionization cross section σ_n is discussed later.

The ionization process would be particularly simple if the electrons were truly independent. Then both initial and final states are given by single electronic configurations and the only nonvanishing transition amplitude $x_q^{(n)}$ is the one for which the ionic configuration differs from the neutral configuration by the occupation of the molecular orbital ϕ_q. The corresponding ionization energy and spectroscopic factor are given by $-\varepsilon_q$

and 1, respectively, where ε_q is the orbital energy. The resulting ionization spectrum now consists of a finite series of lines. Each line corresponds to an orbital ϕ_q occupied in the molecular ground state.

Although the interaction between the electrons is of crucial importance, it is extremely useful to introduce an independent-particle picture which serves as a basis for the interpretation of the observed phenomena. We choose the molecular ground-state Hartree–Fock particles as our independent particles. This choice exhibits distinct conceptual and practical advantages. The ionization energy associated with the ejection of an electron out of Hartree–Fock orbital ϕ_q is described correctly up to first order in the interaction between the Hartree–Fock particles by the corresponding orbital energy ε_q. This well-known result is called Koopman's theorem.[72] Corrections to the ionization energies obtained by applying Koopman's theorem first arise in second-order perturbation theory.

The choice of the independent-particle model has also some impact on the spectroscopic factors. Using an arbitrarily chosen independent-particle model, a given ionic state $|\Psi_n^{N-1}\rangle$ may become accessible in first-order perturbation theory due to the ionization out of any occupied orbital, that is, $x_q^{(n)} \neq 0$ for all occupied orbitals ϕ_q. Hence, in first order, a line in the spectrum may be attributed to several occupied orbitals. The only exception is found when using the ground-state Hartree–Fock orbitals. With this choice, each line in the spectrum is associated exactly with one occupied orbital. Guided by this first-order result we may generally assume that each line in the spectrum is mainly associated with the ionization out of a single orbital. This assumption, which considerably simplifies the interpretation of ionization spectra, is nicely confirmed by the numerical results obtained for many molecules. Exceptions are expected for larger molecules and molecules of low symmetry, where close-lying lines may emerge from different orbitals of the same spatial symmetry. Propiolic acid is a prominent example for a strong mixing of orbitals in the ionic wavefunction.[49]

To proceed with the analysis, the molecular and ionic wavefunctions are expanded in terms of electronic configurations constructed from the ground-state Hartree–Fock orbitals. The expansion of the molecular ground state comprises the Hartree–Fock ground-state configuration as well as singly-excited, doubly-excited, and so on configurations with respect to it. The latter configurations may also be called one-particle–one-hole (1p–1h), two-particle–two-hole (2p–2h), and so on configurations. In analogy, the ionic wavefunction is a superposition of the various single-hole (1h), two-hole–one-particle (2h–1p) and higher excited electron configurations, where holes and particles refer to orbitals occupied and unoccupied in the molecular ground-state configuration, respectively.

For convenience, 1h configurations are often denoted by q^{-1} to specify the orbital ϕ_q out of which the electron has been removed. Analogously, 2h–1p configurations are denoted by $q^{-1}r^{-1}t$, indicating that one electron is ejected from the orbital ϕ_q (or ϕ_r) and another one is simultaneously excited from the orbital ϕ_r (or ϕ_q) to the virtual orbital ϕ_t.

Clearly, on the independent-particle level of approximation there is no access to ionic states which are characterized by 2h–1p or higher-excited configurations; the $x_q^{(n)}$ vanish for these states.‡ The interaction between the Hartree–Fock particles is responsible for the appearance of these kinds of states in the ionization spectrum. As long as the many-body effects resulting from this interaction are weak, the ionic states derived from 1h configurations are much more probable than other states and the corresponding lines will dominate the spectrum. Thus we may call these lines *main* lines and the weaker lines derived from the higher-excited configurations *satellite* lines. In this context it is common to use the illustrative phrase that a satellite line (or state) has *borrowed* its intensity from one or several orbitals and has become visible in the spectrum.

In the language of CI the satellite lines enter the spectrum due to the interaction of the 1h with 2h–1p or higher configurations. Since the 2h–1p configurations usually have higher binding energies than the outer-valence orbitals, one expects a simple structure of the spectrum in the outer-valence energy range. At low binding energy the spectrum is expected to exhibit main lines only. The assignment of these lines and the question whether Koopmans' theorem predicts the correct ordering of lines have been subject to many experimental and theoretical investigations in the literature.[4,5] To the higher-energy side of the outer-valence range one expects more or less weak satellite lines which have borrowed their intensity from the outer-valence orbitals. For small molecules the density of 2h–1p configurations is low in this energy range and the resulting state can be classified by one or a few configurations. A well-studied example is the $C^2\Sigma_u^+$ satellite state of N_2^+ observed at about 25 eV binding energy. This satellite state has borrowed its intensity from the outer-valence $2\sigma_u$ orbital and can be well characterized by the $(3\sigma_g)^{-1}(1\pi_u)^{-1}(1\pi_g)$ configuration.

In the majority of cases of intense lines in the ionization spectrum the numerical calculations exhibit a *single* dominating transition amplitude

‡ There is, of course, a finite probability for the production of such states if we assume different independent-particle approximations for the molecular and ionic states. In principle, one could perform a separate Hartree–Fock calculation for each ionic state, at least for each 1h-derived state. In this way one automatically takes account of the so-called relaxation effects which are especially important in the description of core holes.[73,74] The simplicity of the molecular orbital picture of ionization is lost, however, in such an approach.

associated with each line. Correspondingly, we may view in these cases the ionization process as the ejection of an electron out of a specific orbital. This simple concept of the ionization process also allows for a useful specification of satellite lines in the spectrum associated with final cationic states emerging from a 2h–1p electronic configuration $m^{-1}n^{-1}t$. Let the ionization take place out of the orbital ϕ_q. If q is equal to either m or n, the satellite line mainly originates[31,75] from the relaxation of the molecular orbitals upon creation of a hole in the orbital ϕ_q. Such satellites are often addressed as *shake-up* satellites[1,11] and are particularly important in the ionization of orbitals localized in space, for example, core orbitals, where relaxation effects are substantial. The final state can be viewed as a 1p–1h excitation on top of the 1h configuration from which its intensity has been borrowed.

On the other hand, final-state correlation effects also give rise to satellites for which neither m nor n are equal to q. We denote these satellites as (final-state) *correlation* satellites. The above mentioned $C\,^2\Sigma_u^+$ state of N_2^+ is an example of a correlation satellite. There is a third kind of interesting satellites which cannot appear in the spectrum without ground-state correlation. Neglecting ground-state correlation, it is easily recognized that $x_q^{(n)}$ in Eq. (2.1) is nonzero only for ionic states that transform as the irreducible representations of the orbitals occupied in the molecular ground state. The ionic $^2\Pi_g$ states in N_2^+, for example, cannot appear in the spectrum, since there is no occupied π_g orbital in N_2. Once we include electron correlation effects in the molecular ground state, $a_q|\Psi_0^N\rangle$ does not vanish *a priori* for any virtual orbital ϕ_q and states like the $^2\Pi_g$ states of N_2^+ become accessible. We may term the corresponding lines as *initial-state correlation* satellites.

In general, the appearance of a main line accompanied by weaker satellite lines is not restricted to atomic and molecular systems. In many fields of physics the concept of quasi-particles[76–78] has been introduced to describe formally related phenomena. In these cases many-body effects are relevant, but not strong enough to destroy the appearance of a main structure in the spectrum predicted by the independent-particle model. When discussing molecules we synonymously use the terms "quasi-particle picture" and "molecular orbital picture."

In the energy range of the inner-valence orbitals (typically the $2s$ shell of first row atoms and the $3s$ shell of second row atoms) the density of 2h–1p configurations is often high and satellite states can no longer be expected to correspond to a single configuration. Rather, they are characterized by the superposition of many 2h–1p configurations, and the simple classification scheme discussed above may often fail. Moreover, depending on the molecule in question, several or even many 2h–1p configurations are located energetically close to the q^{-1} configurations of the inner-valence orbitals ϕ_q. The interaction between the 1h and 2h–1p configurations may now be more efficient and distribute a significant portion of the intensity originally confined to a 1h configuration over several satellite states, that is, several states n may acquire considerable spectroscopic factors $|x_q^{(n)}|^2$ from a common orbital ϕ_q. This spread of intensity can lead, in principle, to the situation where it is no longer meaningful to distinguish between satellite and main lines and we encounter a breakdown of the molecular orbital picture of ionization.[14,15] The major part of the present review is devoted to the discussion of this phenomenon and to the general properties of satellite states.

Until now we have discussed the spectrum in terms of transition amplitudes which describe the sudden ejection of electrons and are independent of the experimental conditions. In the following we comment on the relation between these amplitudes and the ionization cross section. Assuming Fermi's golden rule, the partial-channel ionization cross section for production of molecular ions in the state $|\Psi_n^{N-1}\rangle$ is given by

$$\sigma_n(\varepsilon) \propto |T_{n\varepsilon,0}|^2 \equiv |\langle\Psi_{n\varepsilon}^N|\hat{T}|\Psi_0^N\rangle|^2 , \tag{2.2}$$

where $\Psi_{n\varepsilon}^N$ is the N-electron scattering function with the appropriate[79] (incoming waves) boundary condition and $\hat{T}$ is the transition operator describing the photon–molecule interaction. We note that Eq. (2.2) may also apply to ionization experiments other than photoionization and restrict ourselves to those experiments that can be described by transition operators which are one-particle operators, for example, photoionization, dipole $(e, 2e)$, and to some extent also binary[10,27,80] $(e, 2e)$ experiments.

To proceed with the evaluation of Eq. (2.2) we must compute the initial and final states Ψ_0^N and $\Psi_{n\varepsilon}^N$. Whereas the accurate calculation of Ψ_0^N is feasible,[71] the accurate determination of the electron–ion scattering function $\Psi_{n\varepsilon}^N$ is, in general, prohibitively difficult, forcing us to resort to simplifying approximations. An approximation commonly used[12,38,41,81–83] in the present context is to neglect interchannel coupling and put[84]

$$|\Psi_{n\varepsilon}^N\rangle = \tilde{a}_\varepsilon^\dagger|\Psi_n^{N-1}\rangle . \tag{2.3}$$

Here, the creation operator $\tilde{a}_\varepsilon^\dagger$ is associated with the one-electron scattering function $\tilde{\phi}_\varepsilon$ describing the ejected electron. The *tilde* in $\tilde{\phi}_\varepsilon$ indicates that this scattering function is not a member of the molecular orbital basis $\{\phi_p\}$ introduced above. Rather, $\tilde{\phi}_\varepsilon$ is chosen to appropriately describe the ejected electron. For practical reasons it is useful to impose the strong-orthogonality condition $\tilde{a}_\varepsilon|\Psi_n^{N-1}\rangle = 0$ to ensure the normalization of $\Psi_{n\varepsilon}^N$. The state of the art in the computation of the one-electron scattering function and the associated cross sections has recently been reviewed by McKoy et al.[85] We refer the reader to this article for details on this subject not covered here.

Inserting the single-channel approximation (2.3) into expression (2.2) and making use of the well-known commutator relation between annihilation and creation operators defined for nonorthogonal orbitals, one readily obtains the following equation:

$$T_{n\varepsilon,0} = \sum_q \tau_{\varepsilon q} x_q^{(n)} + \langle \Psi_n^{N-1} | \hat{T} \tilde{a}_\varepsilon | \Psi_0^N \rangle, \tag{2.4}$$

where $\tau_{\varepsilon q}$ denotes the one-particle matrix element of $\hat{T}$ performed with $\tilde{\phi}_\varepsilon$ and ϕ_q. This result is a straightforward extension of the Martin–Shirley result[81] and is equivalent to the one derived by Arneberg et al.[41] and Cacelli et al.[83] The first term on the right-hand side (rhs) of Eq. (2.4) represents direct transitions between the orbital ϕ_q and the continuum orbital $\tilde{\phi}_\varepsilon$ and may be addressed as the *direct* term. The second term, on the other hand, vanishes only if $\tilde{\phi}_\varepsilon$ is orthogonal to the one-electron density matrix of the molecular ground state. Hence it is appropriate to follow Cacelli et al. and call this term the *nonorthogonality* correction term. Although a rigorous proof is missing, it is usually conjectured that owing to the fast oscillatory behavior of $\tilde{\phi}_\varepsilon$ at high kinetic energy ε of the ejected electron, the latter term decreases with growing ε faster than the direct term. Consequently, we may neglect the nonorthogonality term at high energies ε.

Very few calculations exist from which we may infer what high energy means in this case. Recent computations[83] on the photoionization of Ne give evidence that for outer-valence electrons the nonorthogonality term falls off rapidly to a minor contribution within 1–2 eV above the ionization threshold. For core electrons, on the other hand, ε must take values of 100 eV and more before this term becomes quantitatively unimportant. This trend can be intuitively understood by considering the state $\tilde{a}_\varepsilon | \Psi_0^N \rangle$ appearing in Eq. (2.4). $\tilde{\phi}_\varepsilon$ calculated in the field of $N-1$ occupied orbitals of the residual ion will be the more different from ϕ_ε calculated in the field of N orbitals the stronger the relaxation effects are due to the removal of an electron. If $\tilde{\phi}_\varepsilon = \phi_\varepsilon$, $\tilde{a}_\varepsilon | \Psi_0^N \rangle$ is essentially zero. Very little is known on the behavior of the nonorthogonality term for satellite states.

For sufficiently high energies ε, which according to the above can be fairly low on an absolute scale, we may discard the nonorthogonality term and determine the partial-channel cross section according to[12,28]

$$\sigma_n(\varepsilon) \propto \left| \sum_q \tau_{\varepsilon q} x_q^{(n)} \right|^2 . \tag{2.5}$$

In many cases of interest mainly a single transition amplitude contributes to the cross section[28] and we can write the cross section as a product of a *spectroscopic factor* (or pole strength) $|x_q^{(n)}|^2$ and an orbital ionization cross section $|\tau_{\varepsilon q}|^2$:

$$\sigma_n(\varepsilon) \propto |\tau_{q\varepsilon}|^2 |x_q^{(n)}|^2 . \tag{2.6a}$$

When the relation (2.6a) is valid, the ionization process may be viewed as the ejection of an electron out of a specific orbital ϕ_q. As should be clear from the discussion following Eq. (2.3), there corresponds a different orbital scattering function $\tilde{\phi}_\varepsilon$ to each ionic state $|\Psi_n^{N-1}\rangle$ even for those states that arise from the ejection of an electron out of the same orbital ϕ_q. Nevertheless, at sufficiently high energy ε, when the limit of the sudden approximation[11] is approached, the ratio of partial cross sections should approach the ratio of spectroscopic factors[86]

$$\frac{\sigma_n(\varepsilon)}{\sigma_{n'}(\varepsilon)} \longrightarrow \left| \frac{x_q^{(n)}}{x_q^{(n')}} \right|^2 . \tag{2.6b}$$

There is little experience on the range of validity of Eq. (2.6b) for photoionization, but, in general, we may expect that ε has to be much larger than that required to fulfill Eq. (2.6a) or, more generally, Eq. (2.5). The available numerous experimental $(e, 2e)$ data provide in many cases good evidence for the validity of Eq. (2.6b).

Once the golden rule (2.2) is accepted, the single-channel ansatz (2.3) comprises the only approximation made to arrive at the final result (2.4) and, therefore, this ansatz deserves more attention. The spherical symmetry of atoms simplifies the calculation of the partial-channel ionization cross section considerably. Indeed, for atoms there is a wealth of literature which goes beyond the single-channel approximation (for a recent review see ref. 87) from which we may draw some conclusions that also apply to molecules. Interchannel coupling exists over the whole spectrum, but its influence is particularly apparent where the bound states of one channel (e.g., Rydberg states) are embedded in the continuum of another channel, thus becoming autoionizing states. By definition, autoionization does not occur within the single-channel approximation scheme. On the other hand, resonances characterized by single-particle quantum numbers, that is, shape resonances, are taken account of in this scheme (e.g., see ref. 88 and the example in Section IV). Shape resonances are common in molecules and, due to their single-particle nature, dominate the overall shape of the cross section, while isolated autoionizing resonances are usually (for an interesting exception see ref. 89) long-lived and perturb the cross section substantially only in narrow spectral regions. Of course, interchannel coupling is also of importance near the threshold of a new channel where the series of Rydberg states accumulate. Calculations on atoms[90] indicate, however, that the coupling at threshold is often small unless one channel is weak.

B. Computational Aspects and Results

The ionization spectrum discussed in the preceding section is characterized by the ionization potentials I_n and the corresponding partial-channel cross sections σ_n. For large kinetic energy of the ejected electron, the latter quantities emerge from the superposition of transition amplitudes $x_q^{(n)}$. There are essentially two different approaches to compute the energies and amplitudes. In the first approach the energies and wavefunctions are separately computed for the molecular ground state and the final ionic states and the desired quantities are subsequently obtained in an obvious way. This approach comprises various types of configuration interaction methods as applied in the field of molecular ionization, for instance, by Müller[39] and Nakatsuji[42] using *ab initio* methods and Koenig and co-workers,[50] Bigelow,[47] and Schulz et al.[44] using semiempirical methods. In the second approach, the ionization potentials and transition amplitudes are calculated directly, usually as the eigenvalues and eigenvector components of a single matrix. The equation-of-motion method[69,91] and the method of Green's functions[28,70,92] are prominent examples of the second approach. The latter method has been used to compute the ionization spectra of a wide range of molecules. An updated list of applications of the Green's function and equation-of-motion methods can be found in ref. 32.

Both ionization potentials and transition amplitudes appear explicitly in the expression for the one-particle Green's function. In energy space this function reads[76]

$$G_{pq}^{-}(\omega) = \sum_n \frac{x_q^{(n)*} x_p^{(n)}}{\omega + I_n - i0^+}, \tag{2.7}$$

where only that part is shown which is analytical in the lower half of the complex energy plane. The other part of the Green's function describes electron attachment and is not relevant here. The ionization potentials appear as poles of the Green's function and the spectroscopic factors as pole strengths, making clear why the latter two names are often used synonymously. Most of the Green's function computations on satellite states published in the literature have been performed using the two-particle-hole Tamm–Dancoff approximation (2ph–TDA).[28,29] More recently,[92] the 2ph–TDA has been extended to yield a theory consistent through third-order perturbation theory. The discussion of these methods is beyond the scope of this review and we refer the interested reader to the above cited original publications. The technical details of the computations are described in a recent comprehensive review article.[32]

In the next section we discuss the various correlation effects present in the ionization of molecules. Results obtained via the 2ph–TDA or its extended version serve as a guideline in this discussion. Although the major conclusions drawn in the following sections are independent of approximations made in these methods, it is useful to assess the reliability of the methods. The calculated data on the first few ionic states of the six molecules explicitly discussed in the next section are collected in Table I. The results for carbonyl sulfide have been taken from the literature.[93] They have been obtained via the standard 2ph–TDA. All the other molecules appearing in the table are investigated using the extended 2ph–TDA. Also shown in the table are the available experimental ionization potentials. The accuracy of the calculated energies is satisfactory. Indeed, for the main lines the extended 2ph–TDA is superior over the 2ph–TDA. For the main ionic states the extended 2ph–TDA is consistent through third-order perturbation theory. As discussed in ref. 92, the extended 2ph–TDA is less accurate for satellite states than for the main ionic states. Experience shows, however, that the first few satellite states of each symmetry are usually predicted to a satisfactory accuracy. The situation is more problematic when a main line lies close to a weak satellite line which has acquired intensity from the same orbital. Then the different levels of accuracy for both lines may lead to an erroneous near-degeneracy of the underlying effective configurations and thus to an inaccurate sharing of intensity between the lines.

The description of high-energy structures is only qualitative in nature owing to the following reasons. The first reason is specific to the present approach in that the coupling of the 2h–1p configurations to higher excited configurations like the 3h–2p ones is neglected. Selective configuration-interaction calculations[38,40] indicate that this deficiency markedly influences the results only at relatively high energies. Of course, this deficiency can systematically be removed.[92] The second reason is common to all computational methods using a finite discrete basis of one-particle states. At higher energies we come close to the first double-ionization threshold to which several Rydberg series of the ion converge. Using a finite discrete basis, only the very first members of the series are described properly. One obtains artificial states and consequently artificial spectral structures at higher excitation energy which can be viewed as an inadequate simulation of higher Rydberg and continuum states. The

TABLE I

The First Few Vertical (Ref. 166) Ionization Potentials and Spectroscopic Factors (Pole Strengths) of CH_4, $ZnCl_2$, COS, C_2H_6, C_2H_4, and C_2H_2 Calculated Via Green's Functions[a]

Molecule	Orbital p	Hartree–Fock $-\varepsilon_p$	Extended 2ph–TDA[b] I_n	$\lvert x_p^{(n)}\rvert^2$	Experiment I_n
CH_4	$1t_2$	14.75	14.25	0.93	14.4[c]
	$1a_1$	25.48	23.14	0.83	23.0[c]
			30.84	0.03	
	$1e^d$		28.79	$<5\times10^{-3}$	
	$1t_1{}^d$		29.13	$<5\times10^{-3}$	
$ZnCl_2$	$2\pi_g$	12.60	11.69	0.93	11.85[e]
	$1\pi_u$	13.02	12.20	0.93	12.41[e]
	$2\sigma_u$	13.68	12.88	0.93	13.09[c]
	$3\sigma_g$	14.58	13.85	0.92	14.13[e]
COS[b]	2π	11.48	10.53	0.92	11.2[f]
	1π	18.09	14.89	0.75	15.5[f]
			17.46	0.02	
			18.70	$<5\times10^{-3}$	
	4σ	17.26	15.31	0.88[h]	16.02[f]
	3σ	21.12	17.29	0.80[h]	18.0[f]
	$1\delta^d$		20.36	$<5\times10^{-3}$	
C_2H_6	$1e_g$	13.08	12.68	0.92	12.5[c]
	$2a_{1g}$	13.88	13.37	0.91	13.3[c]
	$1e_u$	16.02	15.41	0.91	15.4[c]
	$1a_{2u}$	22.61	20.72	0.84	20.4[c]
	$1a_{1g}$	27.39	24.33	0.66	24.0[c]
C_2H_4	$1b_{2u}$	10.21	10.46	0.91	10.51[f]
	$1b_{2g}$	13.82	13.03	0.91	12.85[f]
	$2a_g$	15.89	14.81	0.91	14.66[f]
	$1b_{3u}$	17.48	16.09	0.79	15.87[f]
			17.73	0.06	
	$1b_{3g}{}^d$	−4.02	16.92	0.02	
	$1b_{1u}$	21.47	19.26	0.66	19.23[g]
C_2H_2	$1\pi_u$	11.21	11.37	0.91	11.4[h]
	$2\pi_g{}^d$		17.15	0.006[i]	
			19.46	0.005[i]	
	$2\sigma_g$	18.48	17.29	0.90	17.0[h]
	$1\sigma_u$	20.82	19.26	0.84	19.0[h]

See opposite for footnotes to Table I.

calculation still provides valuable information, since it specifies the various series that couple to the single-hole states. However, the computed specific discrete structures may not be reliable. In such cases we may expect a strong basis-set dependence of the results.[94]

III. CORRELATION EFFECTS IN THE IONIZATION OF MOLECULES

A. Representative Examples

Guided by the results of *ab initio* many-body calculations, we discuss in the present section the various correlation effects that are typically found in the ionization of the valence shell of molecules.

An example where the MO picture is valid over the whole spectral range is methane. The results obtained with the extended 2ph–TDA are depicted in Fig. 1 in the shape of a line spectrum. The position of each line is given by the computed ionization potential and its height by the corresponding spectroscopic factor. The number above each line indicates the orbital out of which ionization takes place. This line spectrum thus represents the ionization spectrum in the sudden approximation [Eq. (2.6)].

In its ground state ($\tilde{X}\,^1A_1$) methane has the electronic configuration (core) $(1a_1)^2(1t_2)^6$. Here and in the following only the valence orbitals are enumerated. The calculation predicts a $1t_2$ main line at 14.25 eV with a large spectroscopic factor of 0.93, a $1a_1$ main line with a spectroscopic factor 0.83 at 23.14 eV, and a few weak satellite lines above 30 eV binding energy. The large spectroscopic factor of the outer-valence ionization line implies that the corresponding ionic state is well described in the in-

---

[a] The basis sets used are at least of double-zeta plus polarization quality. Additional calculated data on the ionization of these molecules are discussed in the text and shown in Figs. 1–5. All energies are in eV.

[b] The results for COS are on the 2ph–TDA level. They are taken from ref. 93.

[c] Estimated centers of gravity from the spectra of refs. 4, 97, 98, and 100.

[d] These orbitals are not occupied in the Hartree–Fock ground-state configuration. Nonvanishing spectroscopic factors for the calculated satellite states arise from ground-state correlation.

[e] Reference 105.

[f] Reference 4.

[g] Reference 125.

[h] Estimated centers of gravity from the spectra of refs. 4 and 98. The $2\sigma_g$ and $1\sigma_u$ bands are broad and asymmetric and the vertical ionization potentials[166] are certainly higher than the band maxima.

[i] For these states the Green's function is nondiagonal.

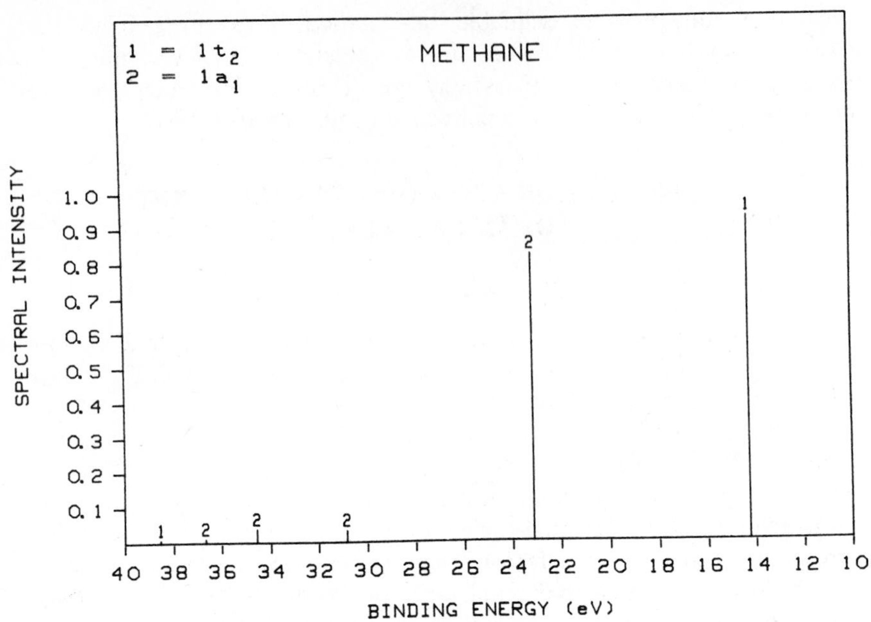

Fig. 1. Ionization spectrum of methane calculated with the extended 2ph–TDA. Shown are the binding energies up to 40 eV and the corresponding spectroscopic factors when larger than 5×10^{-3}. The number on top of each line specifies the orbital ϕ_p from which the ionic state Ψ_n^{N-1} has acquired its spectroscopic factor $|x_p^{(n)}|^2$. Some weak lines at high energy have not been labeled due to space limitations. Additional ionic states with spectroscopic factors smaller than 5×10^{-3} are discussed in the text.

dependent-particle model. The *C2s* inner-valence single-hole state has lost somewhat more of its intensity to the 2h–1p satellite states which can be seen in the figure. This trend that inner-valence single-hole states exhibit smaller spectroscopic factors than outer-valence ones is typical for molecules. Its understanding is facilitated by the obvious fact that the inner-valence 1h configuration is closer in energy to the 2h–1p configurations. Furthermore, the inner-valence orbitals are often more localized, thus giving rise to more pronounced relaxation of the charge density.

Since weak satellite lines are not easily detected in the PE spectrum, one might draw wrong conclusions on the distribution of cationic doublet states from the consideration of PE spectra. This applies also to Fig. 1 where only those states are shown which have spectroscopic factors larger than 5×10^{-3}. According to the computation, the actual number of satellite states in the energy range shown in the figure is 25. Most of these satellites are of ^2E and $^2T_1$ symmetry and, thus, are initial-state correlation satellites according to the classification of the preceding section. In CH_4 these satellite lines acquire only very little spectral intensity because of the weak ground-state correlation effects in this molecule. The lowest ^2E state is predicted at 28.79 eV and that of $^2T_1$ symmetry at 29.13 eV binding energy with the dominant electronic configurations $(1t_2)^{-2}(2a_1)$ and $(1t_2)^{-2}(2t_2)$ in both cases. The satellite lines visible in Fig. 1 can also be classified. The leading configurations associated with these satellites are $(1t_2)^{-2}(p)$ and $(1a_1)^{-1}$ where $p = 2a_1$, $2t_2$, and $3t_2$ for the $^2A_1$ symmetry, and $(1t_2)^{-2}(3t_2)$ for the $^2T_2$ symmetry, respectively. Correspondingly, the satellite lines originating from the ionization out of the $1t_2$ and $1a_1$ orbitals are shake-up and correlation satellites, respectively.

To what extent does the above picture agree with experiment? The ionization spectrum of methane has been investigated experimentally several times via He I,[4] He II,[95–98] ESCA,[99,100] and $(e, 2e)$[101,102] spectroscopy and exhibits two main bands at 14.4 and 23.0 eV binding energy in good agreement with the calculated values of 14.25 and 23.14 eV. Additional weak structures are observed above 28 eV in the $(e, 2e)$ spectrum which correspond to the shake-up processes discussed above. For completeness we mention that other Green's function and configuration-interaction calculations have been reported in the literature.[48,102,103]

Our first example, CH_4, is a typical candidate of weak correlation effects. As our second example we arbitrarily choose zinc dichloride because it exhibits stronger correlation effects, but its spectrum is still simple enough to allow for a straightforward classification. The leading electronic configuration in the ground state of $ZnCl_2$ is (core) $(1\sigma_g)^2(1\sigma_u)^2(1\pi_g)^4(1\delta_g)^4(2\sigma_g)^2(3\sigma_g)^2(2\sigma_u)^2(1\pi_u)^4(2\pi_g)^4$. The spectrum calculated using the extended 2ph–TDA is shown in Fig. 2. Except for some weak satellite lines this spectrum exhibits three groups of lines. The first group at lowest binding energy corresponds to the outer-valence orbitals $2\pi_g$, $1\pi_u$, $2\sigma_u$, and $3\sigma_g$, which are responsible for the bonding of the molecule. As for the outer-valence orbital of CH_4, these main lines also possess large spectroscopic factors around 0.93 and constitute excellent examples for the validity of the MO picture of ionization.

The second group of lines consists of three closely spaced lines. These derive from the $3d$ orbital of the metal atom which, being subject to a cylindrical ligand field, splits into the molecular orbitals $1\delta_g$, $1\pi_g$, and $2\sigma_g$. Similar to the inner-valence orbital of CH_4, the corresponding lines carry somewhat less pole strength than the outer-valence electrons but can be properly described in the MO picture. According to the present calculations, the ejection of an electron out of the outer valence as well as out of the metal d orbitals leads to a number of very weak satellite lines lying 10 eV or more above the corresponding main lines. In contrast to CH_4 these satellite lines can neither be characterized by a few electronic 2h–1p configurations nor be classified as correlation or shake-up satellites.

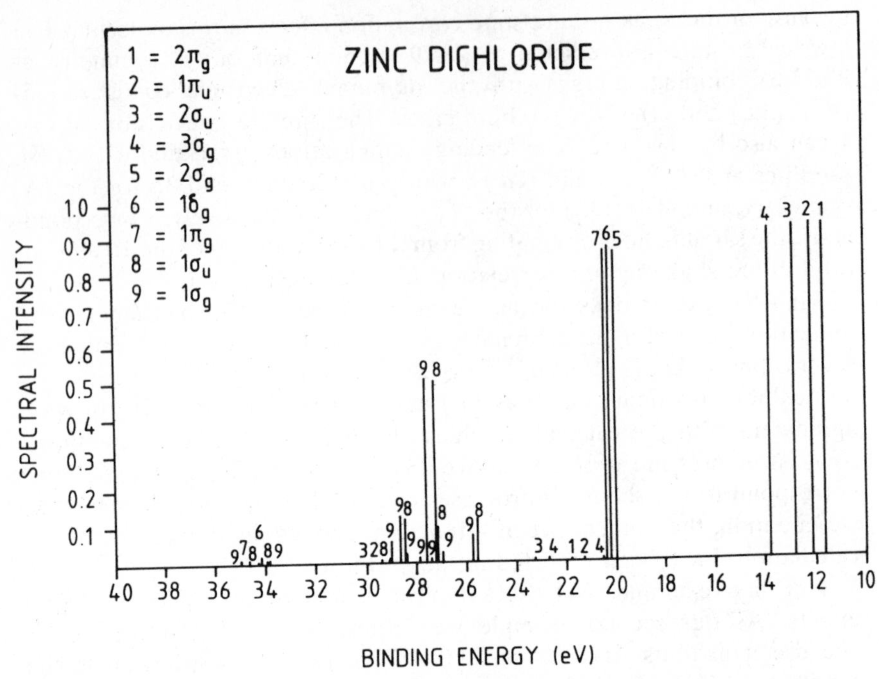

Fig. 2. Ionization spectrum of zinc dichloride calculated with the extended 2ph–TDA. For more details see Fig. 1.

The major difference between the "well-behaved" spectrum of CH_4 and that of $ZnCl_2$ comes to light when the inner-valence orbitals $1\sigma_g$ and $1\sigma_u$ are considered which have mainly chlorine 3*s* character. Ionization out of these orbitals leads to the lines seen in Fig. 2 between 25 and 29 eV binding energy. The spectral intensity associated with the $1\sigma_g$ and $1\sigma_u$ orbitals is no longer confined to a main line and some weak satellite lines, but rather is spread over several intense lines. To a certain extent one can still speak of a main line surrounded by intense satellite lines. The spectroscopic factor of the main lines is 0.5, implying that the satellite states have borrowed 50% of the available total spectral intensity. What leads to this dramatic violation of the MO picture in $ZnCl_2$ in contrast to CH_4? Owing to the heavier atoms in $ZnCl_2$, the manifold of 2h–1p configurations begins at an energy considerably lower than in CH_4. According to our calculations the first 2h–1p satellites of ${}^2\Sigma_g$ and ${}^2\Sigma_u$ symmetries lie at about 18 eV. These lines have not been depicted in Fig. 2 because of their small spectroscopic factors. In CH_4 the first satellite of ${}^2A_1$ symmetry is located above 30 eV, as shown in Fig. 1. Moreover, the large number of electrons in $ZnCl_2$ leads to a high density of such 2h–1p configurations which fall close to the 1h configuration, enlarging significantly the probability for a breakdown of the MO picture of ionization.

The 2h–1p configurations $m^{-1}n^{-1}t$ relevant for the breakdown of the MO picture for the ionization out of the inner-valence orbital ϕ_q are clearly those for which q is neither equal to m nor to n, that is, configurations which are responsible for final-state correlation effects and not for relaxation effects. These configurations describe the ejection of an electron out of the outer-valence orbital ϕ_m accompanied by the excitation of the outer-valence orbital ϕ_n to a virtual orbital ϕ_t. In the case of the $1\sigma_g$ inner-valence orbital of $ZnCl_2$, for instance, the most important 2h–1p configurations are $(n)^{-2}(\sigma_g^*)$, where σ_g^* symbolizes the low-lying virtual orbitals of σ_g symmetry and n the three highest occupied orbitals $2\pi_g$, $1\pi_u$, and $2\sigma_u$. For completeness, we mention that similar results are found also for the related molecule[104] $CdCl_2$. The outer-valence and metal d orbitals have been investigated experimentally[105–109] for both $ZnCl_2$ and $CdCl_2$, but no experimental data are available on the inner-valence region.

We have seen that, compared to CH_4, many-body effects are substantial in $ZnCl_2$. In the following we demonstrate that correlation effects as encountered in $ZnCl_2$ are not uncommon. On the contrary, strong many-body effects govern the ionization spectra of most ordinary molecules. As an example we choose carbonyl sulfide. This linear triatomic molecule possesses the ground-state $(\tilde{X}\,^1\Sigma^+)$ configuration (core) $(1\sigma^2)(2\sigma)^2(3\sigma)^2(1\pi)^4(4\sigma)^2(2\pi)^4$. The states of COS^+ have been investigated many times in the literature. Turner et al.[4], Kovač,[110] Natalis et al.,[111] and Delwiche et al.[112] discuss different aspects of the PE spectrum recorded using He I radiation. The latter authors have also measured a high-resolution spectrum using Ne I radiation. To locate higher-lying ionic states Potts and Williams[113] have used He II radiation and Allan et al.[114] Mg K_α radiation, while Cook et al.[93] report on binary as well as dipole $(e, 2e)$ spectra. Molecular constants of COS^+ have been determined by Frey et al.[115] via threshold PE spectrometry. Utilizing synchrotron radiation, angle-resolved PE spectra measured as a function of photon energy from 21 to 70 eV have been reported by Carlson et al.[116] The valence-shell binding energy spectrum has been investigated theoretically by Cook et al.[93] with the aid of the 2ph–TDA and by Nakatsuji[42] using symmetry-adapted-cluster CI.

The results obtained for COS with the 2ph–TDA calculation of ref. 93 are depicted in Fig. 3. These results are compared in Fig. 4 with several experimental ionization spectra. These figures clearly demonstrate a different behavior of the ionization of outer-valence and inner-valence

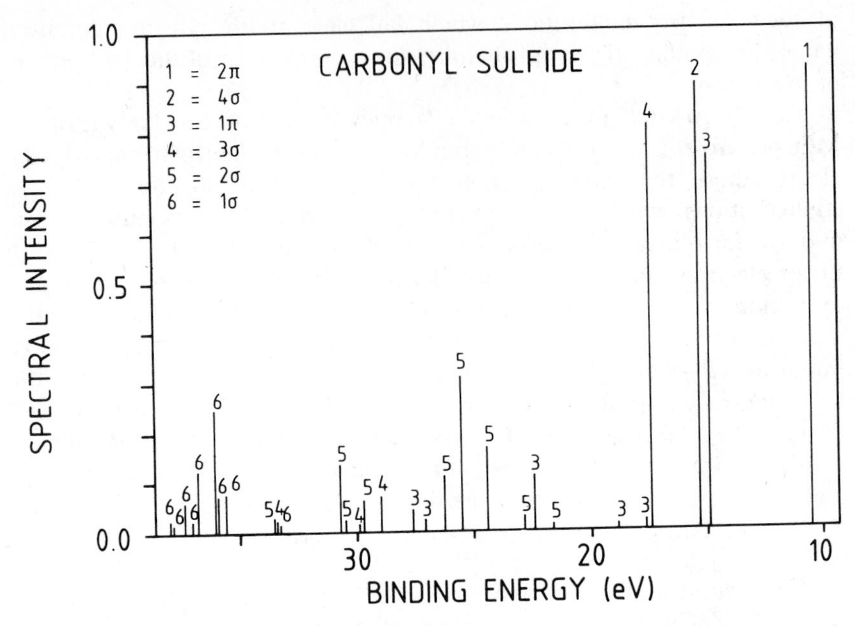

Fig. 3. Ionization spectrum of carbonyl sulfide calculated with the 2ph–TDA and taken from ref. 93. For more details see Fig. 1.

electrons. As typical for many small- and medium-size molecules, the outer-valence ionization lines exhibit relatively large spectroscopic factors, indicating the validity of the MO picture. Contrary to $ZnCl_2$, where the MO picture applies extremely well to all outer-valence electrons (all spectroscopic factors are larger than 0.9), relaxation and correlation effects are important in COS for the ionization of the 3σ and especially of the 1π orbitals. As can be seen in Fig. 3, the missing weight of the 3σ main line has been borrowed by a few high-lying satellite lines. These satellites can be classified as shake-up satellites, since the corresponding states are characterized mainly as $2\pi \to \pi^*$ excitations on top of a vacancy in the 3σ orbital. More interesting is the ionization of the 1π orbital. Application of Koopman's theorem leads to a wrong ordering of the 1π and 2σ main lines. In the 2ph–TDA the 1π main line is shifted to lower binding energies much more than the 2σ main line and the experimental ordering is reproduced. Such a considerable shift due to many-body effects is often an indication for intense satellite structure associated with the orbital in question. Indeed, as can be seen in Fig. 3, several intense satellite lines have borrowed their intensity from the 1π orbital leading to a 1π main line with a spectroscopic factor as small as 0.75.

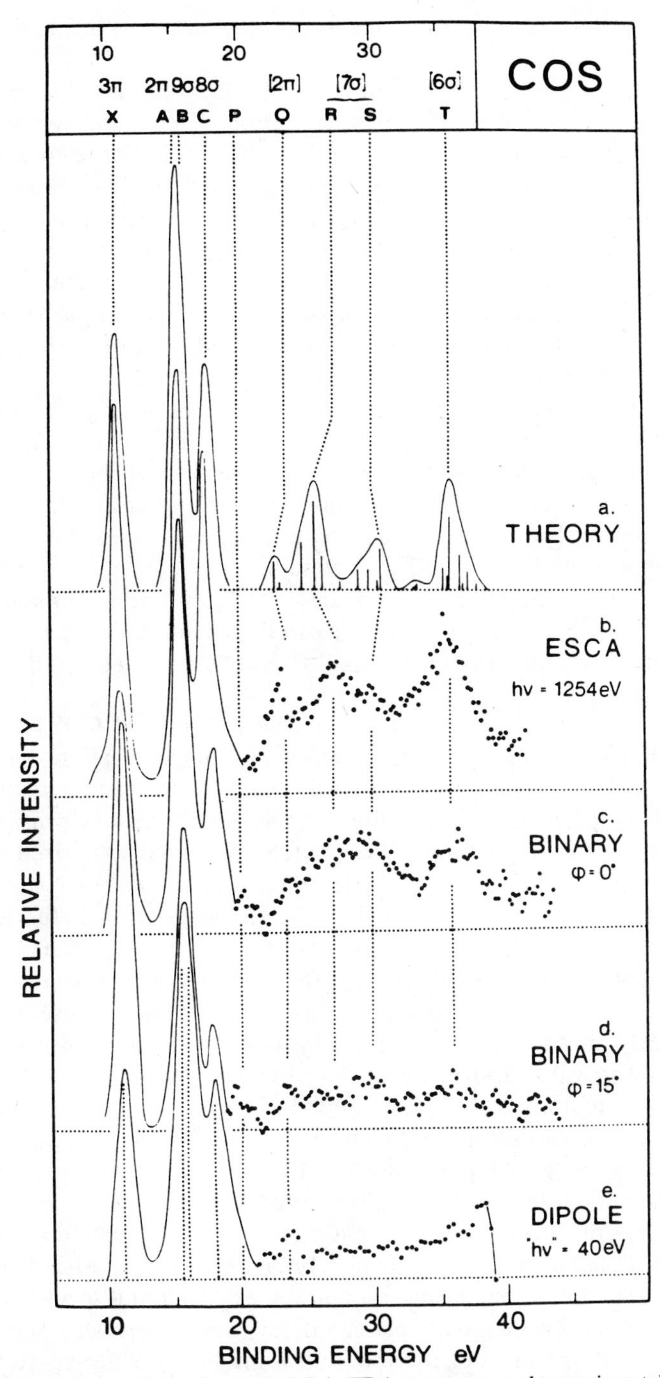

Fig. 4. Comparison of the theoretical 2ph–TDA spectrum and experimental ESCA,[114] binary, and dipole $(e, 2e)$[93] spectra of COS. The figure is taken from ref. 93. To compare with Fig. 3 and the text, note that the energy scale is reversed and the orbitals 2π, 3π, 9σ, 8σ, 7σ, and 6σ should read 1π, 2π, 4σ, 3σ, 2σ, and 1σ.

The origin of the above satellite lines, of which a few are at relatively low binding energy, deserves some attention. The ionization of a 2π electron accompanied by the $2\pi \rightarrow 3\pi$ ($\pi \rightarrow \pi^*$) excitation comprises five symmetry-adapted excitations: three configurations of $^2\Pi$ symmetry, a $^2\Phi$ configuration, and a $^4\Pi$ configuration. Of these configurations only the three $^2\Pi$ configurations can couple to the 1π single-hole configuration due to symmetry restrictions. This coupling gives rise to a 1π main line and three $^2\Pi$ satellite lines which have borrowed intensity from the 1π orbital. According to the classification scheme introduced in Section II, the latter satellites are correlation satellites. The 2ph–TDA indeed predicts the three lowest satellite lines of $^2\Pi$ symmetry shown in Fig. 3 to mainly correspond to the $(2\pi)^{-2}(3\pi)$ configuration. Because of inaccuracies inherent to the 2ph–TDA, the first or even the second $^2\Pi$ satellite could be located in reality between the third and fourth main lines in the spectrum as is the case in the closely related CS_2 molecule. In the latter molecule, two of the predicted[17,42] three $^2\Pi_u$ satellite lines have been unambiguously identified in the experimental spectrum[4,17,116,117] until now. No systematic search for satellite lines has been carried out for COS, but there is experimental evidence for an intense $^2\Pi$ satellite line at about 20.5-eV binding energy.[4,93,113,114]

The existence of a low-lying and localized π^* orbital in COS and CS_2 is responsible for the above low-lying relatively intense satellite lines (see also Section III.B). Contrary to CS_2, the lack of inversion symmetry in COS leads to additional $^2\Pi$ satellites which have borrowed considerable intensity from the 1π orbital. These satellites are characterized by the $(1\pi)^{-1}(2\pi)^{-1}(3\pi)$ configurations and are thus shake-up satellites. They can be seen in Fig. 3 at about 27 eV. We briefly mention here that the 2ph–TDA predicts several low-lying (below and around 20 eV) initial-state correlation satellites of $^2\Delta$ and $^2\Phi$ symmetry not shown in Fig. 3 because of their small spectroscopic factors (the present basis set contains virtual δ orbitals and a more extended basis set will contain also virtual ϕ orbitals to which the 2h–1p configurations can couple).

The rich satellite structure discussed above may serve as an indication for severe limitations of the MO picture of ionization of the inner-valence orbitals 1σ and 2σ. As demonstrated by Fig. 3, the spectral intensity associated with these orbitals is distributed over several or even many lines and the distinction between main and satellite lines becomes useless.[118] We thus encounter a complete breakdown of the MO picture of ionization. The broad bands observed in the experimental spectra and the satisfactory overall agreement between theory and experiment (see Fig. 4) underline the correctness of the theoretical prediction with regard to this breakdown phenomenon. For both inner-valence orbitals of COS, the energy range over which the major portion of the spectral intensity is distributed amounts to about 5 eV and more. This energy is a crude measure for the strength of the interaction between the 1h and 2h–1p or even higher excited electronic configurations and may reach 10 eV and more in some molecules.[119]

So far we have discussed three examples which have been selected out of a large number of calculations to illustrate the range of final-state correlation effects in molecular ionization. While correlation effects were found to be very weak in CH_4, moderate effects are present for the innermost valence orbitals of $ZnCl_2$ and strong effects were predicted in COS affecting even some of the outer-valence orbitals. The breakdown of the MO picture of ionization demonstrated for the inner-valence orbitals of $ZnCl_2$ and COS is a general phenomenon common to the majority of molecules. Of course, the specific features of the breakdown phenomenon depend on various properties of the molecule under consideration, for example, size, symmetry, type of atoms included, and, most of all, its bonding properties.

To illustrate this important fact we compare in what follows the spectra of single-bonded, double-bonded, and triple-bonded hydrocarbons C_2H_n, $n = 6$, 4, and 2, respectively.

In Fig. 5 the calculated spectra of ethane, ethylene, and acetylene are shown. The molecules are highly symmetric and small enough to ensure that the correlation effects are less pronounced than in COS, for example. This choice simplifies the comparison of the spectra.

The spectrum of C_2H_6 is particularly simple. It exhibits five main lines accompanied by a series of weak satellite lines at the high-energy side of the main lines. The spectroscopic factors of the main lines gradually decrease from the value of 0.91 for the outermost valence orbital $1e_g$ to the value of 0.66 for the inner-valence orbital $1a_{1g}$, which is mainly of carbon 2s character. In other words, the MO picture is qualitatively valid for the ionization out of all the valence orbitals, but its quantitative accuracy smoothly diminishes as we go up in binding energy. The situation is typical for small saturated hydrocarbons[48] and, with some modifications, may also prevail in saturated molecules in general.

The overall appearance of the spectra of C_2H_2 and especially of C_2H_4 is in sharp contrast to that of C_2H_6. Whereas the latter spectrum exhibits only weak satellite structures above the innermost main line, the ionization spectra of C_2H_2 and particularly of C_2H_4 are rich in satellite lines beginning at a relatively low binding energy of about 17 eV. The most intense line emerging from the ionization of the inner-valence orbital $1a_g$ of C_2H_4 has a spectroscopic factor smaller than 0.5. The "missing" spectral intensity is spread over several lines, some of which are well

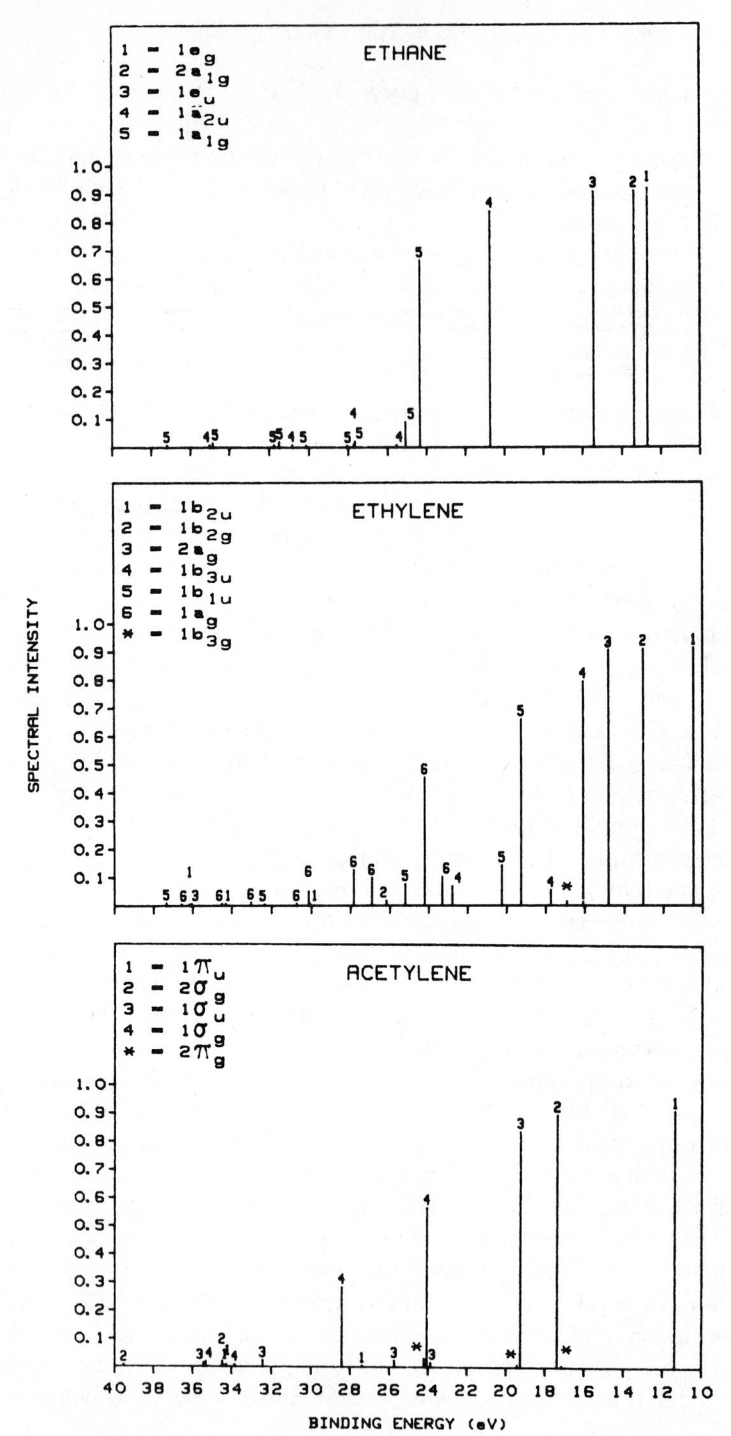

Fig. 5. Ionization spectra of ethane, ethylene, and acetylene calculated with the extended 2ph–TDA. Lines labeled with (*) have gained their spectral intensity from orbitals unoccupied in the Hartree–Fock ground state. For further details see caption of Fig. 1.

separated from this line and have also acquired considerable spectral intensity. The manifestation of correlation effects found for ethylene is typical for small unsaturated hydrocarbons. In larger and less symmetric unsaturated hydrocarbons the breakdown of the MO picture of ionization may become more severe.[48] Hence we recognize a clear-cut correspondence between many-body effects in the ionization spectrum and the chemical bonding. In the next section we present a discussion of the different behaviors of saturated and unsaturated molecules with respect to ionization.

There is a wealth of experimental data on the ionization of the three molecules ethane, ethylene, and acetylene, thus allowing a detailed discussion of the many-body effects present in the spectra. These effects are much more interesting in the unsaturated molecules with which we begin.

Since ethylene is the simplest hydrocarbon with a π-electron bond, its ionization spectrum has been extensively investigated. A series of He I spectra are available in the literature (e.g., see refs. 4, 120–122), including a high-resolution one recorded using a supersonic molecular beam.[122] There exist He II,[95,98,123] ESCA,[52,124,125] and $(e, 2e)$[126,127] spectra as well as an angle-resolved photoelectron spectrum[128] measured using synchrotron radiation. A number of theoretical investigations on C_2H_4 which go beyond the Hartree–Fock level have also been published comprising Green's function and configuration-interaction calculations.[48,126,129–131]

There is general agreement concerning the assignment of the first four observed bands in the ionization spectrum of C_2H_4. In order of increasing binding energy they are associated with the ejection of electrons out of the outer-valence orbitals $1b_{2u}$, $1b_{2g}$, $2a_g$, and $1b_{3u}$ in accordance with the ground-state $(\tilde{X} A_g)$ electronic configuration (core) $(1a_g)^2(1b_{1u})^2(1b_{3u})^2(2a_g)^2(1b_{2g})^2(1b_{2u})^2$ of this molecule. Between the $1b_{3u}$ and $1b_{1u}$ main lines the extended 2ph–TDA calculation predicts the existence of two interesting satellite lines. The first line at 16.92 eV corresponds to the $\tilde{D}\,^2B_{3g}$ state of $C_2H_4^+$. In its ground state ethylene does not possess an occupied orbital which transforms according to the B_{3g} irreducible representation. Following the discussion in Section II, this $^2B_{3g}$ satellite is an initial-state correlation satellite. It has acquired its nonvanishing spectroscopic factor $(= 0.015)$ from the ejection of the electron out of the virtual orbital $1b_{3g}$ which is partially occupied in the ground state of C_2H_4. The leading configuration in the expansion of the $\tilde{D}\,^2B_{3g}$ state is $(1b_{2u})^{-2}(1b_{3g})$. There are many additional initial-state correlation satellite states of $^2B_{3g}$, $^2B_{1g}$, and ^2A_u symmetry between 21.80 and 40 eV, namely, 25, 24, and 27 states, respectively, but all have very small spectroscopic factors. The second satellite line shown in Fig. 5 between the fourth and fifth main lines originates from the ionization of the outer-valence $1b_{3u}$ orbital. Being very close to its main line suggests that it is a correlation satellite. It is predicted to be at 17.73 eV, to have a spectroscopic factor 6×10^{-2}, and to correspond to the leading configuration $(1b_{2g})^{-1}(1b_{2u})^{-1}(1b_{3g})$.

The ionization of the $1b_{1u}$ orbital leads to a main line at 19.26 eV with a spectroscopic factor of 0.66, to a close-by satellite line at 20.25 eV, and to a few weaker satellite lines above 25 eV. The experimental spectra show a band centered at about[125] 19.23 eV which is broad enough to comprise both the main and the close-by satellite line. The ionization of the inner-valence orbital $1a_{1g}$ is more interesting as far as strong final-state correlation effects are concerned. The present calculation predicts that due to final-state correlation effects the $1a_g$ orbital line splits into two pairs of lines centered at 24.04 and 27.44 eV (see Fig. 4). The ratio of spectral intensity of these two groups is 1 : 0.4. The situation is very similar to the one encountered in the 2ph–TDA[48] except that the distribution of spectral intensity within the first pair of lines is different. A number of weaker satellite lines of ^2A_g symmetry appear above 30 eV; the most intense of them is the one at 30.14 eV with a spectroscopic factor 0.05.

The experimental ESCA spectra show in the energy range of the inner-valence $1a_g$ orbital one intense band at 23.7 eV, a less intense band at 27.4 eV, and a weak structure at about 31 eV. The measured ratio of the intensity of the first two bands is[52] 1:0.39. These measurements agree nicely with the results of the extended 2ph–TDA if we attribute the observed intense band to the calculated pair of lines centered at 24.04 eV, the less intense band to the pair of lines centered at 27.44 eV, and the weak structure to the weak satellite line predicted to be at 30.14 eV. Vibrational broadening and the limited experimental resolution prevent the observation of the individual lines and a more detailed comparison with the theoretical predictions. The $(e, 2e)$ measurements[126] provide some evidence for structures above 30 eV binding energy. As can be seen in Fig. 5, a number of weak satellite lines are predicted by the calculation to lie in this energy range.

Martin and Davidson[130] report on a configuration interaction calculation performed within a model orbital space. The calculation predicts three ^2A_g states which originate from the $1a_g$ orbital. With these results it is possible to explain the observed structures. Due to the limited orbital space, one cannot account for the splitting of lines and the rich satellite structure in the inner-valence region predicted by the present calculation A more recent configuration-interaction calculation by Nakatsuji[131] predicts the existence of numerous satellite lines. The calculated intensities, however, are too small to account for the experimental findings discussed above.

Being the simplest example of the triple carbon–carbon bond, acetylene has attracted much attention. The photoelectron spectrum of this molecule has been recorded many times using a variety of photon sources. Several He I,[4,132] He II[98,123,132] and ESCA[40,132] studies have been reported of which we mention only a few. Utilizing synchrotron radiation, Bradshaw et al.,[133] Parr et al.,[134] and more recently Svensson et al.[135] have measured ionization spectra at different photon energies. An $(e, 2e)$ spectrum using different energies of the incident electron together with a simple Green's function calculation are reported by Dixon et al.[136] Other Green's function[137] and 2ph–TDA[48] calculations as well as a recent configuration-interaction[40] calculation on acetylene are also available.

In its $\tilde{X}\,^1\Sigma_g^+$ ground state acetylene is well described by the configuration (core)$(1\sigma_g)^2(1\sigma_u)^2(2\sigma_g)^2(1\pi_u)^4$. The ionization spectrum of C_2H_2 calculated using the extended 2ph–TDA is depicted in Fig. 5. The spectrum is similar to the one of C_2H_4 in that it exhibits low-lying satellite lines and the ionization of the inner-valence orbital leads to a considerable split of lines. The gap between the two major components of the $1\sigma_g$ orbital amounts to nearly 4.4 eV and should be compared with the 3.4 eV in the case of the $1a_g$ orbital of ethylene. The computed ratio of spectral intensity between these components is 1:0.5 in acetylene and 1:0.4 in ethylene. As can be seen in Fig. 5, the calculated spectrum of C_2H_2 exhibits less satellite lines with substantial spectroscopic factors than found for C_2H_4. This finding can be attributed to the higher symmetry of C_2H_2 and to the smaller number of occupied outer-valence orbitals. In particular, the orbitals describing the π-electrons are degenerate which reduces the density of 2h–1p configurations considerably and suppresses the number of satellite lines acquiring relevant intensity. This argument is substantiated by the fact that the gap between the two outermost orbitals is 7.5 eV in C_2H_2 compared to 3.6 eV in C_2H_4.

Already below the second main line in the spectrum of C_2H_2 the calculation predicts another ionization event. The corresponding initial-state correlation satellite state is of $^2\Pi_g$ symmetry and predicted to be at 17.15 eV. This state has the dominant configurations $(1\pi_u)^{-2}(n\pi_g)$ where $n = 2,3$. It is well known that the coupling of three electrons of π symmetry results in three $^2\Pi$ states. Indeed, Fig. 5 shows two additional $^2\Pi_g$ satellite lines at 19.46 and 24.21 eV which acquired spectroscopic factors via ground-state correlation effects. The situation resembles that encountered[17] in the ionization of CS_2, but in this molecule a bound π_g orbital is available. The present calculation predicts the three $^2\Pi_g$ satellite lines to lie in the vicinity of main lines making their detection difficult. High-resolution spectra recorded as a function of photon energy may be helpful in this case. For completeness we mention that numerous initial-state correlation satellites of $C_2H_2^+$ are predicted below 40 eV exhibiting very small spectroscopic factors. Between 24 and 40 eV we find 24 $^2\Pi_g$ states, 12 $^2\Delta_g$ states, and 14 $^2\Delta_u$ states. The $^2\Delta_g$ and $^2\Delta_u$ states lowest in energy appear at 24.98 and 24.54 eV, respectively, and have the leading configurations $(1\pi_u)^{-2}(n\sigma_g)$, where $n = 4$–6, and $(1\pi_u)^{-2}(n\sigma_u)$, where $n = 2$–6.

In agreement with previously obtained 2ph–TDA results,[48] the present calculation predicts the ionization of the $1\sigma_g$ orbital to result in several ionic states, two of which possess significant spectroscopic factors. Figure 5 shows the first line at 24.06 eV with a spectroscopic factor 0.57 and the second line at 28.43 eV associated with a spectroscopic factor 0.28. The most prominent configurations contributing to the corresponding states are $(1\sigma_g)^{-1}$ and $(1\sigma_u)^{-1}(1\pi_u)^{-1}(2\pi_g)$. For the second state the configuration $(1\pi_u)^{-2}(3\sigma_g)$ is also of importance. The ESCA,[40,132] $(e, 2e)$,[136] and synchrotron[133] spectra exhibit an intense band at approximately 23.5 eV and a less intense broad band at approximately 28 eV. The measurement of the electron momentum distribution for these bands indicates[136] that both originate from the ionization of the inner-valence $1\sigma_g$ orbital, thus substantiating the theoretical predictions.

Recent high-resolution spectra recorded using monochromatized Al K_α radiation[40] and synchrotron radiation[135] show that the broad band at about 28 eV exhibits internal structure. This interesting finding has lead Müller et al.[40] to theoretically search for additional states around 28 eV with considerable spectroscopic factors not found by the 2ph–TDA.[48] They have performed a careful study on the problem carrying out several different types of configuration–interaction calculations and a computation of the band profile. In agreement with the 2ph–TDA[48] and the present calculation they do find several $^2\Sigma_g$ states, but only one of them carries a relevant spectroscopic factor. The origin of the observed internal structure is still unclear. A speculative interpretation of this structure is discussed in Section IV where the experimental spectrum is also shown.

Finally, we mention that additional structures are observed in the $(e, 2e)$ spectrum in the energy range above 30 eV. The calculation predicts several weak lines in this energy range (see Fig. 5). The understanding of these structures requires more elaborate experimental and theoretical work.

As mentioned above, many-body effects are less effective in the saturated molecule C_2H_6. The calculation predicts a few weak satellite lines to lie on the high-energy side of the spectrum. The most intense satellite line is close to the $1a_{1g}$ main line and is a correlation satellite. The small spectroscopic factor (0.09) of this satellite implies that the interaction matrix elements between the 2h–1p and 1h configurations are small as well and that this satellite has acquired intensity mainly because of the accidental near-degeneracy of these configurations. A relatively small hypothetical shift of the satellite line away from the main line would rapidly diminish the spectroscopic factor of the former line. Weak interactions of this kind are typical for saturated hydrocarbons and related molecules (see discussion below). There is some evidence in the $(e, 2e)$ spectrum[138] for the calculated weak satellite structures above the $1a_{1g}$ main line. More detailed experiments are necessary to provide additional information on this energy range.

For completeness, we mention that the ionization spectrum of ethane has been investigated using He I,[4,139] He II,[96,98,139–141] ESCA,[99,100] and $(e, 2e)$[138] spectroscopy. There is also considerable theoretical material available on the ionic states of C_2H_6. The Hartree–Fock and configuration-interaction calculations performed until 1977 have been summarized by Richartz et al.[142] Green's function and 2ph–TDA calculations have been published by Cederbaum et al.[48] and computations within the correlation hole model[143] have been carried out by Moscardó et al.[144] More recently, a Green's function calculation has been reported by Cacelli et al.[145]

B. Discussion

This subsection is devoted to the discussion of general aspects of the breakdown of the MO picture of ionization. Several general conclusions have been drawn in the preceding section while discussing explicit examples and are not repeated here. We just summarize that a large number of calculations indicate that the ionization of inner-valence electrons generally cannot be described as the creation of quasi-holes. Instead of finding a main line possibly accompanied by some weak satellite lines, the spectral intensity is distributed over several or even many lines, some of them exhibiting comparable strengths. In general, the high density of 2h–1p and higher ionic configurations found for larger molecules of low symmetry favors the near-degeneracy of these configurations with the configuration of the inner-valence vacancy and thus the breakdown of the MO picture of ionization. The nature of the chemical bonding, on the other hand, controls to a large extent the coupling of these two types of configuration and is thus a major factor influencing the form of the spectrum.

We further note that there is an intimate relationship between the behavior of the inner-valence orbitals with respect to ionization and the satellite structure resulting upon the ionization of outer-valence orbitals. The ionization spectra of molecules exhibiting the breakdown phenomenon often show low-lying satellite lines originating from the outer-valence orbitals. In our examples of the preceding section the first satellite lines appear at about or just below 17 eV for COS, C_2H_4, and C_2H_2, whereas for CH_4 and C_2H_6 they appear at about 31 and 25 eV, respectively. The appearance of intense low-lying satellite lines, on the other hand, is related to the existence of outer-valence main lines which carry a relatively small portion of the intensity originally confined to the orbital in question. The missing intensity is then usually rediscovered in pronounced satellite lines in the neighborhood of the corresponding main line. The ionization of the 1π orbital in COS (see Fig. 3) and the $1\pi_u$ orbital[17] in CS_2 provide nice examples for low-lying intense satellite lines in small molecules.

For a better understanding of the above findings we first discuss which molecular properties favor the breakdown phenomenon to occur in the region of low binding energy. To keep the discussion as simple and transparent as possible we consider the interaction of a single-hole configuration with a set of excited 2h–1p configurations. The interaction among the latter configurations is neglected. It is then clear that a near-degeneracy of the single-hole configuration with the other ionic configurations is prerequisite for the breakdown phenomenon to occur. The energy of a 2h–1p configuration $(k)^{-1}(\ell)^{-1}(j)$ relative to the energy of the ground-state configuration reads

$$E_{k\ell j} = -\varepsilon_k - \varepsilon_\ell + \varepsilon_j + V_{k\ell k\ell} - V_{kjkj} - V_{\ell j\ell j}, \tag{3.1}$$

where V_{mnmn} is the usual Coulomb matrix element and exchange has been suppressed. For low-lying 2h–1p configurations to exist, the energy ε_j of the virtual orbital as well as the absolute values of the energies ε_k and ε_l of the occupied orbitals should be small. However, the quantities $|\varepsilon_k + \varepsilon_\ell - \varepsilon_j|$ are usually much too large to explain the observed low-lying satellite lines and we must expect matrix elements V_{kjkj} and $V_{\ell j\ell j}$ of considerable magnitude. It is worthwhile to note that in first-order perturbation theory $\varepsilon_j - \varepsilon_k - V_{kjkj}$ and $\varepsilon_j - \varepsilon_\ell - V_{\ell j\ell j}$ are the excitation energies of the neutral molecule from the orbital ϕ_k and ϕ_ℓ to the orbital ϕ_j when exchange is neglected. Low-lying ionic 2h–1p configurations are thus expected to occur when the neutral molecule possesses low-lying excited states.

The second factor of relevance in the interpretation of the breakdown phenomenon is obviously the magnitude of the coupling between the single-hole $(p)^{-1}$ configuration and the 2h–1p configurations $(k)^{-1}(\ell)^{-1})(j)$. The absolute value of the corresponding matrix element $V_{pjk\ell}$ is expected to be large only if the virtual orbital ϕ_j is localized in space as the occupied orbitals are. Sufficiently large matrix elements can thus be expected just for those molecules exhibiting low-lying valence-type excited states. Another measure of the localization of the unoccupied orbital ϕ_j is the singlet–triplet splitting of the low-lying excited molecular states. In first-order perturbation theory this splitting is given by $2V_{kjjk}$ for states characterized by the transition of an electron from ϕ_k to ϕ_j. In first-order perturbation theory the electron affinity of the molecule, or, if there are no bound anionic states, the position of a shape resonance in the electron scattering cross section is given by the Hartree–Fock orbital energy ε_j. The presence of a bound anionic state[146] and of a narrow resonance is a clear indication of the localized character of the corresponding unoccupied orbital. The existence of bound anionic states and of low-lying resonances thus favors the occurrence of the breakdown phenomenon. The above criteria are not completely independent, since we may assume intercorrelation between the appearance of low-lying valence-type excited states of the neutral molecule with considerable singlet–triplet splittings and of bound anionic states or low-lying narrow shape resonances.

The lowest excitation energies, the singlet–triplet splitting corresponding to the lowest-lying excited state, and the energies of the lowest resonances in the electron scattering cross section are collected in Fig. 6 for a number of hydrocarbons. The figure shows a clear separation between the saturated and unsaturated hydrocarbons, that is, between molecules which do not and molecules which do exhibit the breakdown of the molecular orbital picture of ionization. The first excitation has lower energy, the single–triplet splitting is more pronounced, and a low-lying resonance is more likely to be found for the unsaturated molecules than for the saturated ones. No long-lived resonance states have been found for the latter molecules. Moreover, saturated molecules lack low-lying valence states at the vertical geometry. The low-lying excited states of these molecules are commonly referred to as Rydberg states owing to their diffuse nature in the vertical geometry.[147] With some care molecules other than hydrocarbons may also be incorporated into the scheme suggested by Fig. 6, but it is more appropriate to compare a set of related molecules rather than very different ones.

Since we have discussed COS in some detail in Section III.A, we comment on its data with respect to Fig. 6. Recent electron impact measurements[148] show that COS possesses several shape resonances. One of these is located at an energy as low as 1 eV above threshold and is associated[149] with the π^* orbital. This result augments the discussion in Section III.A where it has been noted that the π^* orbital plays a central role in the explanation of the calculated low-lying satellite lines in the ionization spectrum. The lowest measured excitation energy of COS is[150]

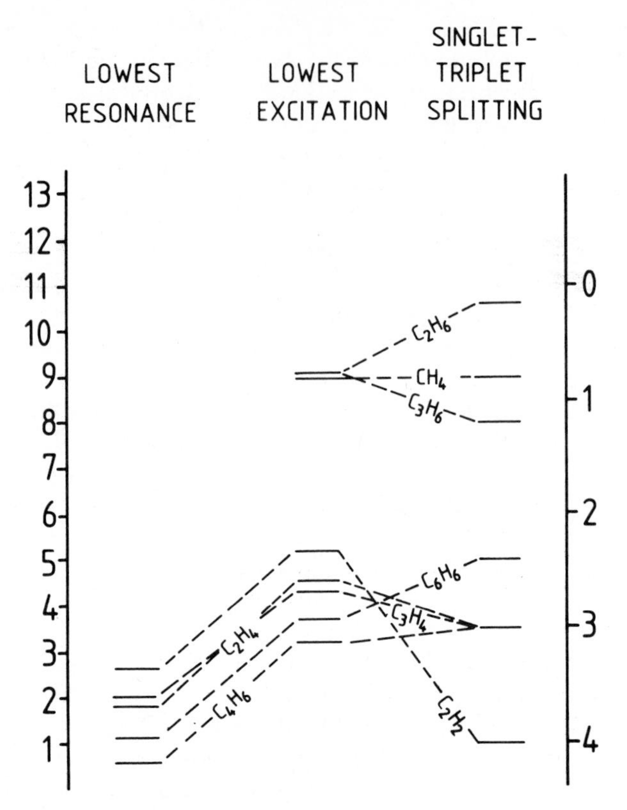

Fig. 6. Energies of the lowest resonances, lowest excitation energies, and singlet–triplet splitting energies of the lowest excited states of several hydrocarbons. The right-hand scale is only for the single–triplet splittings. All energies are in eV.

4.9 eV and falls into the range of energies of the unsaturated hydrocarbons shown in Fig. 6. There is indication[151] that several low-lying excited states of COS exist, being associated with $\pi \rightarrow \pi^*$ transitions. The singlet–triplet splitting of the Σ^+ $(2\pi \rightarrow 3\pi)$ state is around[151] 3 eV and also fits the scheme proposed by Fig. 6.

Very little is known about $ZnCl_2$ which has also been discussed in the preceding section. Since this molecule is saturated, we expect weaker correlation effects than for COS. Indeed, the spectrum of this molecule (Fig. 2) does not show low-lying satellite lines in the vicinity of the outer-valence main lines.

In the high-energy region of the ionization spectrum the density of ionic configurations is generally high and we may *a priori* expect many lines to acquire intensity. However, for pronounced effects to occur some of the matrix elements for the coupling of these configurations with the single-hole vacancy must be large. This requirement is expected to be fulfilled if there are bound anionic states or long-lived single-particle resonance states localized in space. This explains why there is a relationship between correlation effects in the outer- and inner-valence regions. To further elucidate the problem we briefly discuss a specific example. The low-lying excited states of the isoelectronic series CH_4, NH_3, H_2O, HF, and Ne are Rydberg states[152] and these systems do not have bound anionic states or long-lived resonances. Consequently, the ionization spectra do not exhibit intense satellites lines in the outer-valence region. The calculated spectrum of CH_4 has been discussed in the preceding subsection and shown in Fig. 1. Only weak high-energy satellite lines can be seen in this spectrum and the main line associated with the $1a_1$ inner-valence orbital has a large spectroscopic factor. Owing to their N2*s* and O2*s* character, the $1a_1$ orbitals of NH_3 and H_2O lie at deeper energies where the density of ionic configurations is higher. The lowering of symmetry effectively also enhances the density of relevant configurations and the breakdown phenomenon may occur for these orbitals. The lines over which the main portion of the spectral intensity of the $1a_1$ orbital is distributed, however, lie relatively close to each other,[12,34,41,60] reflecting the relatively weak coupling between the relevant configurations and the $1a_1$ vacancy. In contrast to the spectra of unsaturated molecules, the ESCA spectra of NH_3 and H_2O show a single inner-valence band accompanied by some satellite structure on its high-energy wing.[153] However, these bands are broad enough to include several close-lying lines.

Although the Ne2*s* orbital is much deeper in energy than the N2*s* and O2*s* orbitals, the experimental ESCA spectra[154] of the noble gas atoms exhibit a main line associated with the 2*s* vacancy, accompanied with a series of weak higher-lying satellite lines. Qualitatively, the spectrum of Ne rather resembles that of CH_4. This behavior is attributed to the extremely high excitation energy of Ne, to its spherical symmetry, and to the very diffuse nature of its virtual orbitals if computed in an L^2 basis. The available density of ionic configurations in the vicinity of the 2*s* line is thus relatively low, and, more importantly, the coupling of these configurations with the 2*s* single-hole configuration is weak. One has to go as far as to the 4*p* ionization of Xe to find significant final-state correlation effects.[54,55,87] For the breakdown phenomenon to occur, deeper orbitals of many-electron atoms are probably needed, whereas the joint action of two light atoms in a molecule may suffice. Larger molecules built of light atoms and smaller molecules containing heavy atoms may exhibit extremely low-lying satellite structures. Examples are *p*-quinodimethane,[50] *p*-nitroaniline,[47] dimethylisoindene,[44] triacetylene,[155] and nickel dichloride.[104] It should be possible that even the first line in the spectrum is a satellite line.

IV. OTHER EFFECTS AND OUTLOOK

Much work has been devoted to the investigation of the molecular ionization process at fixed energies of the ionizing photon and much valuable information on the residual ionic states has been extracted from experiments and theoretical computations. More recently, the advent of synchrotron radiation has stimulated various investigations on the partial-channel ionization cross sections and asymmetry parameters as functions of the photon energy. Most of this work concentrates on the main outer-valence states, thereby identifying interesting phenomena as, for instance, the appearance of shape resonances[156,157] or core-excited resonances[89,158] in the ionization cross section. Very little work has been carried out on the ionization cross sections of the inner-valence region, where strong correlation effects are commonly present as discussed in the preceding section. We may mention the investigations of Krummacher et al.[19,20] on N_2 and CO and of Bradshaw et al.[133] on C_2H_2. The measurements on C_2H_2 and N_2 will serve in the following as illustrative examples in the discussion of effects present in the inner-valence region and not mentioned in Section III.

Photoelectron spectra of acetylene at several photon energies ω_0 are collected in Fig. 7. The characteristic of the $\omega_0 = 1254\,\text{eV}$ spectrum is a band at approximately 28 eV binding energy with an integrated intensity of about 0.5 relative to the $1\sigma_g$ main band at about 23.5 eV. At lower photon energies the measured relative intensity of the satellite band is reduced. At $\omega_0 = 50\,\text{eV}$ this feature is no longer visible and must therefore have an integrated intensity lower than 0.05 relative to the main band.[133] To interpret the experimental findings we first discuss the intensity ratio of the two lines in the spectrum, assuming they arise from the ionization of the same orbital ϕ_q and adopting the sudden limit.[159] In this high-energy limit the outgoing photoelectron is sufficiently fast that its correlation with the electrons of the residual ion can be neglected and that the variation of the orbital photoionization matrix element $\tau_{\varepsilon q}$ over the energy separation of the two lines can be discarded. Consequently, Eq. (2.6b) can be used to calculate the intensity ratio which is then independent of photon energy. Using Eq. (2.6b) and the spectroscopic factors for the two lines calculated in Section III.A, we obtain the intensity ratio 0.5 in good agreement with the ratio measured at $\omega_0 = 1254\,\text{eV}$ (see also Fig. 5).

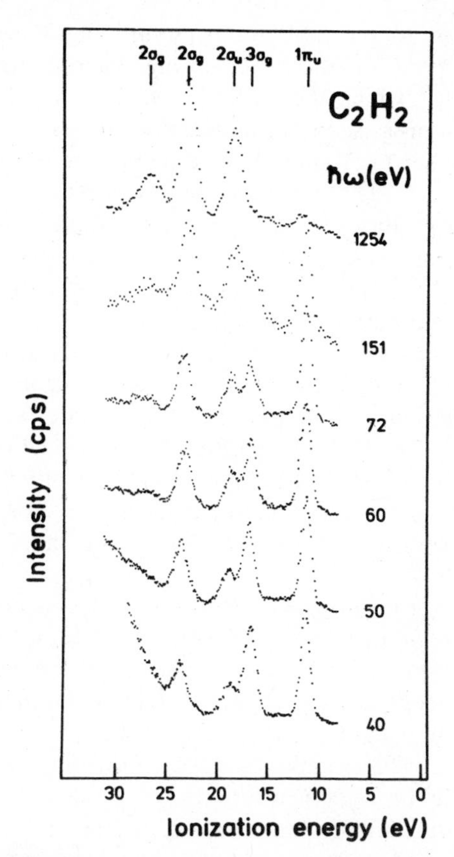

Fig. 7. Photoionization spectra of acetylene for photon energies 40, 50, 60, 72, 151, and 1254 eV. The figure is taken from ref. 133. Note that when comparing with the present text, the orbital indexes $2\sigma_g$, $3\sigma_g$, and $2\sigma_u$ should read $1\sigma_g$, $2\sigma_g$, and $1\sigma_u$, respectively, and the photon energy $\hbar\omega$ should be substituted by ω_0.

Corrections to this simple picture arise if there are lines in the spectrum which derive their intensity from several orbitals. Interferences can then occur and the intensity ratio of satellite lines to main lines may strongly vary with photon energy *even* in the sudden limit. In ref. 133 this effect has been considered as the most likely explanation for the significant photon energy dependence of the relative intensity of the 28 eV satellite band in acetylene. Both orbitals $1\sigma_g$ and $2\sigma_g$ were found to contribute to both the main and satellite lines, but with a different relative sign of the amplitudes $x_{1\sigma_g}^{(n)}$ and $x_{2\sigma_g}^{(n)}$ in each case. In this way destructive and constructive interference may become effective for the satellite and main lines, respectively, explaining the relative decrease of satellite intensity compared to the high-energy limit. This hypothesis is further supported by the experimental results in Fig. 7 and the recent calculations by Machado et al.[160] which show that at low energies the $(2\sigma_g)^{-1}$ partial-channel ionization cross section dominates over that of the $(1\sigma_g)^{-1}$ channel whereas at higher photon energies the opposite is true.

The extended 2ph–TDA calculation on acetylene discussed in the preceding section predicts absolute values for the amplitudes $x^{(n)}_{2\sigma_g}$ which are probably too small to fully explain the observations via Eq. (2.5) assuming unique functions $\tau_{\varepsilon, 1\sigma_g}$ and $\tau_{\varepsilon, 2\sigma_g}$ for both main and satellite lines. What are the additional possible corrections to the above simplified picture? The first obvious correction is to use different orbital scattering wavefunctions for the ejected electron in each channel. Equation (2.5) may then still be used and we encounter again interference effects. The next step is to augment Eq. (2.5) by the nonorthogonality correction term included in the more general expression (2.4). At relatively low energies we expect this term to contribute significantly to the cross sections and to behave differently for a localized single-hole state and a more delocalized satellite state.

The above discussion briefly touches the problems encountered in the quantitative investigation of the intensities of satellite lines. Very little is known on the partial-channel ionization cross section associated with the satellite lines, and considerable experimental and theoretical experience is required before the role of the diverse effects contributing to these cross sections is clarified. The situation is even more complex when investigating the inner-valence region where the breakdown of the molecular orbital picture of ionization commonly occurs. Here, the separated-channel approximation, which considerably simplifies the calculation and the interpretation of the cross sections, most probably becomes inadequate owing to the high density of ionic states that acquire intensity. It is unclear yet whether the separated-channel approximation is capable in even estimating the ionization cross sections in the inner-valence region. The necessity to incorporate interchannel coupling for several channels certainly complicates the problem but, on the other hand, may give rise to interesting new phenomena typical for this region.

Until now we discussed properties of ionization spectra confining ourselves to the electronic degrees of freedom. This restriction must be relaxed for a better understanding of the ionization process. In the other-valence region the density of ionic states is low and one may often successfully apply the adiabatic approximation[161] (often referred to as the Born–Oppenheimer approximation) thus separating the motion of the light electrons from that of the heavy nuclei. In this approximation one can think about the nuclear motion as proceeding on a single potential energy surface. For each electronic band appearing in the spectrum the nuclear energies and wavefunctions can be determined separately and the vibrational progressions associated with this band can be computed in the standard way. The coupling of two or more electronic states via vibrational modes is usually termed vibronic coupling. In a vibronic-coupling situation the electronic and nuclear motions cannot be handled separately. We are rather confronted with a composite system and may speak of vibronic motion. Simply speaking, vibronic coupling becomes important and gives rise to substantial nonadiabatic effects when the potential energy surfaces of two electronic states come close together. In polyatomic molecules this may happen even in the outer-valence region and it is essential to take the vibronic coupling mechanism into consideration in order to interpret the experimental spectra. More details on the subject can be found in a recent review[162] where explicit examples are also given.

The above remarks apply the more to the inner-valence region with its high density of ionic states. As an illustrative example for possible vibronic coupling phenomena we discuss again the inner-valence ionization spectrum of acetylene. Figure 8 shows the photoelectron spectrum of C_2H_2 recently recorded using monochromatized Al K_α radiation.[40] The intense 28 eV satellite band is seen to exhibit internal structure not

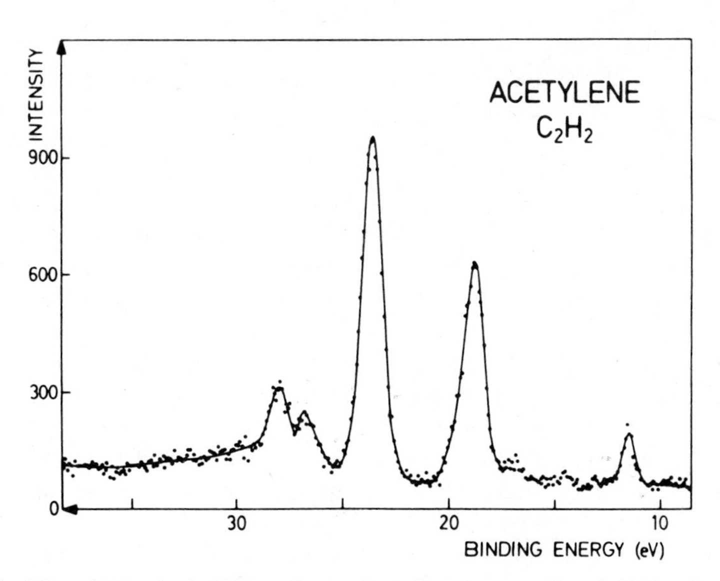

Fig. 8. Monochromatized Al K_α valence photoelectron spectrum of C_2H_2 taken from ref. 40.

observed in the earlier spectra because of lower resolution. It has been discussed in Section III.A that this internal structure cannot be explained from electronic considerations alone. Indeed, Fig. 5 shows just a single intense satellite line at 28.4 eV and the calculations of Müller et al.[40] support this finding.

The observed profile is probably a manifestation of vibronic coupling effects. If so, the question arises which electronic states of $C_2H_2^+$ may be the vibronic counterparts of the state associated with the intense satellite line. In trying to answer this question Müller et al.[40] examined the possibility of vibronic interaction with a close-lying $^2\Sigma_u$ state through the antisymmetric stretching mode ν_3. Their calculations seem to rule out this possibility. In another linear symmetric molecule, namely, C_2N_2, we have found evidence for strong vibronic coupling between the $(1\sigma_g)^{-1}$ inner-valence state and a nearby state of $^2\Pi_u$ symmetry through the *cis*-bending mode.[163] Therefore, we anticipate a bending mode to be relevant also in the present case. This speculation is supported by the extended 2ph–TDA calculation which predicts a weak satellite line at 27.39 eV originating from the $1\pi_u$ orbital (see Fig. 5). As discussed in Section III.A the calculation predicts in addition many initial-state correlation-induced lines. Out of the corresponding states two $^2\Pi_g$ states at 27.93 and 28.85 eV are worth mentioning which may interact vibronically with the relevant $^2\Sigma_g$ state through the *trans*-bending mode. A careful analysis is needed before this interesting problem is settled.

In the outer-valence region as well as in the above example of acetylene we expect that the two- or three-state vibronic coupling mechanism is likely to apply because of the relatively low density of electronic states. Usually the density of states grows rapidly with energy. Consequently, we are faced with the vibronic interaction of several or even many electronic states when investigating highly excited states. A glance at the high-energy side of the computed ionization spectrum of COS shown in Figs. 3 and 4 already facilitates the anticipation of this point. Recent model calculations[164] make clear that we must study the multistate vibronic coupling problem before being able to interpret such spectra correctly. The ionization of the $1\sigma_g$ inner-valence orbital of N_2 provides an interesting, though complicated, example for the interaction of several close-lying electronic states through vibrational modes. The experimental ionization spectrum of the inner-valence orbital in N_2 exhibits a few bands and we concentrate here on the most intense band ranging from approximately 36- to 40-eV binding energy. Figure 9 shows this band as measured by Krummacher et al.[19] using monochromatic synchrotron radiation of 45.8 eV. Several calculations on the ionic states included in this band are available.[14,33,38,94] They clearly indicate that several electronic

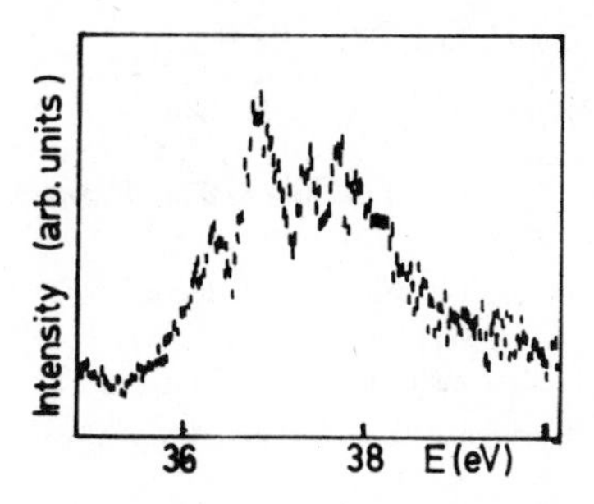

Fig. 9. Inner-valence region of the photoelectron spectrum of N_2 measured by Krummacher et al.[19] using monochromatic synchrotron radiation of 45.8 eV.

states of the ion fall into the 36–40 eV energy range and may thus participate in the vibronic coupling problem leading to the observed structure of the band. To investigate this structure a simple model has been introduced[164] which takes into account 20 interacting ionic states. Using vibronic coupling constants estimated by the 2ph–TDA, the calculation makes clear that the experiment cannot be understood without consulting theory. The multistate vibronic coupling mechanism leads to strong nonadiabatic effects. These tend to smooth out the envelope of the band and make it look like a vibrational progression of a single electronic transition as seen in Fig. 9. High-resolution spectra and more elaborate calculations are necessary to resolve the nature of the underlying nonadiabatic effects.

The above discussion demonstrates the general relevance of vibronic phenomena in ionization spectroscopy of inner-valence electrons. Obviously, the vibronic coupling mechanism has strong impact also on the ionization cross sections as a function of energy and scattering angle. The influence of this mechanism on the asymmetry parameter of some outer-valence orbitals has been investigated by Domcke.[165] This influence is probably amplified in many inner-valence ionization situations. There seems to be a variety of challenging phenomena accompanying the breakdown of the MO picture of ionization of which we have mentioned just a few. Much additional experimental and theoretical work is required in this field which is still in its infancy.

Acknowledgments

Financial support by the Deutsche Forschungsgemeinschaft, Fond der Chemischen Industrie, and the Bundesministerium für Forschung und Technologie is gratefully acknowledged.

References

1. K. Siegbahn, C. Nordling, A. Fahlman, R. Nordberg, K. Hamrin, J. Hedman, G. Johansson, T. Bergmark, S. E. Karlson. I. Lindgren, and B. Lindberg, *ESCA-Atomic Molecular and Solid State Structure Studied by Means of Electron Spectroscopy.* North-Holland, Amsterdam, 1967.
2. K. Siegbahn, C. Nordling, G. Johansson, J. Hedman, P. F. Heden, K. Hamrin, U. Gelius, T. Bergmark, L. O. Werme, R. Manne, and Y. Baer, *ESCA-Applied to Free Molecules.* North-Holland, Amsterdam, 1969.
3. H. Siegbahn and L. Karlsson, in *Handbuch der Physik*, S. Flügge (ed.). Springer, Heidelberg, 1982, p. 215.
4. D. W. Turner, C. Baker, A. D. Baker, and C. R. Brundle, *Molecular Photoelectron Spectroscopy.* Wiley, New York, 1970.
5. (a) J. W. Rabalais, *Principles of Ultraviolet Photoelectron Spectroscopy.* Wiley. New York, 1977. (b) J. Berkowitz, *Photoabsorption, Photoionization and Photoelectron Spectroscopy.* Academic Press, New York, 1979.
6. C. E. Brion and A. Hamnett, *Adv. Chem. Phys.* **45**, 1 (1981).
7. C. Kunz (ed.), *Synchrotron Radiation.* Springer, Heidelberg, 1979.
8. E. E. Koch (ed.), *Handbook of Synchrotron Radiation*, Vol. 1. North-Holland, Amsterdam, 1983.
9. I. E. McCarthy and E. Weigold, *Phys. Rep.* **27c**, 275 (1976).
10. E. Weigold and I. E. McCarthy, *Adv. At. Mol. Phys.* **14**, 127 (1978).
11. T. Åberg, *Ann. Acad. Sci. Fenn. A VI* **308**, 1 (1969); *Phys. Rev. A* **2**, 1726 (1970).
12. L. S. Cederbaum, *Mol. Phys.* **28**, 479 (1974).
13. P. S. Bagus and W.-K. Viinikka, *Phys. Rev. A* **15**, 1486 (1977).
14. J. Schirmer, L. S. Cederbaum, W. Domcke, and W. von Niessen, *Chem. Phys.* **26**, 149 (1977).
15. L. S. Cederbaum, J. Schirmer, W. Domcke, and W. von Niessen, *J. Phys. B* **10**, L549 (1977).
16. W. Domcke, L. S. Cederbaum, J. Schirmer, W. von Niessen, C. E. Brion, and K. H. Tan, *Chem. Phys.* **40**, 171 (1979).
17. J. Schirmer, W. Domcke, L. S. Cederbaum, W. von Niessen and L. Åsbrink, *Chem. Phys. Lett.* **61**, 30 (1979).
18. C. J. Allan, U. Gelius, D. A. Allison, G. Johansson, H. Siegbahn, and K. Siegbahn, *J. Electron Spectrosc.* **1**, 131 (1972).
19. S. Krummacher, V. Schmidt, and F. Wuilleumier, *J. Phys. B* **13**, 3993 (1980).
20. S. Krummacher, V. Schmidt, F. Wuilleumier, J. M. Bizau, and D. Ederer, *J. Phys. B* **16**, 1733 (1983).
21. C. E. Brion, J. P. D. Cook, and K. H. Tau, *Chem. Phys. Lett.* **59**, 241 (1978).
22. C. E. Brion, S. T. Hood, I. H. Suzuki, E. Weigold, and G. R. J. Williams, *J. Electron Spectrosc.* **21**, 71 (1980).
23. R. Fantoni, A. Giardini-Guidoni, R. Tiribelli, R. Camilloni, and G. Stefani, *Chem. Phys. Lett.* **71**, 335 (1980).
24. A. Minchington, I. Fuss, and E. Weigold, *J. Electron Spectrosc.* **27**, 1 (1982).
25. J. P. D. Cook and C. E. Brion, *Chem. Phys.* **69**, 339 (1982).

26. A Minchington, A. Giardini-Guidoni, E. Weigold, F. P. Larkins, and R. M. Wilson, *J. Electron Spectrosc.* **27**, 191 (1982).
27. R. Cambi, C. Ciullo, A. Sgamellotti, F. Tarantelli, R. Fantoni, A. Giardini-Guidoni, I. E. McCarthy, and V. Di Martino, *Chem. Phys. Lett.* **101**, 477 (1983).
28. L. S. Cederbaum and W. Domcke, *Adv. Chem. Phys.* **36**, 205 (1977).
29. J. Schirmer and L. S. Cederbaum, *J. Phys. B* **11**, 1889 (1978).
30. L. S. Cederbaum, J. Schirmer, W. Domcke, and W. von Niessen, *Int. J. Quantum Chem.* **XIV**, 593 (1978).
31. L. S. Cederbaum, W. Domcke, J. Schirmer, and W. von Niessen, *Phys. Scr.* **21**, 481 (1980).
32. W. von Niessen, J. Schirmer, and L. S. Cederbaum, *Comp. Phys. Rep.* **1**, 57 (1984).
33. M. F. Herman, K. F. Freed, and D. L. Yeager, *Chem. Phys.* **32**, 437 (1978).
34. M. Mishra and Y. Öhrn, *Chem. Phys. Lett.* **71**, 549 (1980).
35. I. Cacelli, R. Moccia, and V. Carravetta, *Chem. Phys.* **71**, 199 (1982).
36. J. Baker, *Chem. Phys. Lett.* **101**, 136 (1983); *Chem. Phys.* **79**, 117 (1983).
37. N. Honjou, T. Sasajima, and F. Sasaki, *Chem. Phys.* **57**, 475 (1981).
38. P. W. Langhoff, S. R. Langhoff, J. Schirmer, L. S. Cederbaum, W. Domcke, and W. von Niessen, *Chem. Phys.* **58**, 71 (1981).
39. J. Müller, *Int. J. Quantum Chem.* **XXI**, 465 (1982).
40. J. Müller, R. Arneberg, H. Ågren, R. Manne, P.-Å. Malmqvist, S. Svensson, and U. Gelius, *J. Chem. Phys.* **77**, 4895 (1982).
41. R. Arneberg, J. Müller, and R. Manne, *Chem. Phys.* **64**, 249 (1982).
42. H. Nakatsuji and T. Yonezawa, *Chem. Phys. Lett.* **87**, 426 (1982); H. Nakatsuji, *Chem. Phys.* **75**, 425; **76**, 283 (1983).
43. G. Kluge and M. Scholz, *Int. J. Quantum Chem.* **XX**, 669 (1981).
44. R. Schulz, A. Schweig, and W. Zittlau, *J. Am. Chem. Soc.* **105**, 2980 (1983).
45. D. Saddei, H.-J. Freund, and G. Hohlneicher, *Surf. Sci.* **95**, 527 (1980); D. Saddei, H.-J. Freund, and G. Hohlneicher, *Chem. Phys.* **55**, 339 (1981).
46. D. P. Chong, D. C. Frost, W. M. Lau, and C. A. McDowell, *Chem. Phys. Lett.* **90**, 332 (1982).
47. R. W. Bigelow, *Chem. Phys. Lett.* **100**, 445 (1983); *Chem. Phys.* **80**, 45 (1983).
48. L. S. Cederbaum, W. Domcke, J. Schirmer, W. von Niessen, G. H. F. Diercksen, and W. P. Kraemer, *J. Chem. Phys.* **69**, 1591 (1978).
49. W. von Niessen, G. Bieri, J. Schirmer, and L. S. Cederbaum, *Chem. Phys.* **65**, 157 (1982).
50. T. Koenig and S. Southworth, *J. Am. Chem. Soc.* **99**, 2807 (1977); T. Koenig, C. E. Klopfenstein, S. Southworth, J. A. Hobbler, R. A. Wieselek, T. Balle, W. Snell, and D. Imre, *J. Am. Chem. Soc.* **105**, 2256 (1983).
51. T.-K. Ha, *Mol. Phys.* **49**, 1471 (1983).
52. U. Gelius, *J. Electron Spectrosc.* **5**, 985 (1974).
53. S. Lundqvist and G. Wendin, *J. Electron Spectrosc.* **5**, 513 (1974).
54. G. Wendin and M. Ohno, *Phys. Scr.* **14**, 148 (1976).
55. G. Wendin, *Breakdown of the One-Electron Pictures in Photoelectron Spectra*, Vol. 45, *Structure and Bonding*. Springer, Heidelberg, 1981.

56. H. Ågren, J. Nordgren, L. Selander, C. Nordling, and K. Siegbahn, *Phys. Scr.* **18**, 499 (1978).
57. A. V. Kondratenko, L. N. Mazalov, and D. M. Neiman, *Opt. Spectrosc.* **48**, 587 (1980).
58. V. F. Demekhin, V. L. Sukhorukov, L. A. Demekhina, and V. V. Timoshevskaya, *Opt. Spectrosc.* **51**, 379 (1981).
59. M. Yousif, D. E. Ramaker, and H. Sambe, *Chem. Phys. Lett.* **101**, 472 ((1983).
60. H. Ågren and H. Siegbahn, *Chem. Phys. Lett.* **69**, 424 (1980); **72**, 498 (1980).
61. H. Ågren, *J. Chem. Phys.* **75**, 1267 (1981).
62. O. M. Kvalheim, *Chem. Phys. Lett.* **86**, 159 (1982); **98**, 547 (1983).
63. J. A. Kelber, D. R. Jennison, and R. R. Rye, *J. Chem. Phys.* **75**, 652 (1981).
64. H. Aksela, S. Aksela, M. Hotokka, and M. Jäntti, *Phys. Rev. A* **28**, 287 (1983).
65. C.-M. Liegener, *Phys. Rev. A* **28**, 256 (1983); *J. Chem. Phys.* **79**, 2924 (1983).
66. G. R. Wright, M. J. van der Wiel, and C. E. Brion, *J. Phys. B* **9**, 675 (1976).
67. D. E. Ramaker, *J. Chem. Phys.* **78**, 2998 (1983).
68. R. A. Rosenberg, V. Rehn, A. K. Green, R. R. La Roe, and C. C. Parks, *Phys. Rev. Lett.* **51**, 915 (1983).
69. M. F. Herman, K. F. Freed, and D. L. Yeager, *Adv. Chem. Phys.* **48**, 1 (1981).
70. Y. Öhrn and G. Born, *Adv. Quantum Chem.* **13**, 1 (1981).
71. H. F. Schaefer III (ed.), *Methods of Electronic Structure Theory*, Vol. 3 of *Modern Theoretical Chemistry*. Plenum Press, New York, 1977.
72. T. Koopmans, *Physica* **1**, 104 (1933).
73. P. S. Bagus, *Phys. Rev.* **139**, A619 (1965).
74. M. E. Schwartz, *Chem. Phys. Lett.* **5**, 50 (1970).
75. B. T. Pickup and O. Goscinski, *Mol. Phys.* **26**, 1013 (1973).
76. A. L. Fetter and J. D. Walecka, *Quantum Theory of Many-Particle Systems.* McGraw-Hill, New York, 1971.
77. A. B. Migdal, *Theory of Finite Fermi Systems and Applications to Atomic Nuclei.* Wiley, New York, 1967.
78. D. Pines and P. Nozières, *The Theory of Quantum Liquids.* Benjamin, Reading, Mass., 1966.
79. K. Gottfried, *Quantum Mechanics.* Benjamin, New York, 1966.
80. R. Camilloni, A. Giardini-Guidoni, I. E. McCarthy, and G. Stefani, *Phys. Rev. A* **17**, 1634 (1978).
81. R. L. Martin and D. A. Shirley, *J. Chem. Phys.* **64**, 3685 (1976).
82. B. T. Pickup, *Chem. Phys.* **19**, 193 (1977).
83. I. Cacelli, V. Carravetta, and R. Moccia, *J. Phys. B* **16**, 1895 (1983).
84. For simplicity, closed-shell targets are considered throughout and spin coupling is suppressed.
85. V. McKoy, T. A. Carlson, and R. R. Lucchese, *J. Phys. Chem.* **88**, 3188 (1984).
86. One may slightly improve Eq. (2.6b) by taking account of the different energies of the ionizing projectiles needed to produce two ionic states $|\Phi_n^{N-1}\rangle$ and $|\Phi_{n'}^{N-1}\rangle$ associated with the same energy ε of the ejected electron. In the case of photoionization, for instance, we may write $\sigma_n(\varepsilon)/\sigma_{n'}(\varepsilon) \to \omega_0|x_q^{(n)}|^2/\omega_0'|x_q^{(n')}|^2$.

87. A. F. Starace, *Handbuch der Physik, Vol.* 31. Springer, Heidelberg, 1982, and references therein.
88. J. L. Dehmer, D. Dill, and S. Wallace, *Phys. Rev. Lett.* **43**, 1005 (1979).
89. Z. H. Levine and P. Soven, *Phys. Rev. Lett.* **50**, 2074 (1983).
90. C. D. Lin, *Phys. Rev.* **91**, 171 (1974).
91. J. Simons and W. D. Smith, *J. Chem. Phys.* **58**, 4899 (1973).
92. J. Schirmer, L. S. Cederbaum, and O. Walter, *Phys. Rev. A* **28**, 1237 (1983).
93. J. P. D. Cook, M. G. White, C. E. Brion, W. Domcke, J. Schirmer, L. S. Cederbaum, and W. von Niessen, *J. Electron. Spectrosc.* **22**, 261 (1981).
94. J. Schirmer and O. Walter, *Chem. Phys.* **78**, 201 (1983).
95. E. Lindholm, C. Fridh, and L. Åsbrink, *Faraday Discuss. Chem. Soc.* **54**, 127 (1972).
96. A. W. Potts, T. A. Williams, and W. C. Price, *Faraday Discuss, Chem. Soc.* **54**, 104 (1972).
97. E. Heilbronner, T. B. Jones, and J. P. Maier, *Helv. Chim. Acta* **60**, 1697 (1977).
98. G. Bieri and L. Åsbrink, *J. Electron Spectrosc.* **20**, 149 (1980).
99. K. Hamrin, G. Johansson, U. Gelius, A. Fahlman, C. Nordling, and K. Siegbahn, *Chem. Phys. Lett.* **1**, 613 (1968).
100. J. J. Pireaux, S. Svensson, E. Basilier, P.-Å. Malmqvist, U. Gelius, R. Caudano, and K. Siegbahn, *Phys. Rev. A* **14**, 2133 (1976).
101. S. T. Hood, A. Hammett, and C. E. Brion, *J. Electron Spectrosc.* **11**, 205 (1977).
102. R. Cambi, G. Cuillo, A. Sgamellotti, F. Tarantelli, R. Fantoni, A. Giardini-Guidoni, and A. Sergio, *Chem. Phys. Lett.* **80**, 295 (1981).
103. W. Meyer, *J. Chem. Phys.* **58**, 1017 (1973).
104. W. von Niessen and L. S. Cederbaum, *Mol. Phys.* **43**, 897 (1981).
105. J. Berkowitz, *J. Chem. Phys.* **61**, 407 (1974).
106. G. W. Bogges, J. D. Allan, Jr., and G. K. Schweitzer, *J. Electron Spectrosc.* **2**, 467 (1973).
107. B. G. Cocksey, J. H. D. Eland, and C. J. Danby, *J. Chem. Soc. Faraday Trans. II* **69**, 1558 (1973).
108. A. F. Orchard and N. V. Richardson, *J. Electron Spectrosc.* **6**, 61 (1975).
109. G. M. Bancroft, D. J. Bristow, and L. L. Coatworth, *Chem. Phys. Lett.* **82**, 344 (1981).
110. B. Kovač, *J. Chem. Phys.* **78**, 1684 (1983).
111. P. Natalis, J. Delwiche, and J. E. Collin, *Faraday Discuss. Chem. Soc.* **54**, 98 (1972).
112. J. Delwiche, M. J. Hubin-Franskin, G. Caprace, P. Natalis, and D. Roy, *J. Electron Spectrosc.* **21**, 205 (1980).
113. A. W. Potts and T. A. Williams, *J. Electron Spectrosc.* **3**, 3 (1974).
114. C. J. Allan, U. Gelius, D. A. Allison, G. Johansson, H. Siegbahn, and K. Siegbahn, *J. Electron Spectrosc.* **1**, 131 (1972).
115. R. Frey, B. Gotchev, W. B. Peatman, H. Pollak, and E. W. Schlag, *Int. J. Mass. Spectrom. Ion Phys.* **26**, 137 (1978).
116. T. A. Carlson, M. O. Krause, and F. A. Grimm, *J. Chem. Phys.* **77**, 1701 (1982).
117. I. Reineck, B. Wannberg, H. Veenhuizen, C. Nohre, R. Maripuu, K. E. Norell, L. Mattsson, L. Karlsson, and K. Siegbahn, *J. Electron Spectrosc.* **34**, 235 (1984).

118. The double-ionization energy of COS might lie within the higher-energy part of the spectrum shown in Fig. 3, that is, the states shown above this energy are not discrete states. They partly simulate autoionizing states of COS^+ and partly the underlying double-ionization continuum. This complication should not alter the qualitative conclusions drawn here on the validity of the MO picture as becomes clear from other examples where this complication is not present. See also the discussion at the end of Section II.

119. W. von Niessen, W. Domcke, L. S. Cederbaum, and J. Schirmer, *J. Chem. Soc. Faraday Trans. II* **74**, 1550 (1978).

120. R. Stockbauer and M. G. Ingram, *J. Electron Spectrosc.* **7**, 492 (1975).

121. D. M. Mintz and A. Kuppermann, *J. Chem. Phys.* **71**, 3499 (1979).

122. J. E. Pollard, D. J. Trevor, Y. T. Lee, and D. A. Shirley, *Rev. Sci. Instrum.* **52**, 1837 (1981).

123. D. G. Streets and W. A. Potts, *J. Chem. Soc. Faraday Trans. II* **70**, 1505 (1974).

124. A. Berndtsson, E. Basilier, U. Gelius, J. Hedman, M. Klasson, R. Nilsson, C. Nordling, and S. Svensson, *Phys. Scr.* **12**, 235 (1975).

125. M. S. Banna and D. A. Shirley, *J. Electron Spectrosc.* **8**, 255 (1976).

126. A. J. Dixon, S. T. Hood, E. Weigold, and G. R. J. Williams, *J. Electron Spectrosc.* **14**, 267 (1978).

127. A. Caplan, A. L. Migdall, J. H. Moore, and J. A. Tossell, *J. Am. Chem. Soc.* **100**, 5008 (1978).

128. D. Mehaffy, P. R. Keller, J. W. Taylor, T. A. Carlson, M. O. Krause, F. A. Grimm, and J. D. Allen, Jr., *J. Electron Spectrosc.* **26**, 213 (1982).

129. W. von Niessen, G. H. F. Diercksen, L. S. Cederbaum, and W. Domcke, *Chem. Phys.* **18**, 469 (1976); K. H. Thunemann, R. J. Buenker, S. D. Peyerimhoff, and S. K. Sh'ih, *Chem. Phys.* **35**, 35 (1978); J. Baker, *Int. J. Quant. Chem.* XXVII, 145 (1985).

130. R. L. Martin and E. R. Davidson, *Chem. Phys. Lett.* **51**, 237 (1977).

131. H. Nakatsuji, *J. Chem. Phys.* **80**, 3703 (1984).

132. R. G. Cavell and D. A. Allison, *J. Chem. Phys.* **69**, 159 (1978).

133. A. M. Bradshaw, W. Eberhardt, H. J. Levinson, W. Domcke, and L. S. Cederbaum, *Chem. Phys. Lett.* **70**, 36 (1980).

134. A. C. Parr, D. L. Ederer, J. B. West, D. M. P. Holland, and J. L. Dehmer, *J. Chem. Phys.* **76**, 4349 (1982).

135. S. Svensson, P.-Å. Malmqvist, M. Y. Adam, P. Lablanquie, P. Morin, and I. Nenner, *Chem. Phys. Lett.*, **111**, 574 (1984).

136. A. J. Dixon, I. E. McCarthy, E. Weigold, and G. R. J. Williams, *J. Electron Spectrosc.* **12**, 239 (1977).

137. L. S. Cederbaum, G. Hohlneicher, and W. von Niessen, *Mol. Phys.* **26**, 1405 (1973).

138. S. Dey, A. J. Dixon, I. E. McCarthy, and E. Weigold, *J. Electron Spectrosc.* **9**, 397 (1976).

139. J. W. Rabalais and A. Katrib, *Mol. Phys.* **27**, 923 (1974).

140. L. Åsbrink, E. Lindholm, and O. Edqvist, *Chem. Phys. Lett.* **5**, 609 (1970).

141. A. W. Potts and D. G. Streets, *J. Chem. Soc. Faraday Trans. II* **70**, 875 (1974).

142. A. Richartz, R. J. Beunker, P. J. Bruna, and S. D. Peyerimhoff, *Mol. Phys.* **33**, 1345 (1977).

143. R. Colle and O. Salvetti, Theoret. Chim. Acta **37**, 329 (1975).
144. F. Moscardó, M. Paniagua and E. San-Fabián, *Theor. Chim. Acta* **53**, 377 (1979); **54**, 53 (1979).
145. I. Cacelli, R. Moccia, and V. Carravetta, *Theor. Chim. Acta* **59**, 461 (1981).
146. Anionic states which are bound due to long range forces, for example, dipole, are not considered here. These states are only loosely bound.
147. M. B. Robin, *Higher Excited States of Polyatomic Molecules*, Vol. 1. Academic Press, New York, 1974.
148. C. Szmytkowski, G. Karwasz, and K. Maciag, *Chem. Phys. Lett.* **107**, 481 (1984).
149. M. G. Lynch, D. Dill, J. Siegel, and J. L. Dehmer, *J. Chem. Phys.* **71**, 4249 (1979).
150. W. Lochte-Holtgreven, C. E. H. Bawn, and E. Eastwood, *Nature* **129**, 869 (1932).
151. B. Leclerc, A. Poulin, D. Roy, M. J. Hubin-Franskin, and J. Delwiche, *J. Chem. Phys.* **75**, 5329 (1981).
152. See, for example, R. J. Buenker and S. D. Peyerimhoff, *Chem. Phys. Lett.* **29**, 253 (1974).
153. D. A. Allison and R. G. Cavell, *J. Chem. Phys.* **68**, 593 (1978).
154. A. Wuilleumier and M. O. Krause, *Phys. Rev. A* **10**, 242 (1974).
155. W. von Niessen, L. S. Cederbaum, J. Schirmer, G. H. F. Diercksen, and W. Kraemer, *J. Electron Spectrosc.* **28**, 45 (1982).
156. See, for example, J. L. Dehmer, D. Dill, and A. C. Parr, in *Photophysics and Photochemistry in the Vacuum Ultraviolet*, S. McGlynn, G. Findley, and R. Heubner (eds.). Reidel, Dortrecht, Holland, 1983, and references therein.
157. See, for example, R. R. Lucchese and V. McKoy, *J. Phys. Chem.* **85**, 2166 (1981), and references therein.
158. See, for example, R. Unwin, I. Khan, N. V. Richardson, A. M. Bradshaw, L. S. Cederbaum, and W. Domcke, *Chem. Phys. Lett.* **77**, 242 (1981).
159. H. W. Meldner and J. D. Perez, *Phys. Rev. A* **4**, 1338 (1971).
160. L. E. Machado, E. P. Leal, G. Csanak, B. V. McKoy, and P. W. Langhoff, *J. Electron Spectrosc.* **25**, 1 (1982).
161. M. Born and K. Huang, *Dynamical Theory of Crystal Lattices.* Oxford University Press, Oxford, U.K., 1954.
162. H. Köppel, W. Domcke, and L. S. Cederbaum, *Adv. Chem. Phys.* **57**, 59 (1984).
163. L. S. Cederbaum, W. Domcke, J. Schirmer, and H. Köppel, *J. Chem. Phys.* **72**, 1348 (1980).
164. L. S. Cederbaum and H. Köppel, *Chem. Phys. Lett.* **87**, 14 (1982); L. S. Cederbaum, *J. Chem. Phys.* **78**, 5714 (1983).
165. W. Domcke, *Phys. Scr.* **19**, 11 (1979).
166. The difference of the electronic energies of the ion and parent molecule taken at the equilibrium geometry of the latter is termed *vertical ionization potential.* In the harmonic approximation this energy is given by the center of gravity of a band observed in the spectrum (e.g., see ref. 28). If the potential energy surfaces of the ion and/or molecule are strongly anharmonic it is difficult to assess the vertical ionization potential from the experimental spectrum. In general, care must be taken when comparing calculated fixed-nuclei ionization potentials with measured ones.

SEMICLASSICAL CALCULATION OF QUANTUM MECHANICAL WAVEFUNCTIONS

J. B. DELOS

Department of Physics,
College of William and Mary,
Williamsburg, Virginia 23185

CONTENTS

I. Introduction 162
II. Classical Trajectories and the Lagrange Manifold 165
  A. Hamilton–Jacobi Equation and the Field of Trajectories 165
  B. Lagrange Manifold 167
  C. Transport Equation and the Classical Density 170
III. Local Formal Asymptotic Solution to the Schroedinger Equation 171
IV. Momentum Space and Mixed Spaces 172
  A. Introductory Remarks 172
  B. Definition and Essential Properties of Lagrange Manifolds 175
  C. Classical Functions for General Representations 178
  D. Wavefunctions in Momentum Space and in Mixed Spaces 180
    1. Formal Asymptotic Solution of the Schroedinger Equation 180
    2. Validity of the Asymptotic Expansion 183
V. Global Formal Asymptotic Solutions 185
  A. Local Approximations in Configuration Space 186
  B. Joining Semiclassical Functions Using the Stationary Phase Approximation . . 187
  C. Maslov Index 190
  D. Canonical Formula for the Wavefunction 194
  E. Caustics and Catastrophes 197
VI. Conclusion 198
Appendix A. Formulas Related to Section II 200
  A. Solution to the Hamilton–Jacobi Equation 200
  B. The Manifold Is Lagrangian 202
  C. Solution to the Transport Equation 202
Appendix B. Formulas Related to Section IV 204
  A. Hamilton–Jacobi Equation in Mixed Spaces 204
  B. Density Function in Mixed Spaces 206
  C. Formal Asymptotic Expansion of $\tilde{\Psi}(p_\alpha, q_\beta)$ 208
Appendix C. Summary of Terminology 211
References 212

I. INTRODUCTION

Ever since its discovery by 19th-century mathematicians,[1] the semiclassical or JWKB approximation has been one of the most generally useful methods for constructing solutions to differential equations in one dimension, and for at least half a century there has been interest in generalizing this method to *n* dimensions. Many textbooks on quantum mechanics[2] give some discussion of the semiclassical approximation in two or three dimensions, but these presentations are far from satisfactory. It is all too simple a matter to show a relationship between the Schroedinger equation of quantum mechanics and the Hamilton–Jacobi equation of classical mechanics; it is something else again to show how the solutions to these equations are related—that out of all the solutions to the Schroedinger equation, the particular, desired ones can be constructed approximately by starting from selected solutions to Hamilton–Jacobi equations, and that these, in turn, can be obtained by calculating certain families of trajectories. The textbooks carefully avoid this aspect of the problem, and with good reason: there are many conceptual and mathematical questions that do not have simple answers.

Multidimensional semiclassical approximations can be invaluable to molecular physicists and theoretical chemists, and over the past 20 years this community has invested a lot of effort in the development of such methods.[3] Not surprisingly, physical insight has played a very important role in these developments and has had to be invoked where mathematical proofs were unavailable. Most of these studies[3] have focused on calculating specific properties of systems, such as scattering amplitudes or energy spectra, and the problems associated with calculation of wavefunctions have received less attention.

In contrast, a general mathematical theory would naturally begin with a scheme for calculating wavefunctions, from which all other properties of a system can be found. During this same 20-year period,[4] a community of Russian mathematicians, of whom the principal exponent is V. P. Maslov, has been involved in the development of just such a general theory. Although their papers have rarely been cited, Western physical scientists have known about the existence of this work for a long time.[5] It was widely realized that important developments were taking place, but precisely what had been accomplished was not very well understood, and the formal theory had not been applied in actual calculations.

The publication of an English translation of the monograph by Maslov and Fedoriuk[6](MF) has made possible an integration of Russian mathematical theory with Western calculational methods. This paper is one of a series,[7] the purpose of which is to close the gap between theorems and computations: we shall use the ideas presented by MF[6] to calculate the quantum-mechanical wavefunction for a variety of simple systems. In this first paper, we present the essential parts of the mathematical theory in a simplified form, for, while MF give a very complete and rigorous presentation, much of it is couched in the language of calculus on manifolds, a language that is not part of the training of every chemist. As a result, extraction from MF of the results that are essential for calculations requires a little bit of mathematical fortitude, and a great deal of enthusiasm for the subject. In presenting this material, I hope to make the beautiful work of Maslov & Co. more accessible to theoretical chemists and physicists, for whom it can be most beneficial.

We limit our exposition to the Schroedinger equation, and some of our exposition is limited further to those cases in which the Hamiltonian can be written in the form

$$H = \frac{p^2}{2M} + V(q), \tag{1.1}$$

q and p being n-dimensional coordinates and momenta. This excludes situations in which the Hamiltonian may contain quantities like $f(q)p$, or other powers of p, and so on. The treatment given in MF considers very general types of differential equations and permits these latter possibilities, but thereby it leads to a few additional complications that are not of interest to us at present.

To provide concrete illustrations of the abstract concepts, we discuss two-dimensional scattering on a Lennard–Jones potential:

$$q = (X, Z), \tag{1.2a}$$

$$p = (P_x, P_z), \tag{1.2b}$$

$$R = (X^2 + Z^2), \tag{1.2c}$$

$$\theta = \tan^{-1}(Z/X), \tag{1.2d}$$

and

$$V = \frac{A}{R^{12}} - \frac{B}{R^6}. \tag{1.3}$$

This system is chosen because it is very simple and well understood, and because it provides especially clear illustrations of the ideas involved. In a

TABLE I
Notation

		MF notation (if different)
n	Dimension of configuration space	
$q_i\ (i = 1 \ldots n)$	General configuration-space coordinates, collectively denoted q	x
$p_i\ (i = 1 \ldots n)$	Momenta canonically conjugate to q_i, collectively denoted p	
$p_\alpha q_\beta$	A mixed set of n momenta and coordinates, containing no canonically conjugate pairs, used locally as coordinates for the wavefunction	
$p_\beta q_\alpha$	The mixed set canonically conjugate to $p_\alpha q_\beta$	
$V(q)$	Potential energy in general coordinates	
$H(p, q)$	Hamiltonian	
$S(q)$	Hamilton's characteristic function	
Λ^n	n-Dimensional manifold in $2n$-dimensional phase space representing the field of trajectories	
Λ_0^{n-1}	Initial manifold from which trajectories emanate	
q^0	Points on initial manifold	x^0
$S^0(q^0)$	Characteristic function on initial manifold	
$\int p \cdot dq = \int \sum_{i=1}^{n} p_i\, dq_i$	Integral form of characteristic function	$\int \langle p, dx \rangle$
$q(t), p(t)$	Solutions to Hamilton's equations (2.4)	
$w_i\ (i = 1 \ldots n)$	n Coordinates spanning manifold Λ^n, collectively denoted w	r
$w_i^0\ (i = 1 \ldots n-1)$	$n-1$ Coordinates spanning initial manifold Λ_0^{n-1}, collectively denoted w^0	Various
$\ell[w_I, w_F]$	A path on the manifold going from an initial point w_I to a final point w_F	
Ω	A domain of the manifold	
$X, Z, P_x P_z$	Coordinates and momenta for the two-dimensional scattering problem	
$\mu(\ell[w_I, w_F])$	Index of a path on the manifold from w_I to w_F	ind $\ell[r', r'']$
$\nu(\Omega_1, \ldots, \Omega_s)$	Index of a chain of charts	$\gamma(\Omega_1, \ldots, \Omega_s)$

later paper, we consider the corresponding three-dimensional scattering problem.

A summary of notation is given in Table I, and definitions of some of the less familiar words are given in appendix C.

II. CLASSICAL TRAJECTORIES AND THE LAGRANGE MANIFOLD

A. Hamilton–Jacobi Equation and the Field of Trajectories

The Hamilton–Jacobi equation associated with (1.1) is

$$H(\nabla S, q) = (2M)^{-1}|\nabla_q S|^2 + V(q) = E, \tag{2.1}$$

and, as is typical of nonlinear equations, it has a bewildering variety of solutions. Among them are the *complete integrals*, which are especially emphasized in analytical mechanics, because Jacobi discovered a procedure by which any complete integral can be used to find all the trajectories of the system.[8] In the present case, complete integrals could be calculated by transforming to spherical coordinates and separating variables. However, we are not interested in the complete integrals, but in those solutions to (2.1) that have a relationship to the quantum-mechanical wavefunction that satisfies the standard scattering boundary conditions.

The rules for constructing such solutions to (2.1) are "well-known".[9] One begins by specifying an initial surface Λ_0^{n-1} of dimension $n-1$ in the n-dimensional configuration space. Let points on this initial surface be specified by coordinates q^0. For each initial point q^0, an initial momentum $p^0(q^0)$ and an initial value of the characteristic function $S^0(q^0)$ must be specified. For an arbitrary solution to the Hamilton–Jacobi equation, the initial surface and the initial momentum and characteristic function can be chosen arbitrarily, provided they satisfy certain conditions. The magnitude of the momentum is determined by

$$H(p^0(q^0), q^0) - E = 0 \tag{2.2}$$

and the direction of the momentum must be such that for any arbitrary infinitesimal increment dq^0 on the initial surface

$$dS^0 = p^0(q^0) \cdot dq^0 \equiv \sum_i p_i^0(q^0)\, dq_i^0 \tag{2.3}$$

Starting from each point q^0 on the initial surface and using $p^0(q^0)$ as the initial momentum, one must integrate Hamilton's equations of motion:

$$\dot{p}_i = -\frac{\partial H}{\partial q_i} \quad \text{and} \quad \dot{q}_i = \frac{\partial H}{\partial p_i}. \tag{2.4}$$

The result of this integration is a field of trajectories. *Locally* (i.e., for *some* points sufficiently close to the initial surface, and for t not too large), there is one and only one trajectory connecting the point q with a corresponding point $q^0(q)$ on the initial surface, and for those points $S(q)$

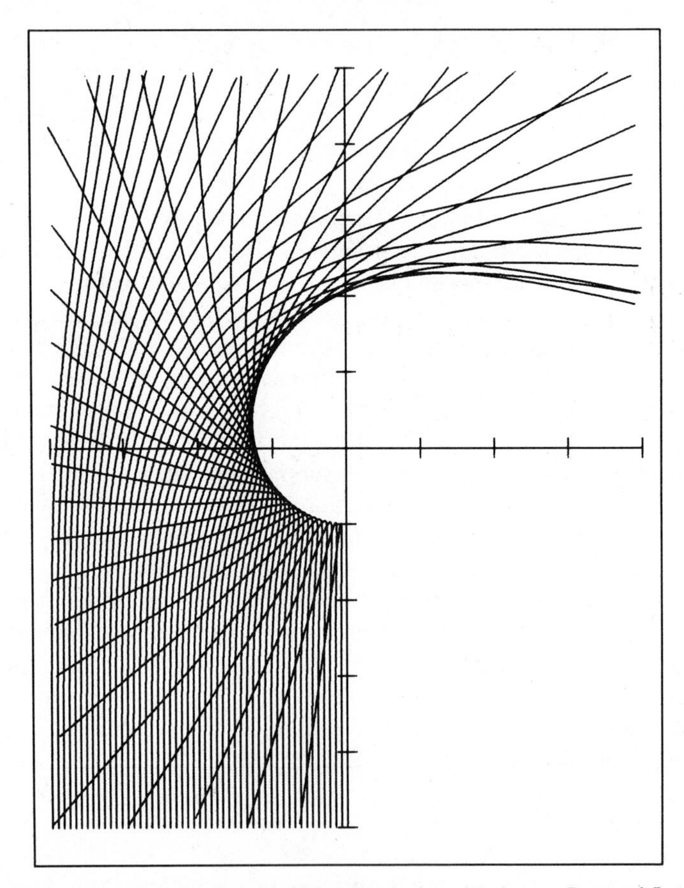

Fig. 1. Family of trajectories for two-dimensional scattering from a Lennard-Jones potential. Each trajectory begins at $Z = Z^0$, with various values of $X_0 \equiv b$. Trajectories with very small impact paremeters are scattered in a backward direction ($\Theta \sim \pi$); as b increases the scattering angle decreases, eventually passing through zero and becoming negative. At a certain value of b there is a maximum negative scattering angle (the *rainbow* angle), and as b continues to increase, the scattering angle approaches zero. Near the center of the figure is a classically forbidden region. The caustic is the boundary between allowed and forbidden regions.

is equal to

$$S(q) = S(q^0(q)) + \int p(t) \cdot \frac{dq(t)}{dt}\, dt, \tag{2.5}$$

where the integral is evaluated on the trajectory connecting q^0 with q. Equation (2.5) can be extended to define S globally (for all t), but in general this can lead to a multivalued function of coordinates q, as indicated in the following example. Henceforth, we refer to S calculated according to (2.5) as the (Hamilton) *characteristic function.*

For the two-dimensional scattering problem, the initial surface is one dimensional, and we take it to be the line $Z = Z^0$, a large negative constant. We have chosen to restrict the initial surface to $X^0 \geq 0$. The direction of the initial momenta may be taken to be perpendicular to this line, and the magnitude of the momenta must satisfy Eq. (2.2). Thus $P_Z^0 = (2ME)^{1/2}$, independent of X^0 for sufficiently large $|Z^0|$. Then from Eq. (2.3) it follows that S^0 is constant on the initial line.

The resulting field of trajectories is shown in Fig. 1. They emerge from $Z_0 = -4$ at the bottom of the figure; in this region $S(X, Z) \simeq S^0 + (2ME)^{1/2}(Z - Z^0)$, and surfaces of constant S are close to the lines $Z =$ constant, corresponding to plane waves approaching the target.[10] Near the center of the figure is a classically forbidden region, and the boundary between allowed and forbidden regions is a caustic curve, at which adjacent trajectories cross over each other. For large t, each point $[X(t), Z(t)]$ is moving away from the force center, and the field of trajectories has a pattern that will correspond in quantum mechanics to an outgoing spherical wave. At every point $q = (X, Z)$ in the classically allowed region, trajectories associated with two different impact parameters intersect each other, and for each trajectory a value of $S(q)$ is specified by the integral (2.5). It follows that $S(q)$ is not a globally single-valued function of q.

B. Lagrange Manifold

It is desirable to find a way of defining S such that it will be a single-valued function of its variables. This can be done by going over to a phase-space description of the trajectories.

In phase space (q, p), when one generates a family of trajectories according to the rules stated above (i.e., integration of Hamilton's equations from a surface with the given initial conditions), then that family of trajectories occupies an n-dimensional collection of points that usually forms a manifold, denoted Λ^n (a manifold is a smooth surface, smoothly embedded in phase-space). A general point on this manifold is denoted by w, and there are several ways to specify it. Obviously w can be specified by $2n$ coordinates and momenta, (q, p). However, since all trajectories in the family have the same energy, manifold points can be represented in a reduced, $2n-1$ dimensional phase space, in which one of the coordinates or momenta can be determined from $H(p, q) = E$. Much more simply, however, an n-dimensional manifold can always be spanned by just n suitably chosen coordinates. Since for each point w on Λ^n there is a unique trajectory passing through w, and a unique point on the initial surface Λ_0^{n-1}. wherefrom that trajectory came, it is possible to span the manifold using the $n-1$ coordinates spanning the initial manifold and using time t as the nth coordinate.

This seemingly abstract statement becomes obvious when the scattering example is considered. As stated above, the initial surface Λ_0^1 is the line $Z = Z^0$, and points on that line are given by the coordinate $X^0 \equiv b$ (impact parameter); if a trajectory is started at some such point and allowed to evolve for a time t, it carries us to a unique point on the manifold, and conversely every point on the manifold Λ^2 is uniquely characterized by the pair (b, t). Henceforth, w may be taken to represent this pair (b, t), and it follows that $S(w)$ or $S(b, t)$ is a single-valued function on the manifold.

A representation of the manifold for the scattering system is shown in Fig. 2. In this case, phase-space is four-dimensional, but because all the trajectories have the same conserved energy, the effective phase-space (the energy shell) is only three-dimensional. Coordinates X, Z, and P_X are convenient for spanning this space. With a bit of thought, one can translate the configuration-space picture of the trajectories into a phase-space picture of the manifold Λ^2 (see Fig. 2).

The coordinates b, t (or more generally the set of coordinates $\{w_i^0, t \mid i = 1 \ldots n-1\}$) may be said to be intrinsic coordinates for the manifold, since they refer only to the manifold itself and not to the phase space in which it is embedded. Alternatively, we may consider an extrinsic characterization, which specifies the embedding of the n-dimensional manifold in the $2n$-dimensional phase space. Most points on the manifold are surrounded by a domain that has a smooth (and smoothly invertible) projection into configuration space. Such domains are said to be *regular*, and for any such regular domain, the embedding of the manifold in phase space is given by n smooth functions $\{p_i(q) \mid i = 1 \ldots n\}$.

The manifold is said to be *Lagrangian*. The general definition of a Lagrangian manifold (MF definition 4.1) is a bit abstract, but for the moment we can specify it this way: a regular domain of the manifold is Lagrangian if and only if the derivatives $\partial p_i/\partial q_j$ and $\partial p_j/\partial q_i$ are equal for all i, j:

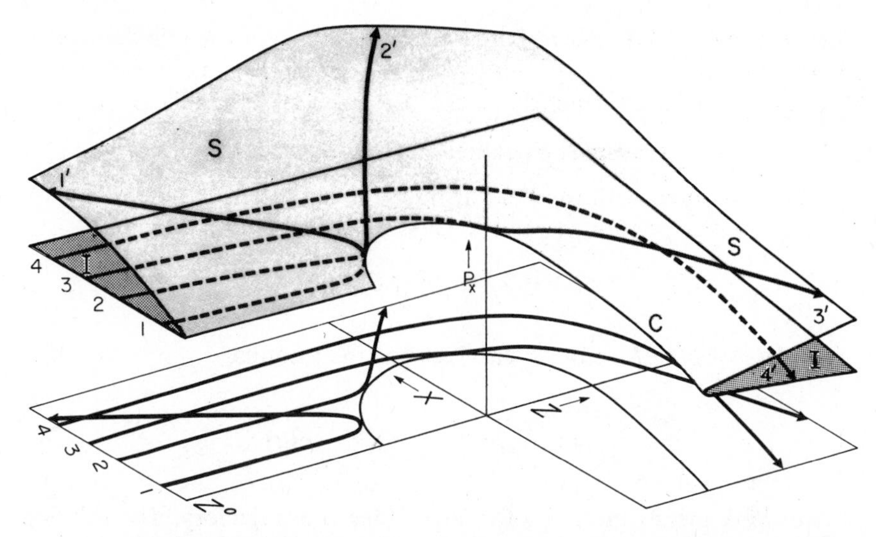

Fig. 2. Manifold Λ^2 in the reduced phase space (XZP_X) and its projection into configuration space. The lower sheet of the manifold contains parts of trajectories that have not touched the caustic; the upper sheet contains trajectories that have passed through the caustic. Labeled trajectories help to define the shape of the surface. The trajectory labeled (1–1′) begins with $P_X^0 = 0$, X^0 small, and $Z = Z^0$. On the incoming part of this trajectory, while Z decreases, X remains nearly constant and P_X remains small. As the trajectory passes through the caustic, P_X increases rather suddenly, then becomes approximately constant on the outgoing part of the trajectory. The trajectory labeled (3–3′) also begins with $P_X^0 = 0$, $Z = Z^0$ but with larger X^0. This trajectory mainly feels the attractive part of the potential energy, and P_X decreases as t increases. After the trajectory passes through the caustic, P_X approaches a negative constant. The trajectory (4–4′) never passes through the caustic, but otherwise is quite similar to (3–3′).

$$\frac{\partial p_i}{\partial q_j} = \frac{\partial p_j}{\partial q_i} \qquad i, j = 1 \ldots n . \tag{2.6a}$$

It follows from this that the integral

$$\int_{w_0}^{w} p \cdot dq = \int_{w_0}^{w} p \cdot \frac{dq}{dt}\, dt \tag{2.6b}$$

is locally independent of the path of integration, and therefore this integral locally defines a function $S(q)$ such that

$$p_i(q) = \frac{\partial S}{\partial q_i} . \tag{2.6c}$$

The function $S(q)$ is said to be a generating function for this domain of the manifold.

Some proofs and further discussion are given in Appendix A.

C. Transport Equation and the Classical Density

Also associated with the field of trajectories or with the manifold is a classical density $\rho(q)$ or $\rho(w)$; it satisfies an equation of continuity

$$\nabla_q(\rho(q)v(q)) = 0\,, \tag{2.7a}$$

where $v(q) = p(q)/M$. Equivalent to this is the formula

$$\frac{d\rho}{dt} \equiv v \cdot \nabla_q \rho(q) = -\rho(q)\nabla_q \cdot v(q)\,, \tag{2.7b}$$

where $d\rho/dt$ means the rate of change along the trajectory. The solution to this equation can be given in several forms, of which the most computationally convenient is

$$\rho(q) = \rho^0(w^0)\frac{J(0, w^0)}{J(t, w^0)}\,. \tag{2.8}$$

Here w^0 is a set of $n-1$ coordinates that specify points on the initial surface, $\rho^0(w^0)$ is an arbitrary "initial" density function, and J is the Jacobian

$$J(t, w^0) = \frac{\partial q(t, w^0)}{\partial(t, w^0)}\,. \tag{2.9}$$

To evaluate this Jacobian, the configuration point q is regarded as a function of t and of the $n-1$ coordinates w^0 spanning the initial surface, $q = q(t, w^0)$. Similarly, (2.8) involves the inverse relationship: q is the independent variable, and $w^0 = w^0(q)$.

Like S, J may be a multivalued function of configuration-space coordinates, but it is single-valued on the manifold Λ^n.

For the two-dimensional scattering problem, the coordinate indicating points on the initial line $Z = Z^0$ is the impact parameter, $w^0 \equiv b$, so to calculate the density we must consider the position X, Z to be a function of t and b through the solution to Hamilton's equations, and we have

$$J = \frac{\partial X}{\partial t}\frac{\partial Z}{\partial b} - \frac{\partial Z}{\partial t}\frac{\partial X}{\partial b}\,. \tag{2.10}$$

$(\partial X/\partial t)_b$ and $(\partial Z/\partial t)_b$ are velocity components that are computed during the integration of (2.4), and $(\partial X/\partial b)_t$ and $(\partial Z/\partial b)_t$ can be obtained by integrating adjacent trajectories.

In some of the work reported in ref. 3, a different Jacobian, sometimes known as the van Vleck determinant, has been used to calculate the classical density function. We believe that the present form, (2.9) and (2.10), is much easier to use.

III. LOCAL FORMAL ASYMPTOTIC SOLUTION TO THE SCHROEDINGER EQUATION

Following standard methods,[2] it is easy to show that the classical functions $S(q)$ and $\rho(q)$, constructed as above by integration from an initial surface, can be used to find formal asymptotic solutions to the Schroedinger equation,

$$[(2M)^{-1}(-i\hbar\nabla_q)^2 + V(q) - E]\Psi(q) = 0 . \tag{3.1}$$

Writing

$$\Psi(q) = A(q)\exp\left[\frac{iS(q)}{\hbar}\right] \tag{3.2}$$

and expanding

$$\Psi(q) = A^{[0]}(q) + \hbar A^{[1]}(q) + \hbar^2 A^{[2]}(q) + \cdots , \tag{3.3}$$

we substitute (3.2) and (3.3) into (3.1) and collect powers of $\hbar$:

$$\begin{aligned}
&[(2M)^{-1}|\nabla_q S(q)|^2 + V(q) - E][A^{[0]} + \hbar A^{[1]} + \hbar^2 A^{[2]} + \cdots] \\
&\quad + (-i\hbar)(2M)^{-1}[A^{[0]}(q)\nabla_q^2 S(q) + 2\nabla_q A^{[0]}(q)\cdot\nabla_q S(q)] \\
&\quad + (-i\hbar^2)(2M)^{-1}[A^{[1]}(q)\nabla_q^2 S(q) + 2\nabla_q A^{[1]}(q)\cdot\nabla_q S(q) - i\nabla_q^2 A^{[0]}(q)] \\
&\quad + \cdots = 0 .
\end{aligned} \tag{3.4}$$

Setting these quantities equal to zero order by order, we obtain: ($\hbar^0$) the Hamilton–Jacobi equation; ($\hbar^1$) the transport equation, with $A^{[0]}(q) = [\rho(q)]^{1/2}$; ($\hbar^2$ and higher) a set of differential equations defining $A^{[j]}$ in terms of $A^{[j-1]}$ for $j \geq 1$.

From Eq. (3.4), it follows that in any region in which $|\nabla^2 A^{[0]}(q)|$ is bounded by a constant K, the function

$$\Psi^{[0]}(q) \equiv A^{[0]}(q) \exp\left[\frac{iS(q)}{\hbar}\right] \tag{3.5}$$

satisfies the Schroedinger equation in the limit as $\hbar \to 0$; that is,

$$|(H-E)\Psi^{[0]}(q)| = \left|\left(\frac{\hbar^2}{2M}\right)\nabla^2 A^{[0]}q\right| \leqslant \frac{\hbar^2 K}{2M}, \tag{3.6}$$

and we can say that $\Psi^{[0]}$ is a formal asymptptic solution of (3.1) "to order $\hbar$."[11] Functions of the form (3.5) are also known as *primitive semiclassical approximations*, to distinguish them from so-called *uniform approximations* that will be developed later.

It is quite clear that (3.5) cannot represent the full solution to the Schroedinger equation anywhere: a glance at Fig. 1 tells us that even arbitrarily close to the initial surface, there must be another part of Ψ corresponding to trajectories that have scattered from the target; (3.5) is strictly valid only for the parts of the trajectory field for which t is sufficiently small. When trajectories touch the caustic, the Jacobian (2.9) passes through zero, the classical density (2.8) becomes infinite, and the asymptotic approximation (3.5) fails. We are then left with the problem of joining different approximations on either side of the caustic, as well as the problem of finding a uniform approximation near the caustic. These are the problems for which Maslov and his co-workers have found "general" solutions. They will be discussed in the next two sections.

IV. MOMENTUM SPACE AND MIXED SPACES

A. Introductory Remarks

Returning to one dimension for a moment, one way to obtain connection formulas and uniform approximations is to transform into momentum space.[12] Let q be the independent variable in the wavefunction $\Psi(q)$, let p be the corresponding independent variable in the Fourier-transformed wavefunction $\tilde{\Psi}(p)$, but let $p(q)$ and $q(p)$ represent the classical correspondence between position and momentum, that is, the solutions to $H(p, q) = E$. The momentum-space Schroedinger equation

$$\left[\frac{p^2}{2M} + V\left(i\hbar\frac{d}{dp}\right) - E\right]\tilde{\Psi}(p) = 0 \tag{4.1a}$$

has a simple semiclassical solution:

$$\tilde{\Psi}(p) = |F(q(p))|^{-1/2} \exp\left\{\left(\frac{i}{\hbar}\right)[S(q(p)) - q(p)p]\right\}, \qquad (4.1b)$$

$$= |F(q(p))|^{-1/2} \exp[i\tilde{S}(p)/\hbar]$$

where $F(q) = -dV/dq$ is the force at the point q, and $\tilde{S}(p) = -\int q(p)\, dp$. Approximation (4.1b) is valid (to order $\hbar$) except in regions of momentum space close to points $\hat{p}$ where $F(q(\hat{p})) = 0$; those are turning points in momentum space. Now in general there is no relationship between turning points $\hat{q}$ in position space [where $p(\hat{q}) \rightarrow 0$] and turning points $\hat{p}$ in momentum space [where $F(q(\hat{p})) \rightarrow 0$]. If (4.1b) is accurate for small p, then its (inverse) Fourier transform can be a good approximation to $\Psi(q)$ near a turning point.

From the above ideas, it is possible to derive WKB connection formulas and the familiar Airy-function uniform approximation near a turning point.[12] We consider next the generalization of this approach to n dimensions.

For the scattering problem, the source of the difficulties at the caustic is shown in Figs. 2 and 3. There we see that above the caustic, the Lagrangian manifold rises vertically—in fact, the caustic is the projection in the XZ plane of the points at which the manifold turns back over itself. We explained earlier that $S(w)$ and $J(w)$ are single-valued functions on the Lagrange manifold, but $S(q)$ and $J(q)$ are two-valued functions of coordinates, and J goes to zero at the caustic. Since $\Psi(q)$ is single valued and finite, the asymptotic approximation fails in this region.

Obviously, the first step is to find a better way to relate points on the Lagrangian manifold to phase-space coordinates. A portion of the manifold Λ^2 near the caustic is shown in Fig. 3, and one can see that in this region, there is a one-to-one relationship between points w on the manifold and the mixed pair of coordinates (P_X, Z). Accordingly, we can hope to construct an asymptotic approximation to the wavefunction in the mixed representation, $\tilde{\Psi}(P_X, Z)$. MF show how to construct such an asymptotic approximation. Just as important, they show that locally there almost always exists such a mixed representation in which the wavefunction has a suitable asymptotic approximant.

The mathematics involved here is quite abstract (ref. 6, pp. 102ff) but the essential ideas are not difficult. Basically, four things are needed: (1) a general definition of a Lagrangian manifold; (2) a demonstration that the manifolds we are discussing have this Lagrangian property; (3) a set of formulas that relate the desired solutions of the (momentum-space)

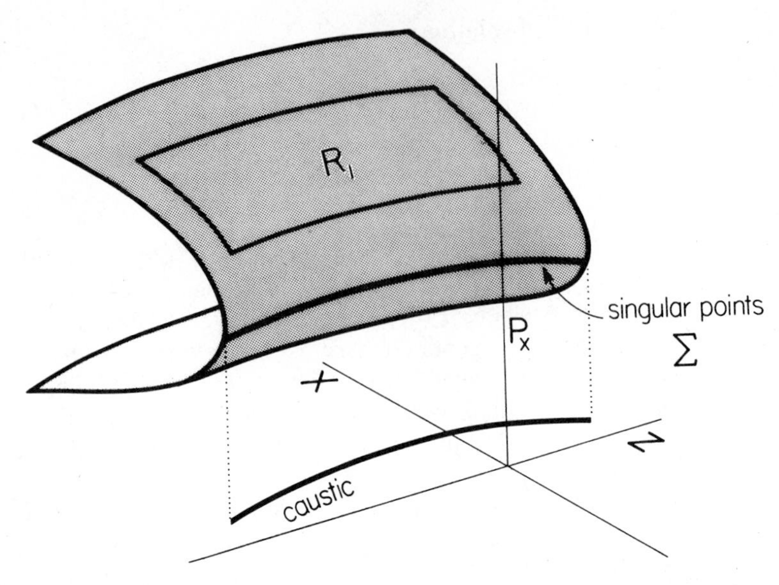

Fig. 3. Portion of the manifold Λ^2. Singular points Σ on the manifold project to the caustic in configuration space. The part of the manifold that is well above the singular points has a smooth projection into configuration space, as does the part that is well below the singular points. Near the singular points, Z and P_X can be used as coordinates. In the region R_1 either set of coordinates can be used.

Hamilton–Jacobi and transport equations to the properties of the manifold; (4) formulas relating the momentum-space wavefunction to the functions found in (3). The first two of these are discussed in Section B, the third in Section C and the last in Section D.

The structure of the argument can be seen most clearly if we momentarily restrict ourselves again to systems with one degree of freedom, for in that case the ideas become quite trivial. In this case, phase space is two dimensional, a trajectory lies on a smooth curve $H(p, q) =$ constant, and any smooth curve in this space is a Lagrangian manifold Λ^1 (Fig. 4). In general, there is not a global one-to-one relationship between coordinates q and points w on the curve, nor is there a global one-to-one relationship between momenta p and points on the curve. Locally, however, in a sufficiently small neighborhood of any point w_0, there is a one-to-one relationship between points w_1 on the curve and points w_1' on the tangent to the curve. Furthermore, for any tangent line, *either* there is a one-to-one relationship between points on the tangent and coordinate points q, *or* there must be such a relationship to momentum points p. For most such tangent lines, both such relationships exist, and for *any* tangent line, at least one of them exists (when the tangent line is perpendicular to the q axis then the only one-to-one relationship is with the p axis and vice

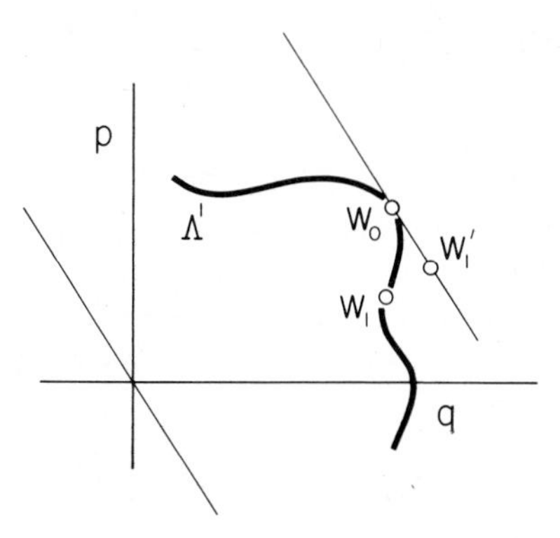

Fig. 4. If $n = 1$, any smooth curve is a Lagrangian manifold. Near any point w_1, we can construct the tangent to the curve and a line passing through the origin parallel to the tangent. That line must pass through the p axis or the x axis transversely (it cannot coincide with both). If it intersects one transversely, then the *other* certainly can be used as a coordinate, and there is a local one-to-one mapping between that coordinate and the points on the manifold Λ^1. MF extend this argument into n dimensions. In their framework, an essential property of an n-dimensional Lagrangian manifold is that its tangent planes intersect at least one Lagrangian coordinate plane transversely; hence the canonically conjugate plane can be used as coordinates for the tangent plane and for a corresponding domain of the manifold. In other words, every small domain of a Lagrangian manifold has a diffeomorphic projection into at least one of the Lagrangian coordinate planes.

versa). It follows, then, that locally we can always use either q or p as the independent variable, and a single-valued function $p(q)$ or $q(p)$ locally describes the curve Λ^1. Furthermore, in any small region, at least one of the functions $S(q) = \int p(q)\,dq$ or $\tilde{S}(p) = -\int q(p)\,dp$ is well defined, and either $p(q) = \partial S/\partial q$ or $q(p) = -\partial\tilde{S}/\partial p$ (or both). It is obvious that $S(q)$ and $\tilde{S}(p)$ satisfy Hamilton–Jacobi equations, respectively, in position space and momentum space; after all if $H(p, q) = E$ and $q = q(p) = -\partial\tilde{S}/\partial p$, then $H(p, -\partial\tilde{S}/\partial p) = E$. With a little more effort, one can show[12] that an associated density function satisfies a transport equation, and that this density function is simply $|F(q(p))|^{-1}$. Once this is done, it is not difficult to show that $\tilde{\Psi}(p)$ defined in (4.1b) gives an asymptotic approximation to the momentum-space wavefunction corresponding to a given region of the manifold. The final step is to transform back into configuration space and to combine the asymptotic approximants from various regions of the manifold into a complete wavefunction $\Psi(q)$.

B. Definition and Essential Properties of Lagrangian Manifolds

Let us now consider the extension of these ideas to systems with n degrees of freedom. We begin with the definition of a Lagrangian manifold.[13]

In $2n$-dimensional phase space, MF define a set of *Lagrangian coordinate planes.* If $n = 1$, these are the q axis ($p = 0$) and the p axis ($q = 0$). If $n = 2$, they are the q_1q_2 plane ($p_1 = 0$, $p_2 = 0$), the q_1p_2 plane, the q_2p_1 plane, and the p_1p_2 plane. For general n, the Lagrangian coordinate planes are a set of 2^n phase-space planes, each of which passes through the origin and contains k coordinate axes and $n - k$ momentum axes ($0 \leq k \leq n$), but which does not contain canonically conjugate coordinate and momentum axes. (For example, in $n = 2$, the q_1p_1 plane is not a Lagrangian coordinate plane.)

An n-dimensional manifold embedded in the $2n$-dimensional phase space is said to be Lagrangian if and only if it has the following two properties:

1. *For every point on the manifold, there exists a small domain of the manifold containing that point, which domain has a diffeomorphic projection into at least one of the Lagrangian coordinate planes.* A diffeomorphic projection is one which is smooth and which possesses a smooth inverse projection from the Lagrangian coordinate plane to the manifold. In Fig. 3, near the singular region of the manifold, there is a diffeomorphic projection into the P_XZ plane but not into the XZ plane.

In general, the selected plane is denoted $(p_\alpha q_\beta)$, where α and β are collective indices. For example, in a three-dimensional problem, we might project a domain into the $p_1q_2q_3$ plane, for which $\alpha = \{1\}$, $\beta = \{2, 3\}$; some other domain might be projected into the $p_1p_2p_3$ plane, for which $\alpha = (1, 2, 3)$ and β is an empty set.

Within that domain, the embedding of the manifold in phase space is given by a set of n smooth functions:

$$\begin{aligned} q_i &= q_i(p_\alpha q_\beta)\,, \qquad i \in \alpha \\ p_j &= p_j(p_\alpha q_\beta)\,, \qquad j \in \beta \end{aligned} \tag{4.2a}$$

specifying the values of the coordinates and momenta that are canonically conjugate to $(p_\alpha q_\beta)$. These functions (or their values on the manifold) are collectively denoted

$$q_\alpha(p_\alpha q_\beta) \quad \text{and} \quad p_\beta(p_\alpha q_\beta)\,. \tag{4.2b}$$

2. *Derivatives of these functions satisfy the following restrictions*:

$$\frac{\partial p_i}{\partial q_j} = \frac{\partial p_j}{\partial q_i}\,, \qquad i \in \beta\,,\ j \in \beta \tag{4.3a}$$

$$\frac{\partial q_i}{\partial p_j} = \frac{\partial q_j}{\partial p_i}, \qquad i \in \alpha, \quad j \in \alpha \tag{4.3b}$$

$$\frac{\partial p_i}{\partial p_j} = -\frac{\partial q_j}{\partial q_i}, \qquad i \in \beta, \quad j \in \alpha . \tag{4.3c}$$

The two conditions stated above may be taken to be the definition of an n-dimensional Lagrangian manifold; the concept can be extended to lower-dimensional manifolds if we say that any k-dimensional submanifold of an n-dimensional Lagrangian manifold is also Lagrangian ($k \leq n$).

From the restrictions (4.3) on the derivatives, it follows that (within the given domain)

$$\begin{aligned} d\tilde{S} &\equiv \sum_{j \in \beta} p_j(p_\alpha q_\beta)\, dq_j - \sum_{i \in \alpha} q_i(p_\alpha q_\beta)\, dp_i \\ &\equiv p_\beta \cdot dq_\beta - q_\alpha \cdot dp_\alpha \end{aligned} \tag{4.4a}$$

is an exact differential, and the integral

$$\tilde{S} = \int p_\beta \cdot dq_\beta - q_\alpha \cdot dp_\alpha \tag{4.4b}$$

is independent of the path of integration. Thus the domain admits a *generating function* $\tilde{S}(p_\alpha q_\beta)$ such that

$$p_\beta(p_\alpha q_\beta) = \frac{\partial \tilde{S}}{\partial q_\beta} \quad \text{and} \quad q_\alpha(p_\alpha q_\beta) = -\frac{\partial \tilde{S}}{\partial p_\alpha} . \tag{4.4c}$$

From the above definition, we see that for a Lagrangian manifold, in any domain in which the configuration-space representation fails, there is always at least one momentum-space representation $(p_\alpha q_\beta)$ in which $\tilde{S}(p_\alpha q_\beta)$ is well defined; the only requirement is that the domain must be sufficiently small. Let us now take up the next issue: do the trajectory manifolds actually have this property? MF establish the important result (their proposition 4.19) that manifolds generated as discussed in Sections II.A and II.B, by integration of Hamilton's equations from a Lagrangian initial surface, are indeed Lagrangian.

This theorem is important enough that it should be stated precisely. Let Λ_0^{n-1} be an $(n-1)$-dimensional initial manifold in phase space, and let it be Lagrangian, that is, let it have the property that when it is projected into *some* Lagrangian coordinate plane $(p_\alpha q_\beta)$ it admits a generating function $\tilde{S}$ such that

$$d\tilde{S} = p_\beta \cdot dq_\beta - q_\alpha \cdot dp_\alpha$$

for any $(dp_\alpha\, dq_\beta)$ restricted to Λ_0^{n-1}. Also let Λ_t^{n-1} be the set of points that result from evolution under the Hamiltonian for time t. [Each phase-space point $w^0 = (p^0, q^0)$ on the initial surface goes to a new point $w^t = (p^t, q^t)$ after time t, and Λ_t^{n-1} is the set of points $(p^t q^t)$ at that specific instant.] MF prove that if the mapping between w^0 and w^t is a diffeomorphism and if the union of the manifolds

$$\Lambda^n = \bigcup_{0 \le t \le T} \Lambda_t^{n-1}$$

is a C^∞ (continuous and infinitely differentiable) manifold of dimension n, then the manifold Λ^n is Lagrangian.

More simply stated, the trajectory fields generated by integration from a suitable (Lagrangian) initial surface do not necessarily form a manifold, but if they do, then the manifold is certainly Lagrangian. It follows that under very general conditions, a semiclassical approximation to the wavefunction in almost any singular region can be obtained by transformation to a suitably chosen momentum-space or mixed-space representation.

C. Classical Functions for General Representations

Let us continue to consider a domain of the manifold having a diffeomorphic projection into the $(p_\alpha q_\beta)$ plane. In this mixed representation, the Hamilton–Jacobi equation is obtained by taking the Hamiltonian $H(p_1 \ldots p_n, q_1 \ldots q_n)$, replacing p_β by $\partial\tilde{S}/\partial q_\beta$ and q_α by $-\partial\tilde{S}/\partial p_\alpha$, and then setting

$$H\left(p_\alpha, \frac{\partial \tilde{S}}{\partial q_\beta}, -\frac{\partial \tilde{S}}{\partial p_\alpha}, q_\beta\right) = E. \tag{4.5}$$

The solution to this equation is constructed by a procedure that is directly analogous to that given in Section II.A and Appendix A, only interchanging the roles of some of the q's and p's and incorporating a minus sign where needed. None of the equations (A.1)–(A.12) in Appendix A depend on the form of the Hamiltonian, and in Hamiltonian mechanics, there is no fundamental distinction between coordinates and momenta (only a minus sign in half of the equations $\dot{p}_i = -\partial H/\partial q_i$, $\dot{q}_i = \partial H/\partial p_i$). It is therefore easy to show that the solution to (4.5) is

$$\tilde{S}(p_\alpha q_\beta) = \int_{w_0}^{w} p_\beta \cdot dq_\beta - \int_{w_0}^{w} q_\alpha \cdot dp_\alpha + C$$

$$= \int_{t_0}^{t(w)} \left(\sum_{i\in\beta} p_i \frac{dq_i}{dt} - \sum_{i\in\alpha} q_i \frac{dp_i}{dt} \right) dt + C, \tag{4.6a}$$

where C is a constant. Thus the solution to the Hamilton–Jacobi equation is the generating function for this domain of the Lagrangian manifold. If the domain is regular, then the function S is also defined, and

$$\tilde{S} = S - \sum_{i\in\alpha} [q_i(t)p_i(t) - q_i(t_0)p_i(t_0)] + C. \tag{4.6b}$$

Formulas that will appear later are simplified if the constant is chosen so that it cancels the terms involving $q_i(t_0)p_i(t_0)$. The quantity $q_i(t)$ means q_i evaluated at the manifold point w, so (4.6) can alternatively be written as

$$\tilde{S}(p_\alpha, q_\beta) = S - \sum_{i\in\alpha} p_i q_i(p_\alpha q_\beta). \tag{4.6c}$$

There is also a density function associated with the trajectories in the mixed representation, and it is given by a formula that is analogous to (2.8). Let an initial surface again be spanned by a set of $n-1$ coordinates w_i^0, and let trajectories emanating from it be given by functions

$$p_\alpha(t, w^0) \quad \text{and} \quad q_\beta(t, w^0). \tag{4.7}$$

Defining a Jacobian

$$\tilde{J}_{\alpha\beta}(p_\alpha, q_\beta) \equiv \tilde{J}_{\alpha\beta}(t, w^0) = \frac{\partial(p_\alpha(t, w^0), q_\beta(t, w^0))}{\partial(t, w^0)}, \tag{4.8}$$

the classical density function is related to its value at the corresponding point on the initial surface by

$$\tilde{\rho}(p_\alpha, q_\beta) = \rho^0(w^0) \frac{\tilde{J}(t=0, w^0)}{\tilde{J}(t, w^0)}. \tag{4.9a}$$

This density function satisfies a transport equation which is analogous to

(2.7), but which is considerably more complicated. Assuming that H has the form (1.1), the equation is

$$\left(\frac{p_\beta}{M}\frac{\partial\tilde{\rho}}{\partial q_\beta}-\frac{\partial V}{\partial q_\alpha}\frac{\partial\tilde{\rho}}{\partial p_\alpha}\right)+\left(\frac{1}{M}\frac{\partial p_\beta}{\partial q_\beta}-\frac{\partial^2 V}{\partial q_\alpha \partial q_{\alpha'}}\frac{\partial q_{\alpha'}}{\partial p_\alpha}\right)\tilde{\rho}=0\,, \qquad (4.9b)$$

where p_β and q_α are manifold functions (4.3). Summation over repeated indices is implied in (4.9b). Proofs of (4.5) and (4.9) are given in Appendix B.

D. Wavefunctions in Momentum Space and in Mixed Spaces

1. *Formal Asymptotic Solution of the Schroedinger Equation*

We now show that the functions $\tilde{S}$ and $\tilde{J}$ defined above can be used to calculate an approximate solution to the corresponding wave equation: we shall show that the semiclassical approximation to the wavefunction in $p_\alpha q_\beta$ space is

$$\tilde{\Psi}(p_\alpha q_\beta)=|\tilde{\rho}(p_\alpha, q_\beta)|^{1/2}\exp[i\tilde{S}(p_\alpha, q_\beta)/\hbar]$$

The analysis is long, but perfectly straightforward.

The Schroedinger equation in momentum space or in mixed spaces is obtained by taking a partial Fourier transform of the configuration-space Schroedinger equation

$$\left[H\left(-i\hbar\frac{\partial}{\partial q}, q\right)-E\right]\Psi(q)=0\,. \qquad (4.10)$$

As in the preceding sections, we select an appropriate mixed set of coordinates and momenta $p_\alpha q_\beta$ (containing no canonically conjugate pairs), multiply (4.10) on the left by $\exp(-ip_\alpha\cdot q_\alpha/\hbar)\equiv\exp(-i\sum_{j\in\alpha}p_j q_j/\hbar)$, and integrate over variables α:

$$(2\pi i\hbar)^{-k/2}\int_{-\infty}^{\infty}\exp\left(-\frac{ip_\alpha\cdot q_\alpha}{\hbar}\right)\left[H\left(-i\hbar\frac{\partial}{\partial q}, q\right)-E\right]\Psi(q)\,dq_\alpha=0\,, \quad (4.11)$$

where k is the number of variables in the set α (the number of variables that are transformed). From well-known properties of the Fourier transform, the resulting Schroedinger equation in the mixed space can be written formally as

$$\left[H\left(p_\alpha, -i\hbar\frac{\partial}{\partial q_\beta}, i\hbar\frac{\partial}{\partial p_\alpha}, q_\beta\right)-E\right]\tilde{\Psi}(p_\alpha, q_\beta)=0\,. \qquad (4.12)$$

The operator can be defined in either of two ways: by its Taylor-series expansion, or by using the fact that the equation arises through Fourier transformation of the (well-defined) configuration-space form.

In general, the Taylor expansion has the form

$$H\left(p_\alpha, -i\hbar\frac{\partial}{\partial q_\beta}, i\hbar\frac{\partial}{\partial p_\alpha}, q_\beta\right) = \sum_{k\ell mn} H_{k\ell mn}\left(i\hbar\frac{\partial}{\partial p_\alpha}\right)^k p_\alpha^\ell q_\beta^m \left(-i\hbar\frac{\partial}{\partial q_\beta}\right)^n, \tag{4.13}$$

where $H_{k\ell mn}$ are constant coefficients, but following our earlier assumption (1.1) that H can be separated into a p-dependent part and a q-dependent part, we can write

$$H = \sum_{k\ell} T_{k\ell} p_\alpha^k \left(-i\hbar\frac{\partial}{\partial q_\beta}\right)^\ell + \sum_{mn} V_{mn}\left(i\hbar\frac{\partial}{\partial p_\alpha}\right)^m q_\beta^n. \tag{4.14}$$

For formal purposes, these Taylor expansions provide everything we need.

To obtain a formal asymptotic expansion of $\tilde{\Psi}(p_\alpha q_\beta)$, let us examine the effect of differential operators on a function of the form

$$\tilde{\Psi}(p_\alpha, q_\beta) = \exp\left[\frac{i\tilde{S}(p_\alpha, q_\beta)}{\hbar}\right]\tilde{A}(p_\alpha, q_\beta). \tag{4.15}$$

By repeated application of the rules

$$\begin{aligned} i\hbar\frac{\partial}{\partial p_\alpha} e^{i\tilde{S}/\hbar}\tilde{A} &= e^{i\tilde{S}/\hbar}\left(q_\alpha + i\hbar\frac{\partial}{\partial p_\alpha}\right)\tilde{A}, \\ -i\hbar\frac{\partial}{\partial q_\beta} e^{i\tilde{S}/\hbar}\tilde{A} &= e^{i\tilde{S}/\hbar}\left(p_\beta - i\hbar\frac{\partial}{\partial q_\beta}\right)\tilde{A}, \end{aligned} \tag{4.16}$$

where in these equations

$$\begin{aligned} q_\alpha &= q_\alpha(p_\alpha, q_\beta) = -\frac{\partial\tilde{S}}{\partial p_\alpha}, \\ p_\beta &= p_\beta(p_\alpha, q_\beta) = \frac{\partial\tilde{S}}{\partial q_\alpha}, \end{aligned} \tag{4.17}$$

one can obtain an expansion of the differential operator in powers of $\hbar$:

$$\left[H\left(p_\alpha, -i\hbar\frac{\partial}{\partial q_\beta}, i\hbar\frac{\partial}{\partial p_\alpha}, q_\beta\right) - E\right]\tilde{\Psi} = e^{i\tilde{S}/\hbar}(\mathscr{H}_0 + \hbar\mathscr{H}_1 + \hbar^2\mathscr{H}_2 + \cdots)\tilde{A}\,, \tag{4.18a}$$

with (Appendix B)

$$\mathscr{H}_0 = H(p_\alpha, p_\beta, q_\alpha, q_\beta) - E \tag{4.18b}$$

and

$$\begin{aligned}\mathscr{H}_1 = &\left(\frac{\partial H}{\partial p_\beta}\right)\left(-i\frac{\partial}{\partial q_\beta}\right) + \left(\frac{\partial H}{\partial q_\alpha}\right)\left(i\frac{\partial}{\partial p_\alpha}\right)\\ &+\frac{1}{2}\left(\frac{\partial^2 H}{\partial p_\beta \partial p_{\beta'}}\right)\left(-i\frac{\partial p_{\beta'}}{\partial q_\beta}\right) + \frac{1}{2}\left(\frac{\partial^2 H}{\partial q_\alpha \partial q_{\alpha'}}\right)\left(i\frac{\partial q_{\alpha'}}{\partial p_\alpha}\right).\end{aligned} \tag{4.18c}$$

Succeeding terms are of higher order in $\hbar$. [In (4.18c) we have assumed that H has the form (1.1); otherwise a few additional terms arise.]

Now, if we also expand $\tilde{A}$ in powers of $\hbar$

$$\tilde{A} = A^{[0]} + \hbar\tilde{A}^{[1]} + \hbar^2\tilde{A}^{[2]} + \cdots, \tag{4.19}$$

apply the expansion to (4.18a), and again collect powers of $\hbar$, the result is obviously

$$\begin{aligned}&\mathscr{H}_0\tilde{A}^{[0]} + \hbar(\mathscr{H}_0\tilde{A}^{[1]} + \mathscr{H}_1\tilde{A}^{[0]})\\ &\quad + \hbar^2(\mathscr{H}_0\tilde{A}^{[2]} + \mathscr{H}_1\tilde{A}^{[1]} + \mathscr{H}_2\tilde{A}^{[0]})\\ &\quad + \cdots = 0\,.\end{aligned} \tag{4.20}$$

Finally, the desired formal asymptotic series is obtained by setting (4.20) to zero line by line. The lowest-order equation is

$$\begin{aligned}0 &= \mathscr{H}_0\\ &= H(p_\alpha, p_\beta, q_\alpha, q_\beta) - E\\ &= H\left(p_\alpha, \frac{\partial\tilde{S}}{\partial q_\beta}, -\frac{\partial\tilde{S}}{\partial p_\alpha}, q_\beta\right) - E\,,\end{aligned} \tag{4.21}$$

which is of course the Hamilton–Jacobi equation, the desired solution to which is given by Eq. (4.5).

The first-order equation

$$\mathcal{H}_1 \tilde{A}_0 = 0 \tag{4.22}$$

is easily seen to be essentially the same as the transport equation (4.9) or (B.22), and its solution is

$$\tilde{A}^{[0]} = [\tilde{\rho}(p_\alpha, q_\beta)]^{-1/2}, \tag{4.23}$$

where $\tilde{\rho}$ is the density function defined in Eq. (4.8). Higher-order terms can also be obtained.

The resulting truncated expansion for the wavefunction,

$$\tilde{\Psi}^{[N]}(p_\alpha, q_\beta) \equiv \sum_{j=0}^{N} \hbar^j \tilde{A}^{[j]} \exp\left(\frac{i\tilde{S}}{\hbar}\right), \tag{4.24}$$

is said to be a formal asymptotic solution to order $\hbar^{N+1}$. We usually limit ourselves to the form

$$\tilde{\Psi}^{[0]}(p_\alpha, q_\beta) = \tilde{A}^{[0]}(p_\alpha, q_\beta) \exp\left[\frac{i\tilde{S}(p_\alpha, q_\beta)}{\hbar}\right], \tag{4.25}$$

where $\tilde{A}^{[0]}(p_\alpha, q_\beta)$ is given in (4.23) and (4.9a), and $\tilde{S}(p_\alpha, q_\beta)$ is defined in (4.6).

2. *Validity of the Asymptotic Expansion*

The derivation given above makes explicit use of the Taylor expansions (4.13) and (4.14), and some of the steps in these derivations are only valid if that expansion converges on the whole real line. In fact, such Taylor series frequently diverge. For example, the potential energy functions $V(R)$ that are relevant in molecular physics almost always diverge at the origin, and their Taylor-series representations usually have very limited utility. For this reason, MF provide alternative proofs of the basic results (their theorems 2.6, 2.9, and 2.10). We do not give any details of their proofs, but let us just mention that instead of using Taylor expansions as above, they define the mixed-space Hamiltonian and derive its properties by referring back to the Fourier transform (4.11), which is clearly more fundamental.

On the other hand, it is worthwhile to state with some care precisely what their theorems tell us. Their proofs make use of assumptions about the Hamiltonian operator, about the resulting classical trajectories, about the solution $\tilde{S}(p_\alpha q_\beta)$ to the Hamilton–Jacobi equation, and about the function $\tilde{A}_0(p_\alpha q_\beta)$, which is related to the solution of the transport equation.

Let us consider a finite, closed region $\mathscr{D}$ of the $p_\alpha q_\beta$ space (Fig. 5). An initial surface spanned by coordinates w_i^0 $(i = 1 \ldots n-1)$ is assumed to lie within this region, and it is assumed that the trajectories provide a mapping between (t, w^0) and $(p_\alpha q_\beta)$ for all points in $\mathscr{D}$. It is further assumed that this mapping is a diffeomorphism, and that $\tilde{S}(p_\alpha q_\beta)$ and $\tilde{A}_0(p_\alpha q_\beta)$ are infinitely differentiable in this region. The range of values of the derivatives

$$p_\beta = \frac{\partial \tilde{S}}{\partial q_\beta} \quad \text{and} \quad q_\alpha = -\frac{\partial \tilde{S}}{\partial p_\alpha}$$

within $\mathscr{D}$ then define a corresponding finite, closed region $\mathscr{D}^*$ in the conjugate $p_\beta q_\alpha$ space. It is necessary to define a larger, open domain in the conjugate space that includes $\mathscr{D}^*$ plus a bit more, and then to include the limit points of this extended domain to get a closed set $\bar{\mathscr{D}}^+$ (Fig. 5). It is assumed that within $\mathscr{D}$ and $\bar{\mathscr{D}}^+$ the Hamiltonian function $H(p, q)$ is real and infinitely differentiable with respect to all its variables. Furthermore, it is assumed that H and all its derivatives with respect to $p_\alpha q_\beta$ are integrable over the region $\mathscr{D}^+$:

$$\int_{\mathscr{D}^+} H(pq)\, dp_\beta\, dq_\alpha < \infty, \quad \int_{\mathscr{D}^+} \frac{\partial H}{\partial p_\alpha}\, dp_\beta\, dq_\alpha < \infty, \quad \int_{\mathscr{D}^+} \frac{\partial H}{\partial q_\beta}\, dp_\beta\, dq_\alpha < \infty, \quad \text{etc.}$$

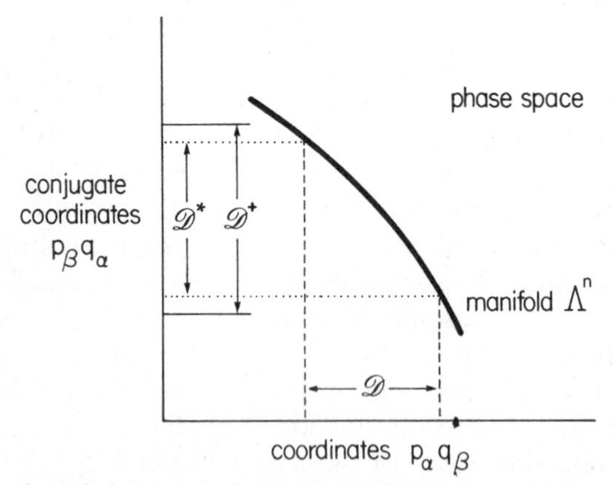

Fig. 5. Illustration of the regions involved in MF's fundamental theorem. $\mathscr{D}$ is a finite, closed region of the p, q space. Corresponding points on the manifold project into a region $\mathscr{D}^*$ of the conjugate space. $\mathscr{D}^+$ is a larger, open domain that contains $\mathscr{D}^*$, and $\bar{\mathscr{D}}^+$ includes $\mathscr{D}^+$ and its limit points. MF prove that the momentum-space semiclassical approximation is valid if the Hamiltonian satisfies certain regularity conditions only in such a finite, closed region of phase space.

Under these conditions, if $\tilde{\Psi}^{[0]}$ is defined as $\tilde{A}^{[0]}(p_\alpha, q_\beta)\times \exp[i\tilde{S}(p_\alpha, q_\beta)/\hbar]$, then within $\mathscr{D}$

$$|(H-E)\tilde{\Psi}^{[0]}| \leq \hbar^2 C,$$

where C is some constant, independent of q, p and $\hbar$. It is then said that the function $\tilde{\Psi}^{[0]}$ represents a *formal asymptotic solution*, which means that as $\hbar$ approaches zero, $(H-E)\tilde{\Psi}$ also approaches zero everywhere in the region $\mathscr{D}$ (in this case, as $\hbar^2$). More generally, for

$$\tilde{\Psi}^{[N]}(p_\alpha, q_\beta) = \exp\left[\frac{i\tilde{S}(p_\alpha, q_\beta)}{\hbar}\right] \sum_{k=0}^{N} \tilde{A}^{[k]}(p_\alpha, q_\beta) \tag{4.26}$$

MF's proposition (2.10) asserts that

$$|(H-E)\Psi^{[N]}| \leq C\hbar^{N+2}, \tag{4.27}$$

so (4.26) is said to represent a formal asymptotic series solution to the Schroedinger equation.

Why is this theorem important? The Fourier transform (4.11) is a global integral over coordinates, and one might expect that the validity of the asymptotic series (4.26) would therefore depend on some global smoothness properties of the Hamiltonian. If this were true, then any singularity in $V(R)$ (e.g., as $R \to 0$) would wreak havoc with the theory. In fact, however, MF have here provided us a theorem based only on *local* smoothness assumptions: only requiring that H be well behaved in a limited region of phase space, they show that (4.26) represents a formal asymptotic series solution to the Schroedinger equation. This theorem therefore establishes that the momentum-space semiclassical approximation (4.26) can be used very generally for the Hamiltonians that describe molecular systems.

V. GLOBAL FORMAL ASYMPTOTIC SOLUTIONS

The arguments developed in the preceding sections tell us that the wavefunction can be described by different formulas in different regions of configuration space. Locally, in some regions, Ψ is written as a combination of primitive semiclassical forms (3.5) while in other regions it is given as a Fourier transform of a momentum-space or mixed-space form (4.25). Each form of Ψ can be multiplied by an arbitrary constant, but all such constants must be chosen consistently in order to obtain a smooth global wavefunction. In the present section we summarize the methods needed to patch together the various semiclassical formulas into a single smooth Ψ.

The essential ideas are quite simple. In the scattering problem, for example, Ψ is given by primitive semiclassical forms throughout most of configuration space, but it would involve an Airy function near the caustic. Sufficiently far (but not too far!) from the caustic, the Airy function can be decomposed into incoming and outgoing parts, and these must join smoothly onto the incoming and outgoing primitive semiclassical wavefunctions. The smooth joining of these functions determines the relative amplitudes and phases of the coefficients of primitive semiclassical and Airy functions.

A. Local Approximations in Configuration Space

The first step is to divide the manifold into various overlapping regions, in each of which an appropriate formula will be applied. For the two-dimensional scattering problem, there are three major divisions (Fig. 2). Region I is most of the lower sheet of the manifold; it represents the *Incident* part of all of the trajectories as well as the outgoing part of those trajectories that never touch the singular points corresponding to the caustic. Region S is most of the upper sheet, representing the *Scattered* parts of trajectories that have passed through the singular points, and region C is the domain connecting I and S near the singular points.

Each of regions I and S are regular, and, for each, a primitive semiclassical form is suitable,

$$\Psi(q) = A(q) \exp\left[\frac{iS(q)}{\hbar}\right]. \tag{5.1}$$

However, as stated above, any such form can be multiplied by an arbitrary complex constant, which we denote by c. Since they both project into the same region of configuration space, the full wavefunction is a linear combination of such forms, which we write as

$$\Psi(q) = \sum_j c_j A_j(q) \exp\left[\frac{iS_j(q)}{\hbar}\right]. \tag{5.2}$$

Normally, in applying this formula we restrict ourselves to the *zero-order* form, with A_j replaced by $A_j^{[0]}$, but we no longer write the superscript$^{[0]}$.

In regions wherein (5.2) is not valid, one selects an alternative representation from the various momentum-space or mixed-space representations. In a two-dimensional system, with phase-space variables

$q_1q_2p_1p_2$, there are four possibilities: q_1q_2 (configuration space), p_1p_2 (momentum space), or q_1p_2 or p_1q_2 (the two mixed spaces). The representation is chosen such that the mapping from the manifold to the selected coordinates is locally a diffeomorphism; the theorem cited in Section IV.B ensures that such a representation generally exists. Denoting the selected representation by $p_\alpha q_\beta$, we have

$$\tilde{\Psi}(p_\alpha q_\beta) = c_j \tilde{A}_j(p_\alpha q_\beta) \exp\left[\frac{i\tilde{S}(p_\alpha q_\beta)}{\hbar}\right]. \tag{5.3}$$

However, before transforming this into the corresponding configuration-space wavefunction, we must note that it is only appropriate for a limited region; elsewhere this function may have singularities (related to mixed-space caustics) that render it unsuitable as an approximation. For this reason, MF multiply (5.3) by a smooth (C^∞) *switching function*, $e(w)$, defined on the manifold such that $e = 1$ in the region in which (4.3) is relevant, and such that e goes smoothly to zero elsewhere. Then the local formal asymptotic approximation to the configuration-space wavefunction is the (inverse) Fourier transform of (5.3):

$$\begin{aligned}\Psi(q) &= F^{-1}_{p_\alpha \to q_\alpha} \tilde{\Psi}(p_\alpha, q_\beta) \\ &= \left(\frac{1}{-2\pi i\hbar}\right)^{k/2} \int \exp\left[i \sum_{j\in\alpha} \frac{p_j q_j}{\hbar}\right] \tilde{\Psi}(p_\alpha, q_\beta) e(p_\alpha, q_\beta)\, dp_\alpha\,, \end{aligned} \tag{5.4}$$

where k is the number of momentum variables in the set α. Often the term *uniform approximation* is used to refer to formulas like (5.4).

B. Joining Semiclassical Functions Using the Stationary Phase Approximation

As stated above, the coefficients c_j are calculated by the requirement that the various forms for Ψ be consistent with each other. To establish this consistency, it is necessary to examine Ψ near the boundaries of the various regions. We mentioned that configuration space was to be divided into overlapping domains; it is in the regions of overlap that the different forms are to be compared. For example, in the two-dimensional scattering problem, we seek a domain (like R_1 in Fig. 3) in which *both* the primitive semiclassical approximation (5.3) and the uniform approximation (5.4) are suitable, and we compare the formulas in that domain. The stationary phase approximation is useful in making that comparison.

Let us begin with a precise statement of the stationary phase approximation in k dimensions. Suppose $a(p)$ is a C^∞ function of k

variables (collectively denoted p), and that $a(p)$ vanishes outside a compact region of p space. Let $\Phi(p)$ also be C^∞ within this region and let $\Phi(p)$ have a single nondegenerate critical point in the region. A nondegenerate critical point is a point p^0 such that

$$\left(\frac{\partial \Phi}{\partial p_i}\right)_{p=p^0} = 0\,, \qquad i = 1 \ldots k \tag{5.5a}$$

but

$$\det\left[\frac{\partial^2 \Phi}{\partial p_i \partial p_j}\right]_{p=p^0} \neq 0\,. \tag{5.5b}$$

Then

$$\int a(p) \exp\left[\frac{i\Phi(p)}{\hbar}\right] dp = (2\pi\hbar)^{k/2} a(p_0) \left| \det\left[\frac{\partial^2 \Phi}{\partial p_i \partial p_j}\right]\right|^{-1/2} \times \exp\left[\frac{i\Phi(p_0)}{\hbar}\right] \exp\left(i\frac{\pi}{4}\operatorname{sgn}\left[\frac{\partial^2 \Phi}{\partial p_i \partial p_j}\right]\right), \tag{5.6}$$

where sgn$[M]$ is the signature of the matrix M and is defined as the number of positive eigenvalues minus the number of negative eigenvalues of the matrix. All quantities on the right-hand side of (5.6) are to be evaluated at the critical point, p^0. It has been proved (MF theorem 1.1) that the error in the stationary phase approximation is bounded by

$$C\hbar^{k/2+1}\,, \tag{5.7}$$

where C is a constant that depends upon $a(p)$. This theorem implies that as $\hbar$ goes to zero the integral in (5.6) approaches the stationary phase value with relative error less than a constant times $\hbar$.

The stationary phase formula is easily applied to the Fourier integral (5.4). For this purpose, we temporarily choose $e(w)$ such that it goes smoothly to zero outside a certain region R_1, which is defined such that (1) the projection of R_1 into configuration space lies within the region of configuration space in which both (5.3) and (5.4) are appropriate, and (2) only one point in R_1 projects to each corresponding configuration-space point. Generally, there are two or more regions having the same properties (Fig. 3), but we choose $e(w)$ so that it is nonzero in only one of them.

It follows that (5.4) can be written in the form (5.6), with [assuming that

(5.3) contains only one term]

$$\begin{aligned}\Phi(p) &\equiv \Phi(p_\alpha; q_\alpha q_\beta) = \tilde{S}(p_\alpha q_\beta) + \sum_{j\in\alpha} p_j q_j , \\ a(p) &= a(p_\alpha; q_\beta) = (2\pi\hbar)^{-k/2} e(p_\alpha, q_\beta) c\tilde{A}(p_\alpha, q_\beta) .\end{aligned} \tag{5.8}$$

In evaluating the integral, $q = (q_\alpha, q_\beta)$ is regarded as a fixed parameter and $p_\alpha = \{p_j \mid j \in \alpha\}$ is the variable of integration. Applying the stationary phase approximation (5.6), several simplifications follow immediately.

The derivative of the phase is

$$\frac{\partial \Phi}{\partial p_j} = \frac{\partial \tilde{S}}{\partial p_j} + q_j , \qquad j \in \alpha \tag{5.9}$$

which vanishes at values of p such that

$$q_j = -\frac{\partial \tilde{S}}{\partial p_j} . \tag{5.10}$$

These are k equations that define the embedding of the manifold Λ^n in the $n+k$ dimensional space $(p_\alpha q_\alpha q_\beta)$. Combining (5.8) and (5.9) with (4.5b), we obtain

$$\Phi(p_\alpha^0(q_\alpha, q_\beta); q_\alpha q_\beta) = S(q) . \tag{5.11}$$

Similarly, the amplitude

$$A(p_\alpha^0; q_\beta) \equiv \tilde{A}(p_\alpha^0, q_\beta) \left| \det \left[\frac{\partial^2 \Phi}{\partial p_i \partial p_j} \right] \right|^{-1/2} e(p_\alpha^0, q_\beta) c \tag{5.12}$$

can be simplified by using (4.23), (4.8), and (4.7),

$$\tilde{A}(p_\alpha^0, q_\beta) = |\tilde{\rho}_0 \tilde{J}_0|^{1/2} \left| \frac{\partial(t, w^0)}{\partial(p_\alpha, q_\beta)} \right|^{1/2} , \tag{5.13}$$

and by evaluating the resulting determinant using (5.10) and (5.11)

$$\begin{aligned}\left| \det \left[\frac{\partial^2 \Phi}{\partial p_i \partial p_j} \right] \right| &= \left| \det \left[\frac{\partial q_i}{\partial p_j} \right] \right| , \qquad i, j \in \alpha \\ &= \left| \frac{\partial(q_\alpha)}{\partial(p_\alpha)} \right| .\end{aligned} \tag{5.14}$$

Then (5.12) reduces to [taking $e(p_\alpha^0, q_\beta) = 1$]

$$A(p_\alpha^0; q_\beta) = c|\tilde{\rho}_0 \tilde{J}_0|^{1/2} \left| \frac{\partial(t, w^0)}{\partial(p_\alpha, q_\beta)} \right|^{-1/2} \left| \frac{\partial(p_\alpha)}{\partial(q_\alpha)} \right|^{1/2} = c|\tilde{\rho}_0 \tilde{J}_0|^{1/2} \left| \frac{\partial(t, w^0)}{\partial(q)} \right|^{1/2} \tag{5.15}$$

which is essentially the formula (2.8) for the configuration-space amplitude.

Finally, the integral contains an additional constant phase $\exp[i(\pi/4)\delta]$, where

$$\begin{aligned} \delta &= \operatorname{sgn} \frac{\partial^2 \Phi}{\partial p_i \partial p_j} + k \\ &= -\operatorname{sgn} \frac{\partial(q_\alpha)}{\partial(p_\alpha)} + k \,. \end{aligned} \tag{5.16}$$

This derivation shows that the stationary phase approximation converts any mixed-space uniform approximation into the primitive configuration-space form, except that there is a constant phase factor $\exp(i\pi\delta/4)$. This means that the coefficients c_j must be chosen in such a manner that the phase factor is correctly incorporated. For this purpose, Maslov invented an index that is defined in the next section.

C. Maslov Index

The constant δ appearing in (5.16) is closely related to the Maslov index. This index is an integer-valued parameter related to the topological structure of the Lagrange manifold Λ^n. Because of its importance, MF devote a chapter to its definition and properties: they give two definitions of the index, and two examples showing how to calculate it; in addition they give a proof that the two definitions are equivalent. In fact, however (at least in the English language edition of their book), the two definitions give two different signs to the index, and so do the two examples, and their proof of the equivalence of the two definitions contains an inconsistency in the sign. The sign is important, because the Maslov index is related to the $\pm\pi/4$ which appears in WKB connection formulas, so we give a discussion of this index here. We follow the sign convention that we think MF intended, and our subsequent development and use of the index will at least be internally consistent.[15]

We have already explained the general idea behind Maslov's method of constructing wavefunctions: configuration space is divided into various regions, and the manifold is also divided accordingly. In certain regions, asymptotic approximations are made in configuration space, while in other regions, asymptotic approximations are made in momentum space or in a mixed space, and then transformed into configuration space. It is helpful to introduce a bit of the language that formalizes this conception.

A *canonical chart* is a simply connected domain Ω of the manifold Λ^n together with a diffeomorphic projection into one of the Lagrangian coordinate planes (p_α, q_β): that is, it is a domain in which (p_α, q_β) are suitable coordinates. A chart is said to be *regular* if it admits a diffeomorphic projection into configuration space $(q_1 \ldots q_n)$; otherwise it is *singular*. In particular, the set of points (denoted Σ) on the manifold which project onto the caustics in configuration space are called *singular points*, and any chart containing a singular point is a singular chart.

Let $\ell[w_I, w_F]$ be a directed curve lying on the manifold Λ^n; its initial point is w_I and its final point is w_F. The endpoints are assumed to be nonsingular (i.e., they do not project onto configuration-space cautics). We now define the Maslov index $\mu(\ell)$ for this curve.

1. Suppose ℓ lies entirely within one chart. Then:

   (a) If the chart is regular, $\mu(\ell) = 0$.

   (b) If the chart is singular, but admits coordinates (p_α, q_β) then

$$\mu(\ell) = \text{inerdex}\left[\frac{\partial(q_\alpha)}{\partial(p_\alpha)}\right]_{w_I} - \text{inerdex}\left[\frac{\partial(q_\alpha)}{\partial(p_\alpha)}\right]_{w_F} . \tag{5.17}$$

Here the manifold is given by n equations of the form (4.2), and $\partial(q_\alpha)/\partial(p_\alpha)$ means the Jacobian matrix

$$\frac{\partial q_j}{\partial p_i} = -\frac{\partial^2 \tilde{S}}{\partial p_i \partial p_j}, \qquad i, j \in \alpha$$

(not the determinant). The inerdex of a symmetric matrix M means the negative inertial index of the matrix, and it is a positive number (or zero) representing the number of negative eigenvalues of the matrix. (If the set α is empty, so that $q_1 \ldots q_n$ are the selected coordinates, the rank of $\partial q_\alpha / \partial p_\alpha$ is zero, and the inerdex vanishes.)

2. If the curve ℓ passes from one chart to another, then it can be divided into pieces, each of which lies in one chart. The index for the whole path is the sum of the indices for each piece. [Changing from one chart to another does not affect $\mu(\ell)$.]

Arnold[14] has shown that for a *typical*[17] Lagrangian manifold, as ℓ passes through a singular point corresponding to a caustic, one of the eigenvalues of the Jacobian matrix changes sign by passing linearly through zero, and then the value of μ is incremented accordingly. If an eigenvalue changes from positive to negative, μ is incremented by -1, while if it changes from negative to positive μ is incremented by $+1$. Figure 6 gives an illustration.

For an n-dimensional manifold, a domain near a singular point might have a diffeomorphic mapping into several different sets of Lagrangian coordinates. For example, in a domain in which $q_1q_2q_3$ are not suitable manifold coordinates, it is possible that *either* $p_1q_2q_3$ *or* $q_1p_2q_3$ might be appropriate. It is a fundamental theorem that under any such conditions, the Maslov index (mod 4) is independent of the selected coordinates.

MF also define another index that we call ν; it is closely related to, but not the same as, index μ. Whereas μ is a property of a path $\ell[w_1, w_2]$, ν is a property of a sequence of charts. We noted that μ changes when a path goes through a singular point, but it does not change when we go from one chart to another. In contrast, ν does not change at singular points,

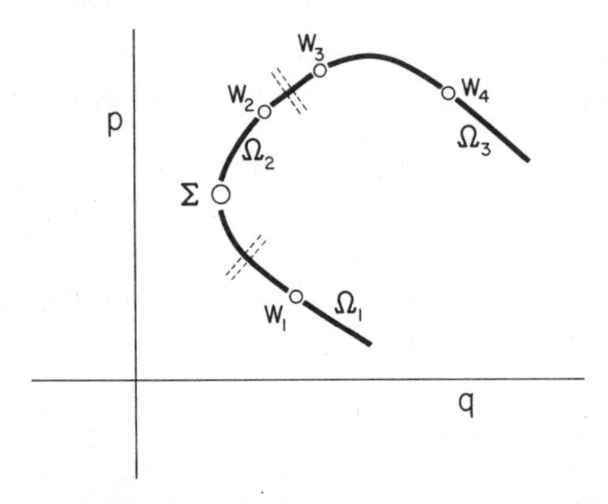

Fig. 6. Maslov index for a path. The smooth curve is a Lagrangian manifold for $n = 1$. The manifold is divided into overlapping domains Ω_1, Ω_2, Ω_3 distinguished by the double dotted lines. A path on the manifold goes from w_1 to w_4. Ω_1 is nonsingular. Ω_2 has a singular point Σ, so p is the appropriate coordinate for Ω_2 and the manifold is specified by a function $q(p)$. We examine $\partial q(p)/\partial p$: below Σ, $\partial q(p)/\partial p < 0$ so inerdex $= 1$, while above Σ, $\partial q(p)/\partial p > 0$ and inerdex $= 0$. Therefore $\mu(\ell[w_1, w_2]) = 1 - 0 = 1$. Continuing on the path to w_3, no further singularities are encountered, and μ remains equal to $+1$. Ω_3 is again a nonsingular chart, and q is its coordinate. Continuing on the path from w_2 to w_4, the sign of $\partial p(q)/\partial q$ changes, and so does the sign of the (locally defined) inverse function $\partial q(p)/\partial p$. This is irrelevant, however. Since Ω_3 is nonsingular, and q is necessarily the independent variable, the set α of independent momentum-space variables is empty, the matrix $\partial q_\alpha/\partial p_\alpha$ has rank zero, and its inerdex is zero. Therefore μ does not change within Ω_3, and $\mu(\ell[w_1, w_4]) = 1$.

but it can change when we go from one chart to another. Consider a pair of charts Ω_1, Ω_2 on the Lagrangian manifold, the first of which has coordinates $p_{\alpha_1} q_{\beta_1}$ and the second of which has coordinates $p_{\alpha_2} q_{\beta_2}$. If in a certain domain the two charts overlap, then either set of coordinates can be used to describe the manifold in that domain. The index $\nu(\Omega_1, \Omega_2)$ of the pair of charts is defined as

$$\nu(\Omega_1, \Omega_2) = \text{inerdex}\left[\frac{\partial q_{\alpha_2}}{\partial p_{\alpha_2}}\right] - \text{inerdex}\left[\frac{\partial q_{\alpha_1}}{\partial p_{\alpha_1}}\right]. \tag{5.18a}$$

[Note that final-minus-initial appears here, but initial-minus-final appears in (5.17).] Now suppose we have a sequence of charts $\Omega_1 \ldots \Omega_s$, with each chart in the sequence overlapping with the preceding and following one. Such a sequence is called a *chain*, and normally this chain will be associated with a path $\ell[w_I, w_F]$ which passes through various domains corresponding to the various charts. The index ν for the sequence or chain is defined as

$$\nu(\Omega_1 \ldots \Omega_s) = \nu(\Omega_1, \Omega_2) + \nu(\Omega_2, \Omega_3) + \cdots + \nu(\Omega_{s-1}, \Omega_s). \tag{5.18b}$$

In making use of this quantity, it will often be convenient to think of ν as being associated with the last chart in the chain; if we continue along a path staying within that chart, then ν does not change, even if the chart is singular, and even if the path goes through a singularity. The value of ν is fixed (by the relative numbers of negative eigenvalues) as soon as we enter a chart and transform to the variables p_α, q_β associated with that chart. It follows that (5.18b) does *not* imply that ν is necessarily the same as

$$\text{inerdex}\left[\frac{\partial q_{\alpha_s}}{\partial p_{\alpha_s}}\right] - \text{inerdex}\left[\frac{\partial q_{\alpha_1}}{\partial p_{\alpha_1}}\right],$$

because if a chart is singular, the inerdex may change within the chart, but ν does not change.

The connection between μ and ν is the following. Suppose Ω_1 and Ω_s are both nonsingular charts having $q_1 \ldots q_n$ as coordinates; suppose w_I and w_F lie in Ω_1 and Ω_s, respectively, and $\ell[w_I, w_F]$ is a path connecting them that goes through the chain of charts $\Omega_1 \ldots \Omega_s$. Then

$$\nu(\Omega_1 \ldots \Omega_s) = \mu(\ell[w_I, w_F]). \tag{5.18c}$$

This is illustrated in Fig. 7.

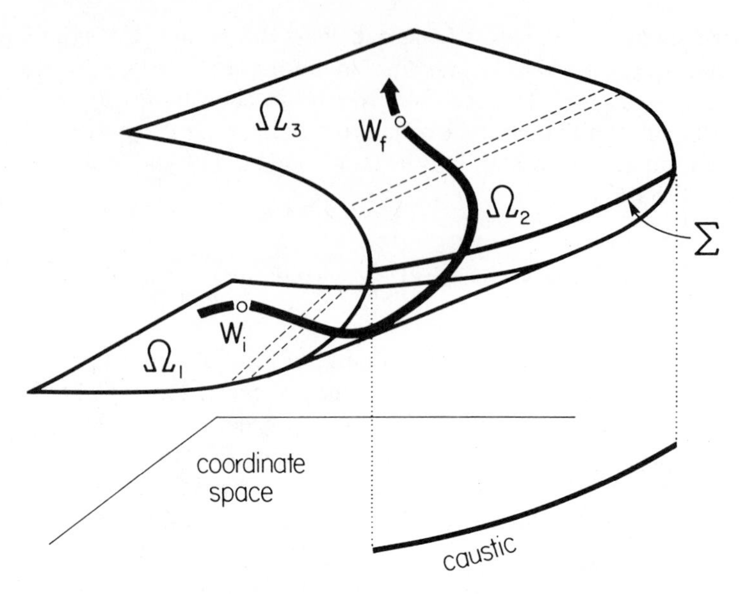

Fig. 7. Connection between μ and ν. Ω_1 is a nonsingular domain and $q_1 \ldots q_n$ are its coordinates. The inerdex for Ω_1 is zero. Ω_2 is a singular domain with coordinates p_α, q_β, and the inerdex has a value i_2^L in the lower part (before the caustic) and i_2^U in the upper part (after the caustic). Ω_3 is again nonsingular, and its inerdex is zero. The index ν is calculated from (5.17):

$$\nu(\Omega_1, \Omega_2) = i_2^L - 0\,,$$

$$\nu(\Omega_2, \Omega_3) = 0 - i_2^U\,,$$

$$\nu(\Omega_1, \Omega_2, \Omega_3) = i_2^L - i_2^U\,.$$

The Maslov index μ for the path $\ell[w_1, w_f]$ which goes through the chain of charts Ω_1, Ω_2, Ω_3 is given by the change (initial − final) of the inerdex in the singular chart Ω_2 [e.g., (5.16)]:

$$\mu(\ell[w_i, w_f]) = i_2^L - i_2^U\,.$$

This type of argument establishes (5.18).

D. Canonical Formula for the Wavefunction

At last we are ready to write down the canonical formula for the wavefunction. One begins by calculating the Lagrangian manifold (by evolution of trajectories from an initial surface), and dividing it into overlapping domains Ω_j, in each of which a suitable set of coordinates p_{α_j}, q_{β_j} has been chosen. To each domain there is assigned a C^∞ switching function $e_j(w)$ such that e_j is nonzero only within the jth domain, but such that

$$\sum_j e_j(w) = 1$$

everywhere on the manifold. Each switching function varies continuously between zero and one in the regions of overlap of two or more domains. (These functions are defined so that we can smoothly combine the different types of asymptotic approximations defined in the various domains.) An arbitrary initial point w_0 on the initial surface is selected (there the phase of the initial term in Ψ is set arbitrarily to zero, and the phase elsewhere is determined relative to the phase at this point).

The appropriate functions $\tilde{S}(p_\alpha, q_\beta)$ and $\tilde{A}(p_\alpha, q_\beta)$ are calculated in the various domains, and these functions are sewn together near the boundaries of domains according to the following rules:

1. At w_0, $\tilde{S} = 0$.
2. Within any domain, $\tilde{S}$ and $\tilde{A}$ are obtained by integration of trajectories using (4.6), (4.9), and (4.23).
3. In the region of overlap of charts m and n (having coordinates $p_{\alpha_m} q_{\beta_m}$ and $p_{\alpha_n} q_{\beta_n}$),

$$\tilde{S}_m - \tilde{S}_n = \sum_{j \in \alpha_n} p_j q_j - \sum_{j \in \alpha_m} p_j q_j \tag{5.19}$$

and

$$\frac{\tilde{A}_m}{\tilde{A}_n} = \left| \frac{\partial(p_{\alpha_n}, q_{\beta_n})}{\partial(p_{\alpha_m}, q_{\beta_m})} \right|^{1/2} \tag{5.20}$$

(by convention, the A's are always positive).

To an arbitrary point in the jth chart Ω_j, we draw a path $\ell[w_0, w_j]$, and we associate with this path the chain of charts through which it passes. Then the index ν for the chain of charts is calculated.

Finally, the wavefunction at a given point $q = (q_\alpha, q_\beta)$ is given by the canonical formula

$$\Psi(q) = \sum_j e^{-i\pi\nu_j/2} F^{-1}_{p_{\alpha_j} \to q_{\alpha_j}} \left(e_j(w) \tilde{A}(w) \exp\left[\frac{i\tilde{S}(w)}{\hbar} \right] \right), \tag{5.21}$$

where $F^{-1}_{p_\alpha \to q_\alpha}$ is the operator of (inverse) Fourier transformation,

$$F^{-1}_{p_\alpha \to q_\alpha} f(p_\alpha q_\beta) = (-2\pi i\hbar)^{-k/2} \int \exp\left[i \sum_{j \in \alpha} \frac{q_j p_j}{\hbar} \right] f(p_\alpha q_\beta)\, dp_\alpha, \qquad (5.22)$$

k being again the number of momentum variables in the set α. In (5.21) there is one term in the sum for each domain Ω_j. However, it is easy to show using the stationary phase formula that if there is no point w within a domain that projects onto the given configuration-space point, then the contribution of that domain is smaller than a constant times $\hbar$. Thus for any point in a classically allowed region of configuration space, it is only necessary to include in the sum those domains which have projections onto that point.

Similarly, if q is in a classically forbidden region, that is, if there is no domain of the mainfold that projects onto q, then Ψ is less than a constant times $\hbar$ at that point. Furthermore, by examining higher-order approximations, MF show that in any classically forbidden region, as $\hbar$ goes to zero, Ψ goes to zero faster than any power of $\hbar$.

It is now also a simple exercise to show that for each nonsingular chart Ω_j having $q_1 \ldots q_n$ for coordinates and having a point w_j that projects to the given configuration-point q, since the set α is empty, the Fourier transform in (5.21) is inoperative, and the contribution to the wavefunction is just

$$A_j(q) \exp i \left[\frac{S_j(q)}{\hbar} - \frac{\pi\mu}{2} \right], \qquad (5.23)$$

where μ is the Maslov index for the path going from w_0 to w_j. Using the stationary phase approximation as in Section V.B above, one can prove that the various asymptotic approximations are all consistent with each other: the indices μ and ν were in fact chosen such that the phases come out correctly!

Equation (5.21) raises one final question. Each term in the sum is defined within a specific chart, and the characteristic function $\tilde{S}(w)$ and the amplitude $\tilde{A}(w)$ depend on the set of coordinates $(p_\alpha q_\beta)$ that were chosen for that chart. Now it frequently happens that there are several different sets of coordinates that might be chosen for the chart; that is, a given region of the manifold might have smooth projections into several different Lagrangian coordinate planes. Furthermore, the switching functions have not been specified. It would therefore seem that the canonical formula (5.21) may lead to ambiguous results. However, MF have proved (their theorem 8.1) that different choices of coordinates or switching functions lead to results that differ by an amount that goes to zero linearly with $\hbar$ as $\hbar$ goes to zero. Since the neglected terms in the expansion of Ψ

are themselves of order $\hbar$, it follows that to within the errors involved in the use of the semiclassical approximation, the canonical formula is independent of the choice of coordinates and switching functions.[16]

E. Caustics and Catastrophes

The formula (5.21) is presumed to be very general and should be valid in the limit $\hbar \to 0$ in a regular region or in any type of singular region that can be associated with an n-dimensional Lagrangian manifold. The formula can be simplified in several special cases, and it turns out that only a few distinctly different types of singular regions occur in "typical" Lagrange manifolds.[17] These are the *elementary catastrophes* that were first defined by Thom[18] and introduced into collision theory and wave mechanics by Connor[3e] and Berry.[3f] In the present case of two-dimensional scattering, the caustic is of the simplest type—a *fold* catastrophe—and the wavefunction near the caustic can be approximated by an Airy function.

To see how the Airy function arises, consider the manifold generated by the function

$$\tilde{S}(p_1, q_2) = \tfrac{1}{3}p_1^3 + \tfrac{1}{2}q_2^2 \tag{5.24}$$

for which

$$q_1 = -p_1^2 \quad \text{and} \quad p_2 = q_2 .$$

For this manifold there is no need for more than one chart, so we can take $\nu = 0$, and Ψ would be given by

$$\Psi(q_1, q_2) = (-2\pi i\hbar)^{-1/2} \int A(p_1, q_2) \exp\left[\frac{i(p_1 q_1 + \frac{1}{3}p_1^3 + \frac{1}{2}q_1^2)}{\hbar}\right] dp_1 \tag{5.25}$$

which, if $A(p_1, q_2)$ is essentially constant, reduces to

$$(-2\pi i\hbar)^{-1/2} e^{iq_2^2/2} A(p_1, q_2) \int \exp\left[\frac{i(p_1 q_1 + \frac{1}{3}p_1^3)}{\hbar}\right] dp_1 , \tag{5.26}$$

and the remaining integral is proportional to an Airy function.

Now, in general, the Taylor expansion for $\tilde{S}(p_1 q_2)$ could contain any number of terms, and we might imagine that arbitrarily complicated caustics could then appear, and each different type of caustic would lead to a different integral representation for Ψ. However, Thom[18] and Arnold[14] established that *stable* caustics (those that retain their structure under small perturbations of the manifold) can be reduced locally to a finite number of elementary catastrophes. For example, in two dimensions, all stable caustics can be locally reduced to just two types: folds, for which Ψ is like an Airy function, and cusps, for which Ψ is given by a Pearcy function.[3e] In the present two-dimensional scattering problem, only a fold is present, and the study of more complicated catastrophes is unnecessary.

One additional remark should be made here. Strictly speaking, a *uniform* approximation ought to be a uniformly accurate approximation to the wavefunction $\Psi(q)$. Now since the primitive semiclassical approximation is valid *to order* $\hbar$ [i.e., the first neglected term in Eq. (3.3) is of order $\hbar$], it would be desirable to obtain a uniform approximation having the same property. Such an approximation could be called *strictly uniform.*

It turns out that such strictly uniform approximations are more complicated than one might expect. In the case of a fold, the leading Airy-function term is proportional to $\hbar^{-1/6}$, and then there is a series of higher-order corrections proportional to $\hbar^{1/6}$, $\hbar^{1/2}$, and $\hbar^{5/6}$. A convenient way of calculating these corrections was worked out by Chester, Friedman and Ursell[21]. The first two terms are easily calculated, and they have been used frequently in the work cited in ref. 3. The way Chester et al. define the series, it turns out that the third term (proportional to $\hbar^{1/2}$) exactly vanishes. To calculate the next term (proportional to $\hbar^{5/6}$), one would need among other things the first correction to the WKB approximation in momentum space (the term denoted $\hbar\tilde{A}_1$ in eq. 4.19). Such a calculation would be a formidable undertaking, so by common consent, the term "uniform approximation" is normally taken to refer to the first two terms in the expansion ($\hbar^{-1/6}$ and $\hbar^{+1/6}$).

VI. CONCLUSION

As stated earlier, the purpose of this series of articles is to close the gap between mathematical theory and computational methods: in this review we have summarized the theory developed by Maslov and Fedoriuk, and in subsequent papers we shall compute global semiclassical wavefunctions for simple systems. For the two-dimensional scattering system considered here, the complete calculation of the wavefunction is given in ref. 7.

The reader who knows about the work represented in ref. 3 will note that we have not presented any startling results, and one might well ask, what precisely is new here? Let us summarize the reasons why the developments made by Maslov and his co-workers are worthy of our attention.

1. The most important reason involves the different orientation of Maslov's work from most of that cited in ref. 3. As we mentioned in the Introduction, most of the "Western" efforts began by examining specific physical properties—the S matrix or the energy spectrum, for example. In contrast, the purpose of Maslov's work is to provide a general semiclassical method for constructing wavefunctions, and from such wavefunctions, all other physical properties can be calculated. This is clearly the more fundamental orientation.

2. Maslov's approach is of course solidly based in sound mathematics. MF give us no speculations and no unprovable assertions.

3. To develop the theory, MF draw our attention away from the behavior of the trajectories in configuration space and direct it to the structure of the Lagrange manifold. Although at first the concept may seem somewhat abstract, it is essential for the mathematical analysis, because many quantities that are multiply-valued or singular in configuration space are all single-valued and well-behaved on the manifold. Furthermore, the manifold provides a very helpful aid in envisioning the structure of singular regions.

4. A Lagrange manifold has the crucial property that any domain can locally be spanned by n coordinates $p_\alpha q_\beta$ containing no canonically conjugate pairs. This means that singular regions can in general be handled by transformation into momentum space. The theorems of Arnold and Thom further restrict the types of singularities that are stable under perturbations.

5. MF give a formula for the density function, $\rho(q)$ or $\rho(p_\alpha q_\beta)$, that is very convenient to apply, but which has not often been used in semiclassical calculations.

6. MF show how to calculate higher-order approximations by including more terms $A^{[n]}$ in Eqs. (3.4) or (4.20). In general, $A^{[n]}$ is determined from $A^{[m]}$ $(m < n)$ by solving a set of inhomogeneous linear differential equations.

7. To connect various uniform approximations smoothly to primitive semiclassical forms, MF use a set of smooth switching functions. The idea is obvious enough, but it is not in common use. (Also by attaching the switching functions to the charts on the manifold, MF circumvent the problems that arise from the fact that each approximation $\tilde{A}(p_\alpha q_\beta)\exp[i\tilde{S}(p_\alpha q_\beta)/\hbar]$ is applicable in only one chart.)

8. The famous Maslov index is essential for connecting primitive semiclassical and uniform approximations with appropriate phases, so that interference patterns are calculated correctly.

9. Finally, MF show that the wavefunction constructed by their method approximately satisfies the Schroedinger equation.

On the other hand, it might also be worthwhile to point out some of the interesting problems that either have not been addressed or have been left unsolved by Maslov and Fedoriuk, because some of these may become the focus of future research. To mathematicians, the most severe limitation in the work of MF is that it provides only *formal* asymptotic series solutions. In Section IV.D we asserted that the function $\Psi^{[M]}$ defined by Eq. (4.26) satisfies the Schroedinger equation with an error that goes to zero as $\hbar^{M+2}$. However, this is not sufficient to prove that such a $\Psi^{[M]}$ is close to an exact solution to the Schroedinger equation. It is one thing to show that $(H-E)\Psi^{[M]}$ is small; it is something quite different to establish that there exists some Ψ such that $(H-E)\Psi$ exactly vanishes in a region and such that $\Psi^{[M]}$ is close to Ψ. About all MF tell us on this subject is that "this question is very important indeed, and extremely hard to answer". Fortunately, from physical considerations we usually know whether or not the solution constructed by these methods makes sense.

Similarly, MF do not tell us much about the wavefunction in classically forbidden regions. For one-dimensional systems, real exponential forms in forbidden regions are quite familiar, and the theory of tunneling between two classically allowed regions is well developed. MF do not give any complete description of wavefunctions in forbidden regions; they tell us only that Ψ is smaller than any power of $\hbar$ in such a region. Physicists and chemists have made many predictions about multidimensional tunneling, but it is worth noting that such predictions are not on such a solid mathematical foundation as are calculations dealing solely with allowed regions.

Finally, while MF tell us what types of singularities are *typical*, they do not give much information about atypical singularities. Important cases exist in which atypical singularities arise. For example, three-dimensional scattering by a spherically symmetric potential leads to a focus on the positive Z axis, and this situation requires additional analysis.

In subsequent papers, we shall use these methods to construct the wavefunction for the two-dimensional scattering problem considered here, then later turn to three-dimensional scattering, a bound-state problem, and (we hope) multidimensional tunneling.

APPENDIX A. FORMULAS RELATED TO SECTION II

A. Solution to the Hamilton–Jacobi Equation

We need to show that the function $S(q)$, constructed according to the rules leading to Eq. (2.5), satisfies the Hamilton–Jacobi equation (2.1)

Actually MF show that *if a solution* to (2.1) *exists* and if it is equal to $S^0(q^0)$ on Λ_0^{n-1}, then that solution is given by Eq. (2.5) (this is their proposition 3.8). Their elegant proof is similar to one given by Pars[19] and is much more direct than the corresponding arguments given by Goldstein.[8] Let $S(q)$ be a solution to (2.1) in a given domain, including the initial surface, on which $S(q) = S^0(q^0)$, and *define* the functions $p_i(q)$ as

$$p_i(q) = \frac{\partial S(q)}{\partial q_i}. \tag{A.1}$$

Differentiating (2.1) with respect to q_i, we have

$$\sum_j \frac{\partial H}{\partial p_j} \frac{\partial^2 S}{\partial q_i \partial q_j} + \frac{\partial H}{\partial q_i} = 0. \tag{A.2}$$

Now starting from a particular point q^0 on the initial surface, *define* a path $q(t)$ such that

$$\frac{dq_i}{dt} = \left.\frac{\partial H(p, q)}{\partial p_i}\right|_{p=\nabla_q S}, \qquad i = 1 \ldots n. \tag{A.3}$$

Here we differentiate $H(p, q)$ with respect to p_i, then substitute $p = \nabla_q S(q)$, so the right-hand side is a function only of q, and (A.3) is a closed set of n equations determining the path $q(t)$ from the initial point q^0. Now consider the function

$$p_i(t) \equiv p_i(q(t)) = \left.\frac{\partial S(q)}{\partial q_i}\right|_{q=q(t)}. \tag{A.4}$$

This satisfies

$$\frac{dp_i}{dt} = \sum_j \frac{\partial p_i}{\partial q_j} \frac{dq_j}{dt} = \sum_j \frac{\partial^2 S}{\partial q_j \partial q_i} \frac{\partial H}{\partial p_j} = -\frac{\partial H}{\partial q_i}, \tag{A.5}$$

where we have used (A.1) and (A.2). Equations (A.3) and (A.5) are Hamilton's equations, which therefore provide an alternative way of specifying the path defined by (A.3). On this path, let us define $S(t) = S(q(t))$; it follows that

$$\frac{dS}{dt} = \sum_i \frac{\partial S}{\partial q_i} \frac{dq_i}{dt} = \sum_i p_i(q(t)) \frac{dq_i}{dt} = p(q(t)) \cdot \frac{dq}{dt}. \tag{A.6}$$

This is a differential form for which the corresponding integral form is (2.5). Hence the function $S(q)$, which was assumed to satisfy the Hamilton–Jacobi equation, also satisfies Eq. (2.5), and so it can be calculated by integration along trajectories.

The above proof is not quite complete: it shows that if a solution to (2.1) exists, then it satisfies (2.5). Separately (their lemma 3.9) MF prove that a solution really does exist, and it is unique if t is restricted to sufficiently small values.

B. The Manifold is Lagrangian

Two theorems, one local and one global, are needed to establish that the manifold defined in Section II has the Lagrangian property. The local theorem is well known and follows trivially from the existence theorem quoted above. If a solution $S(q)$ to the Hamilton–Jacobi equation exists, then according to (A.4)

$$\frac{\partial p_i}{\partial q_j} = \frac{\partial^2 S}{\partial q_j \partial q_i} = \frac{\partial p_j}{\partial q_i}.$$

Obviously, this local theorem applies only in the regular regions of the manifold.

The global theorem is much more difficult. One must first refer to the general definition of a Lagrange manifold, a version of which is given in Section IV.B. The theorem that the trajectory manifold is everywhere Lagrangian then follows from two propositions: (1) the Lagrangian property is preserved under canonical transformations; (2) the time evolution of a system can be described as the continuous development of a time-dependent canonical transformation. The theorem is stated precisely, but not proved, in our section IV.B.

C. Solution to the Transport Equation

Eq. (2.8) can be derived in several ways: MF's method is independent of the form of H or the dimension of the system. First we need a lemma: For any $n \times n$ matrix $\mathbf{J}$, with elements J_{ij},

$$\frac{d}{dt}\ln(\det \mathbf{J}) = \mathrm{Tr}\left[\left(\frac{d\mathbf{J}}{dt}\right)\mathbf{J}^{-1}\right]. \tag{A.7}$$

To prove this, consider the matrix of cofactors C_{ij}, each of which is equal to $(-)^{i+j}$ times the determinant of the matrix obtained by striking out the ith row and jth column of $\mathbf{J}$. From well-known theorems in linear algebra,

$$\det \mathbf{J} = \sum_i J_{ij} C_{ij}$$

so

$$\partial(\det \mathbf{J})/\partial J_{ij} = C_{ij}$$

Furthermore

$$(\mathbf{J}^{-1})_{ij} = C_{ji}/\det \mathbf{J}$$

Therefore

$$\frac{d}{dt}(\det \mathbf{J}) = \sum_{ij} \frac{\partial(\det \mathbf{J})}{\partial J_{ij}} \frac{dJ_{ij}}{dt} = \sum_{ij} C_{ij}\dot{J}_{ij}$$

$$= \sum_{ij} (\det \mathbf{J})(\mathbf{J}^{-1})_{ji}\dot{J}_{ij} = (\det J)\,\mathrm{Tr}(J^{-1}\dot{J})$$

Now let the paths $q(t)$ be generated according to Eq. (A.3), which we write in the form

$$\frac{dq_i}{dt} = f_i(q)\,. \tag{A.8}$$

We will need the matrix

$$F_{ij} = \frac{\partial f_i}{\partial q_j}\,. \tag{A.9}$$

Let the solutions $q(t)$ be regarded as a function of the n variables $\zeta_j = (t, w_i^0)$, with w_i^0 being the $n-1$ variables that specify a point on the initial manifold. Consider the matrix $\mathbf{J}$ having elements

$$J_{ij} = \frac{\partial q_i}{\partial \zeta_j}\,; \tag{A.10}$$

its time derivative is given by

$$\frac{d}{dt} J_{ij} = \frac{d}{dt}\frac{\partial q_i}{\partial \zeta_j} = \frac{\partial}{\partial \zeta_j}\frac{dq_i}{dt} = \frac{\partial}{\partial \zeta_j} f_i(q) = \sum_k \frac{\partial f_i}{\partial q_k}\frac{\partial q_k}{\partial \zeta_j} = (\mathbf{FJ})_{ij}\,. \tag{A.11}$$

Furthermore, according to Eq. (2.9),

$$J(t, w^0) = \det \mathbf{J}\,. \tag{A.12}$$

From (2.8) we find

$$\frac{d\rho(q)}{dt} = -\frac{\rho^0(w^0)J(0, w^0)}{J(t, w^0)^2} \frac{dJ(t, w^0)}{dt} .$$

Using successively (A.12), (A.7), and (A.11), we obtain

$$\frac{d\rho(q)}{dt} = -\frac{\rho^0(q^0)J(0, w^0)}{J(t, w^0)} \operatorname{Tr} \mathbf{F} ,$$

and from (A.9), (A.8), (A.3), and (1.1),

$$\frac{d\rho(q)}{dt} = -\rho(q) \frac{\nabla_q \cdot p(q)}{M} \tag{A.13}$$

so the function $\rho(q)$ defined in (2.8) satisfies the continuity equation (2.7), with $\rho(q) = \rho^0(q^0)$ on the initial surface.

APPENDIX B. FORMULAS RELATED TO SECTION IV

A. Hamilton–Jacobi Equation in Mixed Spaces

Here we verify that the desired solution to the Hamilton–Jacobi equation in the mixed $p_\alpha q_\beta$ space is equal to the integral given in (4.5). The Hamilton–Jacobi equation is

$$H\left(p_\alpha, \frac{\partial \tilde{S}}{\partial q_\beta}, -\frac{\partial \tilde{S}}{\partial p_\alpha}, q_\beta\right) - E = 0 , \tag{B.1}$$

where, as in the main text, α and β are collective indices, so that p_α means "p_i for all i in the set α." The independent variables in this equation are p_α and q_β. Suppose there exists a solution to this equation $\tilde{S}(p_\alpha, q_\beta)$, defined in a domain of n-dimensional $p_\alpha q_\beta$ space that includes an initial surface of dimension $n - 1$. We use the abbreviations

$$\begin{aligned} p_\beta &\equiv p_\beta(p_\alpha, q_\beta) = \frac{\partial \tilde{S}}{\partial q_\beta} , \\ q_\alpha &\equiv q_\alpha(p_\alpha, q_\beta) = -\frac{\partial \tilde{S}}{\partial p_\alpha} . \end{aligned} \tag{B.2}$$

Then differentiating (B.1) with respect to one of the variables p_α (call it $p_{\alpha'}$) gives

$$\frac{\partial H}{\partial p_{\alpha'}} + \frac{\partial H}{\partial p_\beta} \frac{\partial^2 \tilde{S}}{\partial p_{\alpha'} \partial q_\beta} - \frac{\partial H}{\partial q_\alpha} \frac{\partial^2 \tilde{S}}{\partial p_{\alpha'} \partial p_\alpha} = 0 , \tag{B.3a}$$

and differentiating with respect to one of the variables $q_{\beta'}$ gives

$$\frac{\partial H}{\partial p_\beta} \frac{\partial^2 \tilde{S}}{\partial q_{\beta'} \partial q_\beta} - \frac{\partial H}{\partial q_\alpha} \frac{\partial^2 \tilde{S}}{\partial q_{\beta'} \partial p_\alpha} + \frac{\partial H}{\partial q_{\beta'}} = 0 . \tag{B.3b}$$

All equations in this appendix have implied summation over repeated indices.

Now let us define a path $[p_\alpha(t), q_\beta(t)]$ in the n-dimensional space such that at $t = 0$, $p_\alpha(0)$, $q_\beta(0)$ is a point on the initial surface, and such that

$$\frac{dp_\alpha}{dt} = -\frac{\partial H}{\partial q_\alpha} \quad \text{and} \quad \frac{dq_\beta}{dt} = \frac{\partial H}{\partial p_\beta} , \tag{B.4}$$

where, after differentiation, one substitutes $\partial \tilde{S}/\partial q_\beta$ for p_β and $-\partial \tilde{S}/\partial p_\alpha$ for q_α on the right-hand side of these equations. If the solution $\tilde{S}(p_\alpha, q_\beta)$ is known, these equations are a closed set that define a unique path emanating from each point on the initial surface. On this path we define $q_\alpha(t) \equiv q_\alpha(p_\alpha(t), q_\beta(t))$, $p_\beta(t) \equiv p_\beta(p_\alpha(t), q_\beta(t))$; then

$$\begin{aligned} \frac{dq_\alpha}{dt} &= -\frac{\partial^2 \tilde{S}}{\partial p_\alpha \partial p_{\hat{\alpha}}} \frac{dp_{\hat{\alpha}}}{dt} - \frac{\partial^2 \tilde{S}}{\partial p_\alpha \partial q_\beta} \frac{dq_\beta}{dt} \\ &= + \frac{\partial^2 \tilde{S}}{\partial p_\alpha \partial p_{\hat{\alpha}}} \frac{\partial H}{\partial q_{\hat{\alpha}}} - \frac{\partial^2 \tilde{S}}{\partial p_\alpha \partial q_\beta} \frac{\partial H}{\partial p_\beta} = \frac{\partial H}{\partial p_\alpha} \end{aligned} \tag{B.5a}$$

and a similar analysis applied to dp_β/dt leads to

$$\frac{dp_\beta}{dt} = -\frac{\partial H}{\partial q_\beta} . \tag{B.5b}$$

Hence the path specified by the n equations (B.4) is alternatively specified by the standard $2n$ Hamilton equations (B.5). Finally, defining $\tilde{S}(t)$ on each path as $\tilde{S}(t) = \tilde{S}(p_\alpha(t), q_\beta(t))$, we have

$$\frac{d\tilde{S}}{dt} = \frac{\partial \tilde{S}}{dp_\alpha} \frac{dp_\alpha}{dt} + \frac{\partial \tilde{S}}{\partial q_\beta} \frac{dq_\beta}{dt} = -q_\alpha \frac{dp_\alpha}{dt} + p_\beta \frac{dq_\beta}{dt} . \tag{B.6}$$

Thus

$$\tilde{S}(p_\alpha, q_\beta) = \tilde{S}(p_\alpha^0, q_\beta^0) + \int (p_\beta \, dq_\beta - q_\alpha \, dp_\alpha), \tag{B.7}$$

where the integral is evaluated on the path that satisfies Hamilton's equations (B.5) and that goes from the point $p_\alpha^0 q_\beta^0$ on the initial surface to the point $p_\alpha q_\beta$.

As in Appendix A Section A, we have proven that if a solution to (B.1) exists, then it satisfies (B.7); a separate proof of the existence of a solution is also required.

B. Density Function in Mixed Spaces

We now show that the density function given in Eq. (4.8) satisfies a certain "transport" equation (4.9). The derivation is slightly simplified if we assume that the Hamiltonian has the form

$$H(p, q) = T(p) + V(q) \tag{B.8}$$

with no terms containing both coordinates and momenta, such as $q_i p_j$. As discussed in the text, and in Appendix A Section C, there is an $(n-1)$-dimensional initial manifold on which $n-1$ coordinates w_i^0 are defined smoothly. For each point on this initial manifold, Eq. (B.4) specifies a path $p_\alpha(t)$, $q_\beta(t)$ and we regard the path variables as functions of t and the initial point w^0:

$$p_\alpha = p_\alpha(t, w^0) \quad \text{and} \quad q_\beta = q_\beta(t, w^0). \tag{B.9}$$

Since by hypothesis there is locally one and only one such path connecting the initial surface with the point p_α, q_β, we may think of the Jacobian determinant

$$\tilde{J}(p_\alpha, q_\beta) \equiv \tilde{J}(t, w^0) = \frac{\partial(p_\alpha(t, w^0), q_\beta(t, w^0))}{\partial(t, w^0)} \tag{B.10}$$

as being either a function of $p_\alpha q_\beta$ or a function of t, w^0. For each point on the initial surface, there is an arbitrary initial density $\tilde{\rho}^0(w^0)$. Equation (4.8) defines on the Lagrange manifold a density function which can also be regarded either as a function of $p_\alpha q_\beta$ or of t, w^0:

$$\tilde{\rho}(p_\alpha, q_\beta) \equiv \tilde{\rho}(t, w^0) = \tilde{\rho}^0(w^0) \frac{\tilde{J}(0, w^0)}{\tilde{J}(t, w^0)}. \tag{B.11}$$

We wish to obtain a differential equation satisfied by $\tilde{\rho}$. We have

$$\frac{d\tilde{\rho}}{dt} = \frac{\partial\tilde{\rho}}{\partial p_\alpha}\frac{dp_\alpha}{dt} + \frac{\partial\tilde{\rho}}{\partial q_\beta}\frac{dq_\beta}{dt} = -\frac{\partial\tilde{\rho}}{\partial p_\alpha}\frac{\partial H}{\partial q_\alpha} + \frac{\partial\tilde{\rho}}{\partial q_\beta}\frac{\partial H}{\partial p_\beta}. \tag{B.12}$$

But also applying (A.7) to (B.10),

$$\frac{d\tilde{\rho}}{dt} = -\frac{\tilde{\rho}_0\tilde{J}(0, w^0)}{\tilde{J}(t, w^0)^2}\frac{d\tilde{J}(t, w^0)}{dt} = -\tilde{\rho}\frac{1}{\tilde{J}}\frac{d\tilde{J}}{dt} = -\tilde{\rho}\,\mathrm{Tr}\left[\frac{d\tilde{\mathbf{J}}}{dt}\tilde{\mathbf{J}}^{-1}\right], \tag{B.13}$$

where $\tilde{\mathbf{J}}$ is the Jacobian matrix, whose determinant is $\tilde{J}$. As in Appendix A Section C, let the n quantities t, w_i^0 be represented by the symbols ζ_i, and also let $p_\alpha q_\beta$ be represented by the symbols y_i:

$$\begin{aligned} y_i &= p_i\,, && i \in \alpha\,; \\ y_i &= q_i\,, && i \in \beta\,; \\ \zeta_j &= w_j^0\,, && j = 1 \ldots n-1\,; \\ \zeta_n &= t\,. \end{aligned} \tag{B.14}$$

On the trajectory (B.9) defined by (B.4), we write

$$\frac{dy_i}{dt} = f_i(y) \tag{B.15}$$

and define a matrix $\mathbf{F}$ with elements

$$F_{ij} = \frac{\partial f_i}{\partial y_j}. \tag{B.16}$$

[One must remember that p_α and q_β are the only independent variables on the right-hand side of (B.4) and that in Eq. (B.16) $\partial/\partial y_j$ means that only the other y_k are held fixed.] Then, since the matrix elements $\tilde{J}_{ij}$ are

$$\tilde{J}_{ij} = \frac{\partial y_i}{\partial \zeta_j}, \tag{B.17}$$

we have

$$\frac{d\tilde{J}_{ij}}{dt} = \frac{d}{dt}\left(\frac{\partial y_i}{\partial \zeta_j}\right) = \frac{\partial}{\partial \zeta_j} f_i(y) = \sum_k \frac{\partial f_i}{\partial y_k}\frac{\partial y_k}{\partial \zeta_j} = \sum_k F_{ik}\tilde{J}_{kj} \tag{B.18}$$

so

$$\frac{d\tilde{\rho}}{dt} = -\tilde{\rho}\,\mathrm{Tr}\,\mathbf{F}. \tag{B.19}$$

Now from (B.4)

$$\begin{aligned}\mathrm{Tr}\,\mathbf{F} &= \sum_i f_{ii} = \sum_i \frac{\partial f_i}{\partial y_i} = \sum_{i\in\alpha} -\frac{\partial}{\partial y_i}\frac{\partial H}{\partial q_i} + \sum_{i\in\beta}\frac{\partial}{\partial y_i}\frac{\partial H}{\partial p_i} \\ &= -\frac{\partial}{\partial p_\alpha}\frac{\partial H}{\partial q_\alpha} + \frac{\partial}{\partial q_\beta}\frac{\partial H}{\partial p_\beta}.\end{aligned} \tag{B.20}$$

The partial derivatives in this equation have two different meanings. $\partial H/\partial q_\alpha$ arises from the Hamiltonian equations of motion (B.4), and it means $\partial H/\partial q_i$ $(i \in \alpha)$ with all other q_j and all p_j held fixed. However, the subsequent $\partial/\partial p_\alpha$ arises from (B.16), and it means that the other $p_i(i \in \alpha)$ and $q_j(j \in \beta)$ are fixed, but that the p_β and q_α are regarded as functions of $p_\alpha q_\beta$ as specified in (B.2). The same holds, respectively, for $\partial H/\partial p_\beta$ and $\partial/\partial q_\beta$. We repeat that in (B.4), after the derivatives $\partial H/\partial q_\alpha$ and $\partial H/\partial p_\beta$ are evaluated, one substitutes (B.2) into the resulting expression, so that the quantities become functions of only n variables, $p_\alpha q_\beta$. Only after this is done are the derivatives $\partial/\partial p_\alpha$ and $\partial/\partial q_\beta$ in (B.20) evaluated. It follows that

$$\mathrm{Tr}\,\mathbf{F} = -\frac{\partial^2 H}{\partial q_\alpha \partial q_{\hat{\alpha}}}\frac{\partial q_{\hat{\alpha}}}{\partial p_\alpha} + \frac{\partial^2 H}{\partial p_\beta \partial p_{\hat{\beta}}}\frac{\partial p_{\hat{\beta}}}{\partial q_\beta}, \tag{B.21}$$

where now the partial derivatives of H imply that all other p's and q's are held fixed. Only in (B.21) have we used the assumption (B.8). Substituting (B.21) in (B.19) and comparing (B.12), we have two different expressions for $d\tilde{\rho}/dt$, and the two must be equal. Therefore $\tilde{\rho}$ satisfies the equation

$$\left(\frac{\partial H}{\partial p_\beta}\frac{\partial \tilde{\rho}}{\partial q_\beta} - \frac{\partial H}{\partial q_\alpha}\frac{\partial \tilde{\rho}}{\partial p_\alpha}\right) + \tilde{\rho}\left(\frac{\partial^2 H}{\partial p_\beta \partial p_{\hat{\beta}}}\frac{\partial p_{\hat{\beta}}}{\partial q_\beta} - \frac{\partial^2 H}{\partial q_\alpha \partial q_{\hat{\alpha}}}\frac{\partial q_{\hat{\alpha}}}{\partial p_\alpha}\right) = 0 \tag{B.22}$$

and finally making use of (1.1), we obtain Eq. (4.9).

C. Formal Asymptotic Expansion of $\tilde{\Psi}(p_\alpha, q_\beta)$

Let us now derive Eqs. (4.18a)–(4.18c), which give directly the formal asymptotic expansion of Ψ. With the simplifying assumption (1.1), the Hamiltonian contains a potential energy operator, which in this representation is

$$V\left(i\hbar\frac{\partial}{\partial p_\alpha}, q_\beta\right)\tilde{\Psi}(p_\alpha, q_\beta) \tag{B.23}$$

and which we take to be defined by its Taylor series expansion,

$$\sum_{mn} V_{mn}\left(i\hbar\frac{\partial}{\partial p_\alpha}\right)^m q_\beta^n\tilde{\Psi}(p_\alpha, q_\beta)\,. \tag{B.24}$$

Here m and n are composite indices: for example, if there were four coordinates $(q_1 \ldots q_4)$, of which the first two were transformed to momentum space, the operator would be written out completely as

$$V\left(i\hbar\frac{\partial}{\partial p_1}, i\hbar\frac{\partial}{\partial p_2}, q_3, q_4\right) = \sum_{m_1m_2n_1n_2} V_{m_1m_2n_1n_2}\left(i\hbar\frac{\partial}{\partial p_1}\right)^{m_1}\left(i\hbar\frac{\partial}{\partial p_2}\right)^{m_2} q_3^{n_1}q_4^{n_2}\,. \tag{B.25}$$

Because of assumption (1.1), V contains no noncommuting operators, and when it acts on a function of the form

$$\tilde{\Psi}(p_\alpha, q_\beta) = \exp\left[\frac{i\tilde{S}(p_\alpha, q_\beta)}{\hbar}\right]\tilde{A}(p_\alpha, q_\beta)\,,$$

it gives

$$e^{i\tilde{S}/\hbar}\sum_{mn} V_{mn}q_\beta^n\left(q_\alpha + i\hbar\frac{\partial}{\partial p_\alpha}\right)^m\tilde{A}\,, \tag{B.26}$$

where, as before,

$$q_\alpha = q_\alpha(p_\alpha, q_\beta) = -\frac{\partial\tilde{S}}{\partial p_\alpha}\,.$$

Expanding the binomial and collecting powers of $\hbar$, for a single variable we would obtain

$$\left(q_1 + i\hbar\frac{\partial}{\partial p_1}\right)^{m_1} = q_1^{m_1}\left[1 + \left(\frac{m_1}{q_1}\right)\left(i\hbar\frac{\partial}{\partial p_1}\right) + \frac{m_1(m_1-1)}{2q_1^2}\left(i\hbar\frac{\partial q_1}{\partial p_1}\right) + \mathcal{O}(\hbar^2)\right]$$

so for several variables

$$\prod_{i\in\alpha}\left(q_i + i\hbar\frac{\partial}{\partial p_i}\right)^{m_i} = \left(\prod_{i\in\alpha} q_i^{m_i}\right)$$

$$\times\left\{1 + \sum_{j\in\alpha}\left(\frac{m_j}{q_j}\right)\left(i\hbar\frac{\partial}{\partial p_j}\right) + \frac{m_j(m_j-1)}{2q_j^2}\left(i\hbar\frac{\partial q_j}{\partial p_j}\right) + \sum_{\substack{j\in\alpha\\ \ell\in\alpha\\ j\neq\ell}}\frac{m_j m_\ell}{q_j q_\ell}\left(i\hbar\frac{\partial q_\ell}{\partial p_j}\right)\right\}$$

where we have used the fact that $\partial q_l/\partial p_j = \partial q_j/\partial p_l$. Using these results in (B.26) and condensing the notation using composite indices, the result is

$$e^{i\tilde{S}/\hbar}\left\{\left[\sum_{mn} V_{mn} q_\beta^n q_\alpha^m\right] + \left[\sum_{mn} V_{mn} q_\beta^n m q_\alpha^{m-1}\left(i\hbar\frac{\partial}{\partial p_\alpha}\right)\right]\right.$$

$$+\sum_{mn} V_{mn} q_\beta^n\left[\tfrac{1}{2}m(m-1)q_\alpha^{m-2}\left(i\hbar\frac{\partial q_\alpha}{\partial p_\alpha}\right) + \tfrac{1}{2}mm' q_\alpha^{m-1} q_{\alpha'}^{m'-1}\right.$$

$$\left.\left.\times\left(i\hbar\frac{\partial q_{\alpha'}}{\partial p_\alpha} + i\hbar\frac{\partial q_\alpha}{\partial p_{\alpha'}}\right)\right] + \cdots\right\}\tilde{A} \tag{B.27}$$

and we can perform the summations to find

$$e^{i\tilde{S}/\hbar}\left[V(q_\alpha, q_\beta) + \frac{\partial V}{\partial q_\alpha} i\hbar\frac{\partial}{\partial p_\alpha} + \frac{1}{2}\frac{\partial^2 V}{\partial q_\alpha \partial q_{\alpha'}} i\hbar\left(\frac{\partial q_{\alpha'}}{\partial p_\alpha}\right) + \cdots\right]\tilde{A}. \tag{B.28}$$

Finally, a similar development applied to the kinetic energy operators leads directly to (4.18).

Some questions about convergence are raised by the above development. Equation (B.24) does not make any sense unless the Taylor series has a wide domain in which it converges, and the rearrangement involved in going from (B.26) to (B.27) normally requires absolute convergence. Furthermore, the convergence properties are related not only to the magnitudes of q_α and q_β, but also to the rate of change of $\tilde{A}$. For this reason, MF give more general and rigorous derivations, which establish the essential results under quite general conditions.

APPENDIX C. SUMMARY OF TERMINOLOGY

The reader who wishes to pursue multidimensional semiclassical theory, but who has better things to do than to learn the theory of calculus on manifolds, might find the few words below to be helpful.[22]

Mapping. A function or a set of functions.

Smooth. Continuous and infinitely differentiable (C^∞).

Diffeomorphism. A smooth mapping with a smooth (unique) inverse.

Manifold. A smooth surface (a concept that is surprisingly difficult to define). Locally, a k-dimensional manifold has the properties of k-dimensional Euclidean space $\mathbb{R}^k$; in addition, it must be smoothly embedded in the $2n$ dimensional phase apace. One should examine Figs. 4.9–4.11 of ref. 20.

Chart. A local coordinate system for a portion of the manifold, that is, a domain of the manifold together with the smooth functions that relate it to fundamental coordinates. A collection of overlapping charts is called (of course) an atlas.

Regular/Singular. A domain of the manifold Λ^n is called regular if it has a diffeomorphic projection into a domain of configuration space. Otherwise it is singular.

Lagrangian. The definitions are all oriented toward one central idea: every small domain of a Lagrangian surface admits a generating function with properties given by Eq. (4.4).

Lagrangian Plane. (Geometrical definition) A hyperplane in phase space is said to be Lagrangian if for any two points on the plane, specified by $(q_1^a \ldots q_n^a, p_1^a \ldots p_n^a)$ and $(q_1^b \ldots q_n^b, p_1^b \ldots p_n^b)$,

$$\sum_i q_i^a p_i^b - p_i^a q_i^b = 0 \,.$$

Lagrangian Coordinate Planes. Any of 2^n planes in phase space, each characterized by n equations $\{p_i q_i = 0 \mid i = 1 \ldots n\}$; that is, for each i, either $p_i = 0$ or $q_i = 0$.

Lagrangian Manifold. (Geometrical definition) Every tangent plane is Lagrangian.

Properties:

1. The dimension does not exceed n.
2. If the dimension is n, then every (sufficiently small) domain has

a diffeomorphic projection into (at least) one of the Lagrangian coordinate planes.

3. In a regular domain with coordinates $q = \{q_i \mid i = 1 \ldots n\}$ and embedding $p_i = p_i(q)$, the cross-derivative equation (2.6a) holds, and

$$dS = \sum_i p_i \, dq_i$$

is an exact differential.

4. In a singular domain with coordinates $p_\alpha q_\beta$, and embedding $p_\beta = p_\beta(p_\alpha q_\beta)$, $q_\alpha = q_\alpha(p_\alpha q_\beta)$, the cross-derivative equations (4.3) hold, and

$$d\tilde{S} = \sum_{i \in \beta} p_i \, dq_i - \sum_{j \in \alpha} q_j \, dp_j$$

is an exact differential.

5. This definition and these properties are invariant under canonical transformations.
6. An n-dimensional manifold obtained by integration of Hamilton's equations starting from an $(n-1)$ dimensional Lagrangian manifold is Lagrangian.

Acknowledgments

This article was written during the period in which the author was supported by a fellowship sponsored by the J. Bruce Bredin Foundation. Additional support for portions of this work came from the National Science Foundation and the Jeffress Foundation. I especially thank G. Rublein for his unflagging attempts to make me understand some of the mathematics underlying this theory, S. K. Knudson for his heroic efforts to improve the presentation, P. Pechukas for helpful suggestions, D. Fannin for typing and retyping, and L. Menges for the drawings.

References

1. A short historical survey is given by J. Heading, *An Introduction to Phase-Integral Methods.* Wiley, New York, 1962. He credits Carlini (1817), Liouville (1837), and Green (1837) with the discovery of phase-integral approximations, and Stokes (1857–1889), Rayleigh (1912), Gans (1915), and Jeffreys (1923) with important further developments.
2. For example, A. Messiah, *Quantum Mechanics,* Vol. I. North-Holland, Amsterdam, 1964, pp. 222–228. L. D. Landau and E. M. Lifschitz, *Quantum Mechanics,* 2nd ed. Pergamon Press, New York, 1965, p. 158.
3. A complete bibliography will not be attempted here, but some seminal papers can be listed:

(a) R. A. Marcus, *Chem. Phys. Lett.* **7**, 525 (1970); *J. Chem. Phys.* **54**, 3935 (1971); **56**, 311, 3548 (1972); *Discuss. Faraday Soc.* **55**, 34 (1973). J. N. L. Connor and R. A. Marcus, *J. Chem. Phys.* **55**, 5636 (1971). D. W. Noid and R. A. Marcus, *J. Chem. Phys.* **62**, 2119 (1975).
(b) W. H. Miller, *J. Chem. Phys.* **53**, 1949, 3578 (1970); **54**, 5386 (1971); **57**, 2458 (1972). W. H. Miller and T. F. George, *J. Chem. Phys.* **56**, 5668 (1972).
(c) J. B. Keller, *Ann. Phys.* **4**, 180 (1958). J. B. Keller and S. I. Rubinow, *Ann. Phys.* **9**, 24 (1960). D. J. Vezzetti and S. I. Rubinow, *Ann. Phys.* **35**, 373 (1965).
(d) J. C. Y. Chen and K. M. Watson, *Phys. Rev.* **174**, 152 (1968); **188**, 236 (1969); J. C. Y. Chen, C.-S. Wang, and K. M. Watson, *Phys. Rev. A* **1**, 1150 (1970).
(e) J. N. L. Connor, *Mol. Phys.* **31**, 33 (1976); also, *Mol. Phys.* **26**, 1217 (1973) and **27**, 853 (1974).
(f) M. V. Berry, *Adv. Phys.* **25**, 1 (1976).
(g) E. Heller, *J. Chem. Phys.* **62**, 1544 (1975); **75**, 2923 (1981). M. J. Davis and E. Heller, *J. Chem. Phys.* **71**, 3383 (1979); **75**, 3916 (1981).

4. MF (ref. 6) cite, for example, papers written in Russian by Maslov (1961–1963), Arnold (1967), Kicherenko (1969), Fedoriuk (1971), and Weinberg (1975). Somewhat more accessible is V. P. Maslov, *Théorie des Perturbations et Methodes Asymptotiques.* Dunod, Paris, 1972.
5. For example, Maslov's work is mentioned by P. Pechukas, *J. Chem. Phys.* **57**, 5577 (1972) and discussed somewhat by I. Percival, in *Advances in Chemical Physics*, Vol. XXXVI, I. Prigogine and S. A. Rice (eds.). Wiley, New York, 1977, p. 1.
6. V. P. Maslov and M. V. Fedoriuk, *Semiclassical Approximation in Quantum Mechanics.* D. Reidel, Boston, 1981.
7. S. K. Knudson, J. B. Delos, and B. Bloom, *J. Chem. Phys.* **XX**, XX (1985).
8. H. Goldstein, *Classical Mechanics*, 2nd ed. Addison-Wesley, Reading, MA, 1980. Again we emphasize that Goldstein mainly discusses calculation of separable complete integrals and that these are not the solutions desired.
9. R. Courant and D. Hilbert, *Methods of Mathematical Physics*, Vol. II. Interscience, New York, 1966. M. Born and E. Wolf, *Principles of Optics*, 6th ed. Pergamon, New York, 1980.
10. If we were being more strict about scattering theory, we would note that we are supposed to take the limit as $Z^0 \to -\infty$, and then we would discover that $\lim_{Z^0 \to -\infty} S(X, Z)$ does not exist. The quantity that has a limit is $S(X, Z) - S^*(X, Z)$, where $S^*(X, Z)$ is the function obtained by the stated rules from trajectories of the *unperturbed* Hamiltonian, $(P_X^2 + P_Z^2)/2M$.
11. The statement that Ψ is a solution "to order $\hbar$" means that the term proportional to $\hbar$ in (3.3) has been neglected.
12. The complete derivation for the one-dimensional case is given by D. ter Haar, *Selected Problems in Quantum Mechanics*, 2nd. Academic Press, New York, 1965.
13. Here our presentation differs somewhat from that given by MF. They give a (quite abstract) geometrical definition of a Lagrangian manifold, and then they establish as theorems the properties that we call definitions.
14. V. I. Arnold, *Funct. Anal. Appl.* **1**, 1 (1976).
15. Our definition of μ is consistent with MF (definition 7.7) and their example given on p. 146. It has the opposite sign from their definition 7.4 and their example on p. 144. MF also define another index that they call γ, but its definition (MF 7.2) is inconsistent with its later usage (MF 8.2). Our index ν corresponds to the γ that appears in (MF 8.2) and differs by a sign from that in (MF 7.2).

16. This theorem holds if the manifold is connected in such a way that for any closed path ℓ_c on the manifold, $\int_{\ell_c} p \cdot dq = 0$ and $\mu(\ell_c) = 0$. The manifolds involved in this review have this property.

17. The word "typical" can be given a precise definition, and it has the same meaning here that it has in catastrophe theory. Not all Lagrange manifolds are typical, but the ones that are not typical arise from some unusual aspect of dynamics (such as a special symmetry property), and in general a small perturbation applied to an unusual manifold will convert it into a typical one.

18. R. Thom, *Structural Stability and Morphogenesis.* Benjamin, Reading, MA, 1975.

19. L. A. Pars, *A Treatise on Analytical Dynamics.* Heinemann, London, 1965, Sec. 16.2, pp. 269–271.

20. T. I. Poston and I. Stewart, *Catastrophe Theory and Its Applications.* Pitman, London, 1978, p. 65.

21. C. Chester, B. Friedman and F. Ursell, *Proc. Camb. Phil. Soc.* **53**, 599 (1957).

22. An especially clear presentation of the mathematical formalism is given by M. Spivak, *Calculus on Manifolds*, Benjamin, N.Y. (1965). Its application to dynamics is given by V. I. Arnold, *Math. Methods of Classical Mechanics*, Springer-Verlag, N.Y. (1978).

CORRELATION FUNCTIONS IN SUBCRITICAL FLUID

JOHN KERINS, L. E. SCRIVEN, and H. TED DAVIS

Department of Chemical Engineering and Materials Science, University of Minnesota, Minneapolis, Minnesota 55455

CONTENTS

I. Introduction . . . 215
II. Background . . . 221
III. Method of Solution . . . 225
IV. High Temperature . . . 234
V. Low Temperature . . . 242
A. PY Case . . . 243
B. HNC Case . . . 250
C. BGYK Case . . . 255
VI. Summary of Results . . . 262
VII. Discussion . . . 264
Appendix A . . . 270
Appendix B . . . 273
References . . . 275

I. INTRODUCTION

In an isotropic fluid of density n, the pair correlation function $g(r)$ contains information on the microscopic structure of the fluid.[1–4] If we focus our attention on a particular particle in the fluid, then $ng(r)$ is the local density of particles a distance r away. That is, the local density of particles at a distance r from a reference particle would differ by $n[g(r)-1]$ from the macroscopic fluid density n, due to the influence of the reference particle. When the fluid particles interact only through a short-ranged pair potential, the range of influence ξ of the reference particle is usually finite, and the pair correlation function satisfies the physically sensible condition that $g(r)\to 1$ as r increases to values larger than ξ. In fluids with pair potentials that are strongly repulsive at short distance, the reference particle excludes other particles from its immediate neighborhood and thus $g(r) \to 0$ as $r \to 0$. Typically, $g(r)$ rises sharply from a small value to a maximum value greater than unity at a distance r of roughly one particle diameter σ, and then decays, sometimes in an oscillatory fashion, to its asymptotic value of unity (see Fig. 2); usually $g(r)$ has approached very close to its asymptotic limit within several multiples of σ. Closely related to the pair correlation function $g(r)$ is the direct correlation function $c(r)$ defined through the Ornstein–Zernike (OZ) equation,[5]

$$c(r) = g(r) - 1 - n \int d\mathbf{s}[g(|\mathbf{s}|) - 1]c(|\mathbf{s} - \mathbf{r}|) . \tag{1.1}$$

By design, the direct correlation function has a shorter range than $g(r)$ and a simpler structure; $c(r)$ typically has one positive peak at a distance r of about σ, approaches a negative value as $r \to 0$, and decays rapidly and monotonically to 0 as r increases beyond σ. One of the main objectives in the theory of the fluid state is characterization of the microscopic fluid structure through $g(r)$ or $c(r)$.

The correlation functions also provide various routes to the thermodynamic properties of the fluid. The isothermal bulk modulus B [$\equiv (kT)^{-1} (\partial P/\partial n)_T$ with T absolute temperature, P pressure, and k Boltzmann's constant] of the fluid can be calculated either from the direct correlation function,

$$B_c = 1 - n \int d\mathbf{r}\, c(r) , \tag{1.2}$$

or from the pair correlation function

$$B_g = \left[1 + n \int d\mathbf{r}(g(r) - 1)\right]^{-1} . \tag{1.3}$$

The equivalence of B_c and B_g follows from the OZ equation (1.1). The isothermal compressibility K_T is simply related to the isothermal bulk modulus: $B = (nkTK_T)^{-1}$. The pressure P^c of the fluid could be obtained by integrating the bulk modulus with respect to the density, or, in the present case where fluid particles interact through a pair potential $u(r)$, the pressure P^v can be obtained from the virial equation,

$$P^v = nkT - \frac{n^2}{6} \int d\mathbf{r}\, r g(r) \frac{du(r)}{dr} . \tag{1.4}$$

The thermodynamics of a fluid of structureless particles having only pair interactions is also accessible through the internal energy per particle e,

$$e = \frac{3kT}{2} + \frac{n}{2} \int d\mathbf{r}\, g(r) u(r) . \tag{1.5}$$

If the correlation functions were known exactly, then all the above routes (1.2)–(1.5) to the fluid thermodynamics would be consistent and equivalent.

Since an exact statistical mechanical calculation of $g(r)$ or $c(r)$ is not possible, a number of approximate schemes have been proposed. Two widely studied approximations,[1–4] which result in nonlinear integral equations for $g(r)$ or $c(r)$, are the Percus–Yevick (PY) approximation,[6]

$$c(r) = g(r)[1 - e^{\beta u(r)}] , \tag{1.6}$$

where $\beta = 1/kT$, and the hypernetted chain (HNC) approximation,

$$c(r) = g(r) - 1 - \ln g(r) - \beta u(r) , \tag{1.7}$$

which was derived independently by several authors.[7–12] A third well-known approximation is the Kirkwood superposition approximation[13] as applied to the second member of the Born–Green–Yvon (BGY) hierarchy[14,15] for a uniform fluid. The BGY equation under the superposition approximation (BGYK) can be written in the form

$$\begin{aligned} \ln g(r) + \beta u(r) = {} & \pi n \int_0^\infty dt\, t[g(t) - 1] \int_{|r-t|}^{r+t} g(s) \frac{du(s)}{ds} \frac{s^2 - (r-t)^2}{r} \\ & + \pi n \int_0^\infty dt\, 4t^2[g(t) - 1] \int_{r+t}^\infty ds\, g(s) \frac{du}{ds} , \end{aligned} \tag{1.8}$$

where the condition $g(r) \to 1$ as $r \to \infty$ has been used to convert the usual BGY integrodifferential equation to the above BGYK equation.[16] Each of the three approximations (1.6), (1.7), and (1.8) essentially defines a closure relation which, when coupled with the OZ equation (1.1), leads to two functional equations for the two unknown functions $g(r)$ and $c(r)$. Because the correlation functions calculated under the PY, HNC, or BGYK approximation are not exact, the alternative routes—bulk modulus B (1.2), (1.3), virial pressure (1.4), or internal energy (1.5)—to the fluid thermodynamics are not equivalent for the approximate $g(r)$ and $c(r)$. The degree of consistency among the alternative routes is one simple measure of the goodness of the approximation.

Given a homogeneous fluid at density n and temperature T in which the pair potential is a (truncated and shifted) Lennard-Jones potential, there is a region in the (n, T) parameter plane associated with a region of liquid–vapor coexistence, and a point (n_c, T_c) associated with the liquid–vapor critical point. The thermodynamic functions of the fluid are expected to be singular at the critical point, and such singularities have been seen in experiment.[17] At subcritical temperatures metastable states, in which the fluid may persist for a long time in a homogeneous state even though it lies inside the coexistence region, are often observed experimentally.[18] The region of metastability lies between the coexistence curve and the spinodal curve (Fig. 1); the spinodal curve bounds the spinodal region where homogeneous fluid states are unstable and are never observed to persist for very long, but rather separate into two coexisting phases. One might expect that the thermodynamic properties of metastable states could be calculated by analytically continuing the stable fluid thermodynamic functions across the coexistence curve and into the metastable region. There are doubts, however, about whether the stable equilibrium thermodynamic functions can be continued into the metastable region. In the droplet model, a simple statistical mechanical model for first-order phase transitions studied by Fisher[19] and Langer,[20] and in its later modifications,[21,22] the thermodynamic functions are found to be singular (although only very weakly) at the coexistence curve. Several series analyses[23,24] for Ising models also indicate an essential singularity in the thermodynamic functions at the coexistence curve. If such singularities are present at the phase boundary (notwithstanding unphysical

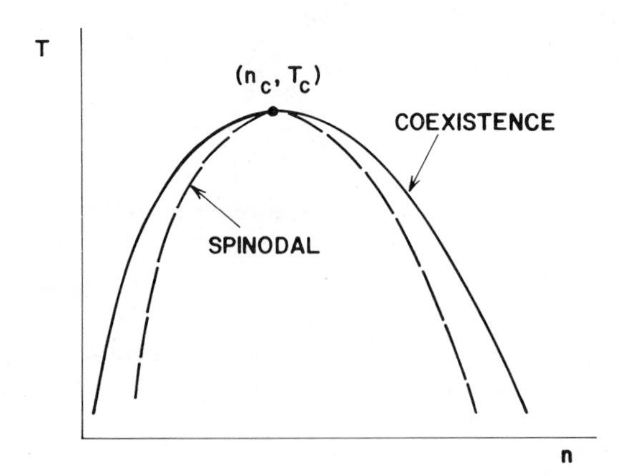

Fig. 1. In the density–temperature or (n, T) plane, the spinodal curve (– –) lies inside the liquid–vapor coexistence curve (—) except at the critical point (n_c, T_c), where the two curves touch.

features[25] of the droplet model or the difficulties in the analysis of a divergent series), then unique analytic continuation of thermodynamic functions across the coexistence curve is not possible. An appealing alternative to the notion of analytic continuation for metastable states is the method of restricted equilibrium discussed by Penrose and Lebowitz.[26] In this approach metastable states are possible, but only in the context of a restricted equilibrium state (i.e., in terms of a restricted statistical mechanical ensemble) rather than of a full thermodynamic equilibrium state. One of our purposes here is to examine the singularities or metastable states found in the thermodynamic functions of a Lennard-Jones fluid under the PY, HNC, or BGYK closure conditions at subcritical temperatures. Accordingly, although we do give some results for supercritical temperatures $T > T_c$, where the PY,[27–33] HNC,[39–41] and BGYK[45–48] approximations have already been studied, the majority of our results are for subcritical temperatures, $T < T_c$, where only limited attention has been directed.[34–38,42–44]

Our interest in the correlation functions near the coexistence curve is also motivated by several questions which arise in theories of nonuniform fluid systems; these questions are distinct from, but related to, those above about phase–boundary singularities and metastability. In an inhomogeneous fluid, the density is not constant but depends on position, $n = n_I(\mathbf{r})$. Moreover, the pair correlation function no longer depends only on the separation distance r, but rather on the particle positions $\mathbf{r}'$ and $\mathbf{r}' + \mathbf{r}$; that is, $g = g_I(\mathbf{r}', \mathbf{r}' + \mathbf{r})$, and similarly $c = c_I(\mathbf{r}', \mathbf{r}' + \mathbf{r})$. The inhomogeneous correlation function g_I appears in the equations for stress in an inhomogeneous fluid[49] and in approximate theories has often been replaced[49] by the pair correlation function g of a homogeneous fluid at some local density or densities. For example, g_I might be approximated by the homogeneous-fluid correlation function g for the density at mean location,

$$g_I(\mathbf{r}', \mathbf{r}' + \mathbf{r}) = g\left(r; n_I\left(\frac{2\mathbf{r}' + \mathbf{r}}{2}\right)\right), \tag{1.9a}$$

by g for the mean density,

$$g_I = g\left(r; \frac{n_I(\mathbf{r}') + n_I(\mathbf{r}' + \mathbf{r})}{2}\right), \tag{1.9b}$$

or by the mean correlation function

$$g_I = \tfrac{1}{2}[g(r; n_I(\mathbf{r}')) + g(r; n_I(\mathbf{r}' + \mathbf{r}))]\,. \tag{1.9c}$$

Such schemes as (1.9a)–(1.9c) for g_I are also useful in a modified van der Waals theory of inhomogeneous fluids,[50] where the free energy of the fluid is written as a functional of the density $n_I(\mathbf{r}')$, the pair potential $u(r)$, and the inhomogeneous correlation function $g_I(\mathbf{r}', \mathbf{r}' + \mathbf{r})$. In each case (1.9a)–(1.9c), g is required at a local density which may lie inside the coexistence or spinodal curve, and hence information on the correlation functions at metastable or spinodal densities is required. The same information is required, in both the pressure-tensor formalism[49] and the density functional formalism,[51,52] in order to develop an expansion in terms of the density $n_I(\mathbf{r}')$ and its gradients. The leading coefficients in such an expansion involve integrals over either the pair correlation function $g(r)$[49] or the direct correlation function $c(r)$[51] of a homogeneous fluid of density $n_I(\mathbf{r}')$, where again $n_I(\mathbf{r}')$ may lie inside the coexistence curve. Alternatively, if the free-energy density functional is expanded in terms of the deviations of $n_I(\mathbf{r}')$ from some mean value n_m, then the leading coefficients depend on $c(r; n_m)$, where n_m is inside the coexistence curve.[52] The same type $c(r; n_I)$ is also required in the approximate-density-functional formalism[53] in which a partial summation of the gradient expansion is effected. Our main purpose in this study is to examine the type of correlation functions $g(r; n, T)$ and $c(r; n, T)$ predicted under the PY, HNC, or BGYK closure condition when (n, T) lies inside the coexistence curve.

In Section II we give a brief outline of the theory for the three approximate closure conditions for the correlation functions. We describe in Section III the techniques for finding solutions and analyzing the solution sensitivity for the three approximate integral equations which result from the PY, HNC, and BGYK closures. The finite-element formulation used for our numerical solution of the integral equations by a Newton method is discussed in some detail because we feel the approach has much to recommend it over the traditional technique of successive substitution. The behavior of $g(r)$, $c(r)$, and the associated thermodynamic functions for supercritical temperatures is summarized in Section IV. In this section we also illustrate the changes in $g(r)$ and $c(r)$ induced by the truncation of the Lennard-Jones potential. Results at subcritical temperatures are presented in Section V. When $T < T_c$ the correlation functions depend sensitively on the numerical approximations as illustrated by several examples. A summary of our results is presented in Section VI. In the concluding paragraphs, Section VII, we discuss our results in relation to the questions of thermodynamic singularities, of metastability, and of correlation functions inside the spinodal region. Appendix A contains some further details and remarks on the theory of correlation functions outlined in Section II. Additional comments on the level of error in our numerical approximation are contained in Appendix B.

II. BACKGROUND

Although the original derivations of the PY,[6] HNC,[7–12] and BGYK[13–15] equations are rather different, there are several common threads that run through each approximation. The PY, HNC, and BGYK equations may each be generated by truncating functional expansions (in the density) of $n\exp(\beta u_{ext})$, $\ln n + \beta u_{ext}$, and $\beta n \nabla u_{ext}$, respectively;[54] here $u_{ext}(\mathbf{r}')$ is an arbitrary external field which should be set to 0 after carrying out the necessary functional differentiations. Cluster expansions for the PY and HNC equations are well known,[55] and the BGYK equation can be viewed as the lowest-order approximation in an exact cluster expansion.[56–58] Since the nature and limitations of the PY, HNC, and BGYK closure conditions have been reviewed elsewhere,[3,4] we do not discuss the detailed background of these approximations. We would, however, like to set an exact reference point from which we can view the approximations, and we choose potential distribution theory[59–61] to establish this vantage point. In potential distribution theory the number density n and the pair correlation function $g(r)$ are expressed as ensemble averages of the Boltzmann factor for the interaction of a test particle with the rest of the fluid. For a fluid of N particles in which any two, say one at $\mathbf{r}_1$ and the second at $\mathbf{r}_2$, interact pairwise with a potential $u(r_{12})$, $r_{12} = |\mathbf{r}_1 - \mathbf{r}_2|$, the interaction energy of a test particle at $\mathbf{r}'$ with the fixed N particles is $\psi(\mathbf{r}')$,

$$\psi(\mathbf{r}') = \sum_{i=1}^{N} u(|\mathbf{r}' - \mathbf{r}_i|)\,. \tag{2.1}$$

(Note that we assume no external field u_{ext} is present.) In the thermodynamic limit the canonical average of $\exp(-\psi(\mathbf{r}')/kT)$ equals the density $n(\mathbf{r}')$ divided by the fugacity z;

$$\frac{n(\mathbf{r}')}{z} = \langle e^{-\beta\psi(\mathbf{r}')}\rangle\,. \tag{2.2}$$

The fugacity $z = \lambda^{-3}\exp(\beta\mu)$, where μ is the chemical potential and $\lambda = (h^2/2\pi mkT)^{3/2}$, the thermal de Broglie wavelength for particles of mass m. Equation (2.2) is derived by expressing the configuration integral Q_{N+1} in terms of Q_N and $\psi(\mathbf{r}')$; the chemical potential μ arises through the ratio Q_{N+1}/Q_N in the limit as N becomes very large. The pair correlation function $g(\mathbf{r}', \mathbf{r}' + \mathbf{r})$ is similarly expressed as an average related to $\psi(\mathbf{r}')$ and $\psi(\mathbf{r}' + \mathbf{r})$,

$$g(\mathbf{r}', \mathbf{r}' + \mathbf{r}) = \frac{z^2 e^{-\beta u(r)}}{n(\mathbf{r}')n(\mathbf{r}' + \mathbf{r})}\langle e^{-\beta\psi(\mathbf{r}')}e^{-\beta\psi(\mathbf{r}'+\mathbf{r})}\rangle\,. \tag{2.3}$$

In an isotropic fluid the density is independent of position, and g depends only on the magnitude r of the separation vector $\mathbf{r}$. In that case the fugacity and density can be eliminated from Eq. (2.3) through Eq. (2.2), to yield the equation

$$y(r) \equiv g(r)e^{\beta u(r)} = 1 + K(r)\,, \tag{2.4a}$$

where $K(r)$ is a normalized covariance for $e^{-\beta\psi}$:

$$K(r) = \frac{\langle e^{-\beta\psi(0)}e^{-\beta\psi(\mathbf{r})}\rangle - \langle e^{-\beta\psi(0)}\rangle^2}{\langle e^{-\beta\psi(0)}\rangle^2}\,. \tag{2.4b}$$

Several features of (2.4) are noteworthy. First, the function $y(r)$, defined in (2.4a), arises naturally as the correlation function of interest. It represents the pair correlation function for a pair of fluid particles whose pair interaction with each other has been turned off. We find below that each of the three closure approximations leads directly to an integral equation for $y(r)$. Second, we note that $y(r)$ is a measure of the correlation in $e^{-\beta\psi}$ between the points 0 and $\mathbf{r}$. As r increases to values much greater than the correlation length ξ, $y(r)$ differs negligibly from the value of unity; this in turn implies that $g(r) \sim e^{-\beta u(r)} \sim 1$ as $r \to \infty$ in accord with our earlier remarks on the asymptotic behavior of g. It is important to emphasize, however, that this asymptotic behavior for $g(r)$ is expected only at distance r much larger than the correlation length or range of influence ξ (e.g., $r > 10\xi$). Once $r > \xi$, we believe that the correlation effects are small and $g(r)$ is close to unity, but at intermediate distances (e.g., $\xi < r < 10\xi$) we expect to observe either the classical Ornstein–Zernike decay ($\eta = 0$) or nonclassical decay ($\eta \neq 0$) in $g(r)$[62] rather than the actual asymptotic behavior $g(r) \sim e^{-\beta u(r)}$ as $r \to \infty$. Finally, we note that $g(r)$ is also proportional to $e^{-\beta u(r)}$ as $r \to 0$,[59] but in this case the proportionality constant is $\langle e^{-2\beta\psi(0)}\rangle / \langle e^{-\beta\psi(0)}\rangle^2$.

When the PY approximation (1.6) is used in the Ornstein–Zernike equation (1.1), the result is conveniently written as a nonlinear integral equation for $y(r)$,

$$y(r) = 1 + \frac{2\pi n}{r}\int_0^\infty ds\, s[(f(s)+1)y(s) - 1]\int_{|r-s|}^{r+s} dt\, t f(t)y(t)\,, \tag{2.5}$$

with $c(r) = f(r)y(r)$ and $f(r)$ the well-known Mayer function $(\equiv e^{-\beta u(r)} - 1)$. In a like manner, use of the HNC approximation (1.7) in OZ (1.1) gives

$$\ln y(r) = \frac{2\pi n}{r} \int_0^\infty ds\, s[(f(s)+1)y(s)] \times \int_{|r-s|}^{r+s} dt\, t[(f(t)+1)y(t) - 1 - \ln y(t)]\,, \qquad (2.6)$$

with $c(r) = (f(r)+1)y(r) - 1 - \ln g(r)$. Finally, the BGYK equation (1.8) can be recast as a nonlinear integral equation for $y(r)$:

$$\ln y(r) = \frac{2\pi n}{r} \int_0^\infty ds\, s[(f(s)+1)y(s) - 1] \times \left\{ \int_{|r-s|}^{r+s} dt \left[\frac{t^2 - (r-s)^2}{2}\right] \frac{du}{dt} [(f(t)+1)y(t)] + \int_{r+s}^{\infty} dt\, 2rt \frac{du}{dt} [(f(t)+1)y(t)] \right\}; \qquad (2.7)$$

in this case $c(r)$ must be determined by using the function $(f(r)+1)y(r)$ for $g(r)$ in the OZ equation (1.1).

We take Eqs. (2.5), (2.6), and (2.7) as our working PY, HNC, and BGYK equations, respectively. In this study, the pair potential $u(r)$ is a Lennard-Jones potential which, for the majority of our results, has been truncated and shifted:

$$u(r) = \begin{cases} 4\varepsilon[(\sigma/r)^{12} - (\sigma/r)^6] - u_0\,, & r < \ell_0\,, \\ 0\,, & r > \ell_0\,, \end{cases} \qquad (2.8)$$

where $u_0 = 4\varepsilon[(\sigma/\ell_0)^{12} - (\sigma/\ell_0)^6]$. The potential $u(r)$ is continuous and strictly finite-ranged. For this potential the direct correlation function is necessarily of finite range in the PY approximation, while $c(r) \sim [g(r)-1]^2/2$ as $r \to \infty$ in the HNC approximation; no such asymptotic behavior is apparent for $c(r)$ in the BGYK case. Note that the similarity among the PY, HNC, and BGYK equations is emphasized when $r > \ell_0$, since then the second t-integral in the BGYK equation (2.7) vanishes.

Upon comparison of the exact equation (2.4) for $y(r)$ with each of the approximate equations (2.5)–(2.7), we see that the right side of the PY equation (2.5) is a direct approximation $K_{PY}(r)$ to the normalized covariance $K(r)$, whereas both the HNC (2.6) and BGYK (2.7) approximations amount to approximations for $\ln[1 + K(r)]$. In these terms the HNC and BGYK approximations are similar. On the other hand, suppose that $\ln(1 + K(r))$ is expanded,

$$\ln(1 + K(r)) = K(r) - \frac{K(r)^2}{2} + \frac{K(r)^2}{3} + \cdots, \tag{2.9a}$$

and all terms nonlinear in $K(r)$ are collected in a remainder $\mathscr{S}(K(r))$ so that

$$\ln(1 + K(r)) = K(r) + \mathscr{S}(K(r)). \tag{2.9b}$$

Then the HNC approximation amounts to replacing the linear term by $K_{\mathrm{PY}}(r)$ and approximating the remainder $\mathscr{S}(K(r))$ as

$$\mathscr{S}(K(r)) = \frac{2\pi n}{r} \int_0^\infty ds\, s[(f(s) + 1)y(s) - 1] \int_{|r-s|}^{r+s} dt\, t[y(t) - 1 - \ln y(t)]. \tag{2.9c}$$

(This is analogous, in a sense, to the usual statement in the cluster expansion formalism that the HNC includes more graphs than PY.) Since both the PY and HNC equations use $K_{\mathrm{PY}}(r)$, these two approximations can be viewed as similar.

If the pair potential is short-ranged (i.e., negligible beyond some finite microscopic length) and hence fixes a microscopic length scale, then only fluid particles in a local microscopic neighborhood of the test particle at $\mathbf{r}$ contribute to $\psi(\mathbf{r})$ given in (2.1). Thus the normalized covariance $K(r)$ of (2.4b) is nonzero when the arrangement of particles in the local neighborhood of the origin influences the arrangement of particles in a local neighborhood at a distance r away. At most temperatures and densities the predominant fluctuations in local density occur on a microscopic scale. Consider a length ℓ greater than the range of the pair potential but still submacroscopic, and for a fixed microscopic configuration let $N_\ell(\mathbf{r})$ be the number of particles in a cube of volume ℓ^3 centered at $\mathbf{r}$. Those configurations in which $N_\ell(\mathbf{r})$ differs from $n\ell^3$ for some $\mathbf{r}$ are either rare (i.e., few in number) or improbable (i.e., have a very small Boltzmann probability factor) relative to the large number of probable configurations in which $N_\ell(\mathbf{r})$ is virtually equal to $n\ell^3$ for all $\mathbf{r}$. Accordingly, the influence of particles in the neighborhood of the origin on particles in a neighborhood a distance $\mathbf{r}$ away is usually small if the microscopic neighborhoods do not overlap; hence the correlation length ξ measured by $K(r)$ is usually on the order of the microscopic length characteristic of the pair potential. If configurations that show a large local density variation, away from the overall density n, become important (as at the critical point), then a more pronounced correlation between the microscopic neighborhoods at the origin and at a distance r away reflects the role of large fluctuations, especially for r greater than the pair-potential range. As large density fluctuations become important, the range of $K(r)$ increases. In any case we see that the three approximations, PY, HNC, and BGYK, amount to different estimates for the normalized covariance, and hence to different estimates for the effects of fluctuations in the fluid. [See Appendix A for a more formal discussion of $K(r)$.]

III. METHOD OF SOLUTION

To put equations in simplest form, lengths are scaled by the Lennard-Jones particle diameter σ, temperature is measured relative to the Lennard-Jones interaction ε, energies are scaled by the thermal energy, and a dimensionless pressure is defined using σ^3 and ε:

$$\begin{aligned} r^* &\equiv \frac{r}{\sigma}, & n^* &\equiv n\sigma^3, \\ T^* &\equiv \frac{kT}{\varepsilon}, & u^*(r) &\equiv \frac{u(r)}{kT}, \\ e^* &\equiv \frac{e}{kT}, & p^* &\equiv \frac{p\sigma^3}{\varepsilon}. \end{aligned} \tag{3.1}$$

From this point on we suppress the asterisks with the understanding that all parameters and functions are dimensionless. Note that the $y(r)$ calculated from any one of the approximate equations (2.5)–(2.7) depends parametrically on the temperature T, the density n, and the distance ℓ_0 at which the Lennard-Jones potential is truncated.

We discuss the numerical scheme in the context of the PY equation (2.5); the same analysis applies to the HNC or BGYK equation. The difference between the right and left sides of the PY equation (2.5) defines the functional residual $R[y(r); r]$ which must be 0 at all r for an acceptable solution $y(r)$. Our first assumption is that $y(r) = 1$ for $r > \ell_1 > \ell_0$; consequently, the limits for the integrals in the residual $R[y; r]$ are bounded by ℓ_1. In reduced units we then have from (2.5)

$$\begin{aligned} 0 = R[y(r); r] = 1 - y(r) + \frac{2\pi n}{r} \int_0^{\ell_1} ds\, s[(f(s)+1)y(s) - 1] \\ \times \int_{a(s,r)}^{b(s,r)} dt\, t f(t) y(t), \end{aligned} \tag{3.2}$$

where $a(s, r) \equiv \min(|r - s|, \ell_1)$ and $b(s, r) \equiv \min(r + s, \ell_1)$. The functional equation (3.2) for $y(r)$ is solved by Newton iteration. Given an initial estimate $y_{i=0}(r)$ for which $R[y_i] \neq 0$, the next iterate $y_{i+1}(r) = y_i(r) + \delta y_i$ is treated as if it were exactly the solution with $R[y_{i+1}] = 0$; δy_i is found by expanding $R[y_{i+1}]$ about $R[y_i]$ to first order in δy_i:

$$0 = R[y_{i+1}] \cong R[y_i] + \int ds \frac{\delta R[y_i ; r]}{\delta y_i(s)} \delta y_i(s) . \tag{3.3}$$

$\delta R/\delta y_i$ denotes the functional derivative of $R[y(r); r]$ with respect to $y_i(r)$. This linear integral equation for δy_i is solved, y_{i+1} calculated, and the process repeated, $i = 1, 2, \ldots$, until a specified convergence criterion is satisfied, at which point we have an acceptable solution $y(r)$. The convergence criterion might be based on the norm of the residual $\|R[y_i; r]\|$, where the norm is with respect to R as a function of r, or on the norm of the update difference $\|\delta y_i(r)\|$, or on both. The kernel $\delta R(r)/\delta y(s)$[63] of the integral equation (3.3) for δy_i is known as the Jacobian $J(r, s)$ for the Newton method.[64]

$$J(r, s) = \frac{\delta R(r)}{\delta y(s)} = -\delta(r-s) + \frac{2\pi n}{r}\Big\{ s(f(s)+1) \int_a^b dt\, t f(t) y(t) + sf(s) \int_0^{\ell_0} dt\, t[(f(t)+1)y(t) - 1]\theta(r, s, t)\Big\} , \tag{3.4a}$$

where $\delta(r)$ is the Dirac delta function and

$$\theta(r, s, t) = \begin{cases} 1 , & a(r, t) \leq s \leq b(r, t) \\ 0 , & \text{otherwise} . \end{cases} \tag{3.4b}$$

Note that $J(r, s)$ is a functional of $y(r)$ and thus must be updated at each iteration.

Although the residual $R[y(r); r]$ is identically zero for $2\ell_1 \leq r$ under the assumption $y(r) = 1$ for $r > \ell_1$, we can check the consistency of this assumption by calculating $R(r)$ for $\ell_1 < r < 2\ell_1$. If in that range $R(r)$ differs appreciably from 0, then ℓ_1 is increased. The difference between the two compressibilities B_c and B_g, from Eqs. (1.2) and (1.3), respectively, serves as a further check on the integral-cutoff assumption since $B_c - B_g = 0$ when $\ell_1 = \infty$; if the difference is larger than some preset value, then ℓ_1 is increased.

We also examine the sensitivity or precision of the solution $y(r)$ for the given ℓ_1, as distinct from the question of consistency or accuracy in the solution $y(r)$. Differentiating $R(r)$, $0 < r < \ell_1$, with respect to ℓ_1 at fixed density n, temperature T, and potential-cutoff ℓ_0, we find

$$0 = \int ds\, J[y; r, s] \frac{\partial y(s)}{\partial \ell_1} + \frac{2\pi n}{r} \left\{ \lim_{s \to \ell_1^-} s[(f(s)+1)y(s)-1] \int_{\ell_1 - r}^{\ell_1} dt\, t f(t) y(t) + \lim_{s \to \ell_1^-} s f(s) y(s) \int_{\ell_1 - r}^{\ell_1} dt\, t[(f(t)+1)y(t)-1] \right\}. \quad (3.5)$$

This inhomogeneous linear integral equation for $\partial y(r)/\partial \ell_1$ requires the Jacobian $J(r, s)$ evaluated at the final iterate $y(r)$. Unlike the Newton equation (3.3), in which the inhomogeneous term $R[y_i; r]$ converges to 0, the inhomogeneous term in the sensitivity equation (3.5) is fixed for the given $y(r)$ and ℓ_1 and depends strongly on the value of $y(r)$ as $r \to \ell_1^-$. Obviously, if the inhomogeneous term in (3.5) vanishes and the kernel $J(r, s)$ is nonsingular, then $\partial y(s)/\partial \ell_1 = 0$ and the solution $y(r)$ is independent of the cutoff ℓ_1.

In addition to ℓ_1, the solution $y(r)$ depends parametrically on n, T, and ℓ_0. A very good initial guess $\bar{y}(r)$ for the solution at $n + \delta n$ with the other parameters fixed is generated from $J(r, s)$ and $y(r)$ at density n. An expansion of $\bar{y}(r)$ in δn about $y(r)$ yields, to first order,

$$\bar{y}(r; n + \delta n) \cong y(r; n) + \frac{\partial y(r)}{\partial n} \delta n. \quad (3.6)$$

(This is known as the technique of first-order continuation. The case in which $\bar{y}$ is approximated by y at n is known as zero-order continuation.[65]) The partial derivative $\partial y(r)/\partial n$ is determined by solving the linear integral equation generated by differentiating $R[y; n, r]$ with respect to n:

$$0 = \int ds\, J(r, s) \frac{\partial y(s)}{\partial n} + \left(\frac{\partial R}{\partial n} \right)_y . \quad (3.7a)$$

For the PY residual (3.2) the explicit dependence of $R[y; r]$ on n is linear and we obtain the particularly simple form

$$\left(\frac{\partial R}{\partial n} \right)_y = \frac{R[y; r] - 1 - y(r)}{n}. \quad (3.7b)$$

Similar manipulations are possible to determine $(\partial y/\partial T)$ or $(\partial y/\partial \ell_0)$, but

$(\partial R/\partial T)_y$ or $(\partial R/\partial \ell_0)_y$ are complicated enough that zero-order continuation with a small step δT or $\delta \ell_0$ is usually more practical.

Although up to this point we have treated y as a function of r which depends on the scalar parameters ℓ_0, ℓ_1, T, and n, we sometimes find it advantageous to parameterize $y(r)$ by the potential cutoff ℓ_0, integral cutoff ℓ_1, the temperature T, and the value y_0 of $y(r)$ at a particular point r_0.[66–68] Then n becomes a variable which depends on the parameters ℓ_0, ℓ_1, T, and y_0. Using such an alternative set of parameters, we can track the solution branch for $y(r)$ around turning points in the density n.[34,44] If we had rigidly fixed n as a continuation parameter, the method of Newton iteration for calculating $y(r)$ would break down at a turning point in the density because the Jacobian $J(r, s)$ would be singular. By first-order continuation we can still find, at fixed ℓ_0, ℓ_1, and T, an initial guess $\bar{n}$ and $\bar{y}(r)$, $0 < r < \ell_1$ and $r \neq r_0$, for a change δy_0 in the parameter y_0:

$$\bar{y}(r; y_0 + \delta y_0) \cong y(r; y_0) + \left(\frac{\partial y(r)}{\partial y_0}\right) \delta y_0 , \qquad r \neq r_0 \tag{3.8a}$$

and

$$\bar{n}(y_0 + \delta y_0) \cong n(y_0) + \left(\frac{\partial n}{\partial y_0}\right) \delta y_0 . \tag{3.8b}$$

The partial derivatives with respect to y_0 are found as solutions of the linear integral equation given by differentiating $R[y]$ with respect to y_0,

$$0 = \left(\frac{\partial R}{\partial n}\right)\left(\frac{\partial n}{\partial y_0}\right) + \int ds\, J(r, s) \left(\frac{\partial y(s)}{\partial y_0}\right), \tag{3.9a}$$

subject to the normalization condition

$$\frac{\partial y(r = r_0)}{\partial y_0} = 1 . \tag{3.9b}$$

By allowing for flexibility in our choice of continuation parameter, we can ensure a very good initial estimate $\bar{y}(r)$. For example, in the PY case, n is not usually a good choice for the continuation parameter because at several points r_i the solution $y(r_i)$ is extremely sensitive to n (i.e., $\partial y(r_i)/\partial n > 1$). Thus even for small δn the first-order continuation estimate $\bar{y}(r; n + \delta n)$ from equation (3.6) may be very crude at points $r = r_i$. (At turning points in the density n, it is certain that $\partial y(r_i)/\partial n \gg 1$ for some

r_i, but an extreme sensitivity of $y(r_i)$ with respect to n can also occur away from any turning point in n.) On the other hand, if the continuation parameter $y_0 = y(r_0)$, where r_0 is the point where $|\partial y(r)/\partial n|$ is maximal, then $\partial y(r)/\partial y_0$ and $\partial n/\partial y_0$ are less than one and $\bar{n}(y_0 + \delta y_0)$ and $\bar{y}(r; y_0 + \delta y_0)$ from (3.8) are close to the true solutions $n(y_0 + \delta y_0)$ and $y(r; y_0 + \delta y_0)$.

In addition to the crucial role of $J(r, s)$ in iteration (3.3), sensitivity (3.5), and continuation [(3.7) and (3.9)] calculations, the Jacobian also determines whether the solution is locally unique with respect to the specified parameters. Given a solution $y(r)$ of the PY equation (3.2) at fixed parameters ℓ_0, ℓ_1, T, and n, suppose there is a nearby but distinct solution $\tilde{y}(r)$ of (3.2) at the same parameter values which can be written

$$\tilde{y}(r) = y(r) + \omega p(r)\,, \qquad \omega \ll 1\,. \tag{3.10}$$

[Note that as part of the assumption that $\tilde{y}$ is close to y we restrict $p(r)$ to vanish for $r > \ell_1$.] When terms of order ω^2 are neglected, $p(r)$ satisfies the linear integral equation

$$0 = \int ds\, J(r, s) p(s)\,, \tag{3.11}$$

inasmuch as both $\tilde{y}$ and y are solutions to (3.2). This equation has nontrivial solutions $p(r)$ if and only if the kernel $J[y; r, s]$ is singular (i.e., J defines a singular operator). Thus the singularities of the integral operator J locate, for the specified set of parameters, the points where multiple solutions are to be found; when the only solution to (3.11) is $p(s) \equiv 0$, then the solution $y(r)$ is locally unique with respect to the specified parameters.

The PY equation (3.2) applies to a homogeneous fluid in which the macroscopic density n is identical to the average microscopic density. So we have really only established that, when $J(r, s)$ is nonsingular, $y(r)$ is unique with respect to perturbations $p(r)$ which preserve the homogeneous structure of the fluid. The question of uniqueness with respect to perturbations in which the local density is allowed to vary is much more involved. Under the approximation that y is a function only of the magnitude r, Fulinski[69] solved the (sticky-sphere) PY equation for an assumed form $n_l(\mathbf{r})$ of the local density which maintains the value n for the macroscopic density. Although his solution does seem to produce nonclassical critical exponents, the use of $y(r)$ rather than $y_l(\mathbf{r}', \mathbf{r}' + \mathbf{r})$ is somewhat inconsistent. When the local density $n_l(\mathbf{r})$ is variable, it must be related consistently to either the inhomogeneous direct correlation function $c_I(\mathbf{r}, \mathbf{s})$,[70,71]

$$\nabla \ln n_I(\mathbf{r}) = \int d\mathbf{s}\, c_I(\mathbf{r}, \mathbf{s}) \nabla n_I(\mathbf{s}) , \tag{3.12a}$$

or the inhomogeneous pair correlation function $g_I(\mathbf{r}, \mathbf{s})$,

$$\nabla n_I(\mathbf{r}) = -\int d\mathbf{s}\, \nabla \phi(|\mathbf{r}-\mathbf{s}|) n_I(\mathbf{r}) n_I(\mathbf{s}) g_I(\mathbf{r}, \mathbf{s}) . \tag{3.12b}$$

[Equation (3.12b) is simply the first member of the well-known BGY hierarchy.[1,2]] The functions n_I, g_I, and c_I are also related through the generalized OZ equation for inhomogeneous fluids. [Note that under an inexact closure condition Eqs. (3.12a) and (3.12b) are not equivalent, and different triples (n_I, g_I, c_I) can result from coupling the OZ equation under an approximate closure with either (3.12a) or (3.12b). This is the analog in an inhomogeneous fluid to the inequivalence of the various thermodynamic routes under approximate closure.] Since an analysis of the integral equations connecting $n_I(\mathbf{r})$ and $y_I(\mathbf{r}, \mathbf{s})$ would take us too far afield from our present study of homogeneous fluids, we do not attempt any further discussion of the stability or uniqueness of solutions $y(r)$ with respect to inhomogeneous perturbations.

Having established a formal framework of integral equations for Newton iteration, sensitivity, and continuation, we need only pick a discretization scheme to transform a continuous integral equation into a set of coupled algebraic equations. Traditionally, this discretization[72] is accomplised by choosing a specific quadrature rule for a given uniform mesh $r_i = (i-1)h$, $i = 1, 2, \ldots, N$, to get a set of equations for $y(r_i)$, or by choosing a particular orthonormal basis for an expansion to get a set of equations for the expansion coefficients. As our discretization scheme we choose the finite-element method[73] which is an effective hybrid of the two classical approaches. The unknown solution $y(r)$ is approximated by an expansion with coefficients y^k in a finite set of N linearly independent basis functions $H^k(r)$:

$$y(r) \cong \sum_{i=1}^{N} y^i H^i(r) = \mathbf{y} \cdot \mathbf{H}(\mathbf{r}) . \tag{3.13}$$

Once a mesh or set of node positions r_i, $i = 1, 2, \ldots, N$, is specified in the domain $0 < r < \ell_1$, the intent in the finite-element approach is to approximate the function $y(r)$ locally over small subdomains centered at r_i. The key features of the finite-element expansion are (1) that the basis function $H^i(r)$ takes the value 1 at node r^i and the value 0 at all other nodes and (2) that $H^i(r)$ is nonzero over only a small subdomain of the entire domain $0 < r < \ell_1$. Because of the first feature, the coefficients y^i are simply approximations to the value $y(r^i)$; that is, in contrast to the coefficients in an orthonormal series expansion, the finite-element coefficients provide clear and direct information on the approximate value of the unknown function $y(r)$ at $r = r_i$. Because of the second feature, the finite-element expansion retains the convenience of local finite-difference or collocation methods. Most importantly, although not an orthonormal series expansion, the finite-element expansion is in terms of linearly independent basis functions, and the techniques of functional analysis can be used to address the questions of the distance between $y(r)$ and $\mathbf{y} \cdot \mathbf{H}$ or of the rate of convergence of $\mathbf{y} \cdot \mathbf{H}$ to $y(r)$ as the basis functions are refined.

Given a set of N nodes, $0 = r_1 < r_2 < \cdots < r_N = \ell_1$, the simplest basis functions are the linear polynomials or *Chapeau* functions defined by

$$H^i(r) = \begin{cases} \dfrac{r - r_{i-1}}{r_i - r_{i-1}}, & r_{i-1} \leq r < r_i, \\ \dfrac{r_{i+1} - r}{r_{i+1} - r_i}, & r_i \leq r < r_{i+1}, \\ 0, & \text{elsewhere}. \end{cases} \tag{3.14}$$

With this set (3.14) of basis functions, the approximation $\mathbf{y} \cdot \mathbf{H}$ of (3.13) to $y(r)$ is continuous and piecewise linear. Rather than Chapeau functions, piecewise quadratic or cubic functions could be used as basis functions.[73] Just as a given segment of a curve is better approximated by quadratic or cubic interpolation than by linear interpolation, so can the finite-element expansion $\mathbf{y} \cdot \mathbf{H}$ better approximate $y(r)$ when basis functions of higher order than linear are used. The improvement in accuracy with higher-order basis functions is not unlike the gain in accuracy produced when a larger number of basis functions are used in a truncated orthonormal series expansion. (Such truncation is unavoidable in a numerical solution for an orthonormal series expansion.) For any particular set of basis functions, if the number N of nodes is increased, then the size of subdomains over which the function is locally approximated is decreased, and the approximation error is decreased.[73] This is analogous to the decrease of error on grid refinement in finite-difference or collocation schemes, but now in the framework of a formal expansion in independent basis functions. The coefficients y^i in the finite-element expansion (3.13) always are approximations to the values of $y(r)$ at the nodes r_i, independent of the details of node placement, subdomain size, and basis-function complexity.

For our correlation function calculations, we divide the domain of interest, $0<r<\ell_1$, into $N-1$ subdomains of size $h=\ell_1/(N-1)$ by a uniform mesh of N points $r_i=(i-1)h$, $i=1,2,\ldots,N$. As basis functions we adopt the linear polynomials defined in (3.14). When used in an integral equation for the correlation function $y(r)$, the finite-element approximation (3.13) essentially amounts to specifying a simple quadrature rule. Since we have restricted ourselves to piecewise linear basis functions, the error between the finite-element solution and the exact solution is controlled by the size h of the subdomain.

The next numerical approximation we make is the Swartz–Wendroff approximation.[74] Any function $F(r)$ that involves the unknown $y(r)$ is expanded in the basis functions as $F(r)=\Sigma F^i H^i(r)$, with coefficients $F^k=F(r^k)$. For example, a nonlinear function $F(y(r))=\Sigma F(y^i)H^i(r)$. The integral $\int_a^b dt\, t f(t)y(t)$ which appears in the PY equation (3.2) would take the form $\Sigma\, y^k \int_a^b dt\, t f(t)H^k(t)$ if the finite-element expansion (3.13) was simply substituted for $y(r)$; even though the functions f and H^k are known, the resulting t-integral cannot be evaluated in closed form. When the Swartz–Wendroff approximation is used, the original integral is replaced by the form $\Sigma_k\,[y^k t_k f(t_k)] \int_a^b dt\, H^k(t)$ in which the t-integral can be immediately evaluated. The Swartz–Wendroff approximation, like the finite-element expansion itself, is expected to introduce errors proportional to the square of the subdomain size h (see Appendix B). Clearly, the approximation emphasizes the quadrature-rule aspect of the finite-element approach and is in the same spirit as the Nyström method[72] for linear integral equations.

When the expansion $\mathbf{y}\cdot\mathbf{H}$ is substituted into the PY residual equation (3.1) and the Swartz–Wendroff approximation invoked, no choice of expansion coefficients y^k can satisfy the condition that $R[\mathbf{y};\mathbf{r}]=0$ for all $0<r<\ell_1$. As our final step in the discretization of the residual equation, we require that the residuals $R[\mathbf{y};r_k]$ be 0 at each of the nodes r_k, $k=1,2,\ldots,N$. This is the collocation version of the weighted-residual technique in finite elements and results in N coupled, nonlinear algebraic equations for the N expansion coefficients y^k:

$$0=R[\mathbf{y};r_k]=1-y^k+\frac{2\pi n}{r_k}\sum_{i=1}^{N}\sum_{j=1}^{N} s_i[(f(s_i)+1)y^i-1]t_j f(t_j)y^j A^{ijk}\,, \tag{3.15a}$$

where

$$A^{ijk} = \int_0^{\ell_1} ds\, H^i(s) \int_{a(s_i, r_k)}^{b(s_i, r_k)} dt\, H^j(t)\,, \qquad i, j, k = 1, 2, \ldots, N. \tag{3.15b}$$

This set of N residual equations is solved by using the appropriately discretized version of the Newton method embodied in Eq. (3.3). The kernel $J(r, s)$ is now a matrix $J[\mathbf{y}; r_i, s_j]$ in which the elements J_{ij} depend on the expansion coefficients $\mathbf{y}$. The iteration is carried on until the Euclidean norm $\|\delta \mathbf{y}_i\|$ of the update vector for the ith iteration or the Euclidean norm of residuals $\|\mathbf{R}[\mathbf{y}_i; r_k]\|$ is less than $10^{-10}\sqrt{N}$. We monitor the accuracy of the truncation approximation $[y(r) = 1 \text{ for } r \geq \ell_1]$ by calculating the extra residuals $R[\mathbf{y}; \bar{r}_k]$ at the N nodes $\bar{r}_k = \ell_1 + r_k$, $k = 1, 2, \ldots, N$, outside the domain $0 < r < \ell_1$. If each of these extra residuals is less than 10^{-6}, then we regard the assumption $y = 1$ for $r \geq \ell_1$ as valid.

Discrete versions of the sensitivity (3.5) and continuation [(3.7) or (3.9)] equations are produced by using the same scheme of finite-element expansion, Swartz–Wendroff approximation, and collocation. In Eq. (3.5) for the sensitivity of $y(r)$ to ℓ_1, the value $F(r_{N-1})$ is used for the $\lim_{s \to \ell_1^-} F(s)$ of the appropriate functions F. The sensitivity vector $(\partial \mathbf{y}/\partial \ell_1)$ provides information on the precision of the solution with respect to ℓ_1 while the continuation vector $(\partial \mathbf{y}/\partial n)$ is used to generate an initial guess $\bar{\mathbf{y}}(n + \delta n)$.

We have used the procedure described above for the PY equation (2.5) to derive and solve discretized versions of the HNC (2.6) and BGYK (2.7) equations. Since the BGYK equation does not involve any explicit approximation for the direct correlation function $c(r)$, an appropriately discretized form of the OZ equation (1.1) was used to fix the coefficients $\mathbf{c}$ in the expansion of $c(r)$, given the BGYK coefficients $\mathbf{y}$. Whereas we have based our numerical algorithm on a simple version of the finite-element approach, a variety of refinements could be introduced. For example, we could avoid the Swartz–Wendroff approximation, choose higher-order basis functions,[73] or use a nonuniform, adaptive mesh.[75] In our implementation of the Newton method the practical limit to N, the number of nodes, is dictated by the computer time necessary to calculate the $N \times N$ Jacobian J_{ij}; the solution of the linear equations involving J_{ij} by Gauss elimination took slightly less time. Most of our calculations are for a domain spacing $h = 1/15$ with N between 90 and 270 depending on the choice of ℓ_1, $6 < \ell_1 < 18$. Our algorithm was designed to optimize vectorization, and, although mostly written in standard FORTRAN, the code did implement some special syntax peculiar to vector supercomputers. Production runs were carried out either on a CRAY 1B computer at the University of Minnesota or on a Cyber 205 computer owned by Control Data Corporation.

We should contrast our numerical method for the correlation-function integral equations[76] with those that have been reported in the literature.[27–48] With a few notable exceptions for the PY,[31–34,38] HNC,[44] and BGYK[48] equations, the standard technique is to use a successive substitution method. While this method often works well, there are invariably convergence difficulties as the coexistence curve is approached (e.g., see refs. 30, 35, or 47). Moreover, as the Jacobian is not calculated, there is no basis to judge the stability or sensitivity of the solution. In the exceptional cases noted above (see also ref. 77), a Newton method was used and no such convergence problems appeared; in no case, however, where the questions of accuracy or sensitivity in the solution addressed as completely as we do here.

The values of $g(r)$ and $c(r)$ at the node points r_k, $k = 1, \ldots, N$, are fixed by the expansion coefficients $\mathbf{y}$. In the PY case

$$g(r_k) = (f(r_k) + 1)y^k \quad \text{and} \quad c(r_k) = f(r_k)y^k . \tag{3.16}$$

The thermodynamic integrals—bulk modulus B_c or B_g, virial pressure P^v, or internal energy e—are computed by first using a cubic spline interpolation based on the nodal values of g or c (3.16) to determine the values of g or c on a mesh of $3N$ uniformly spaced points in $0 \leq r \leq \ell_1$, and then using Simpson's rule on the mesh of $3N$ points to evaluate the integrals. Although a subdomain size of h is sufficient to compute accurately the expansion coefficient y^k, $k = 1, 2, \ldots, N$, a mesh size of $h/3$ is better to calculate accurately the thermodynamic integrals.[30] This is because the thermodynamic integrands contain factors that become large (viz. r^2) or change rapidly near $r = 0$ or $r = 1$ [viz. $\phi(r)$ and $d\phi/dr$]. In the next section we report results for the PY, HNC, and BGYK correlation functions and themodynamic functions at a supercritical temperature and in Section V, at subcritical temperatures.

IV. HIGH TEMPERATURE

Our primary purpose in this section is to provide a common high-temperature baseline for the low-temperature results to be presented in Section V. We should emphasize that at the high temperatures considered in this section we easily (i.e., with $N \leq 120$) reproduced results reported in the literature to within 1%. Since no results had been reported for the truncated and shifted Lennard-Jones potential, our high-temperature calibrations were made using potentials that matched those in the literature. A more detailed comparison of our work with published results at the lower temperatures is given in Section V. We do discuss here the effects of the potential truncation ℓ_1, the integral cutoff ℓ_0 and the subdomain size h.

Because the correlation functions which we calculated are not exact, the location of the liquid–vapor critical point, (n_c, T_c), depends on the particular thermodynamic function used to specify the condition of criticality. We adopt the bulk modulus B as the thermodynamic standard for locating the critical point. In practice, B_c (1.2) is superior to B_g (1.3) since $c(r)$ is less sensitive to the cutoff ℓ_1 than is $g(r)$, although in principle B_c and B_g are equivalent. The critical point (n_c, T_c) is defined by the condition that, on the critical isotherm, $B_c > 0$ for $n \neq n_c$ $(n_c < 0.65)$ whereas $B_c = 0$ at $n = n_c$. Accordingly, at high (supercritical) temperatures B is positive for all $0.05 < n < 0.65$. The temperature $T = 1.65$ is supercritical for all three closure approximations. The correlation functions from the PY, HNC, and BGYK equations at $T = 1.65$ are shown in Fig. 2 for low $(n = 0.10)$, intermediate $(n = 0.30)$, and high $(n = 0.50)$ densities. In each case the direct correlation function is short-ranged; only the structure of $c(r)$ for $0 < r < 1$ changes appreciably as density increases. In Table I the bulk modulus B for $T = 1.65$ and $0.05 \leq n \leq 0.65$ is tabulated. The difference between the PY, HNC, and BGYK results for B reflect the differences in the critical temperature T_c for each approximation. In accord with our definition of the critical point, the closer the positive minimum of B on the isotherm $T = 1.65$ is to 0, the closer the isotherm is to the critical isotherm $(T = T_c)$. Thus from Table I we have $T_c(\text{BGYK}) > T_c(\text{HNC}) \gtrsim T_c(\text{PY})$, in agreement with the literature results[4] that $T_c(\text{BGYK}) \cong 1.6$, $T_c(\text{HNC}) \cong 1.4$, and $T_c(\text{PY}) \cong 1.3$. (We cannot quote exact values for the critical temperatures because we use a pair potential that differs slightly from those used in the literature.)

For the PY and HNC results in Table I we also note that $B_g(\ell_1 = 12)$ differs by less than 0.5% from $B_c(\ell_1 = 12)$ which itself is equal to $B_c(\ell_1 = 6)$. On the other hand, $B_g(\ell_1 = 6)$ is larger than $B_g(\ell_1 = 12)$, although $g(r; \ell_1 = 12)$ is virtually identical to $g(r; \ell_1 = 6)$, $0 < r < 6$. Thus, even though the cutoff approximation that $y(r) = 1$ for $r \geq \ell_1 = 6$ is not highly accurate, inasmuch as the tail of $y(r)$ is still significant at $r = 6$, the solution $y(r)$, $0 \leq r \leq 6$, is rather insensitive to the cutoff approximation. For example, in the PY results at $\ell_1 = 6$ and $n = 0.30$ we find the first few extra residuals $R(r)$, $6 < r < 7$, are of order 10^{-3} and hence we judge the cutoff approximation to be insufficiently accurate; at the same time the sensitivity of the solution $\partial y(r)/\partial \ell_1$, $0 \leq r \leq 6$, is at most of order 10^{-5}. When the cutoff ℓ_1 is changed from 6 to 12, $y(r)$ is of order 10^{-4}–10^{-6} over the range $6 < r < 12$. The resulting changes in the short-ranged direct correlation function are much less pronounced [e.g., for PY, $c(r) = f(r)y(r)$] and not even evident in B_c, while the changes in the tail of $g(r)$

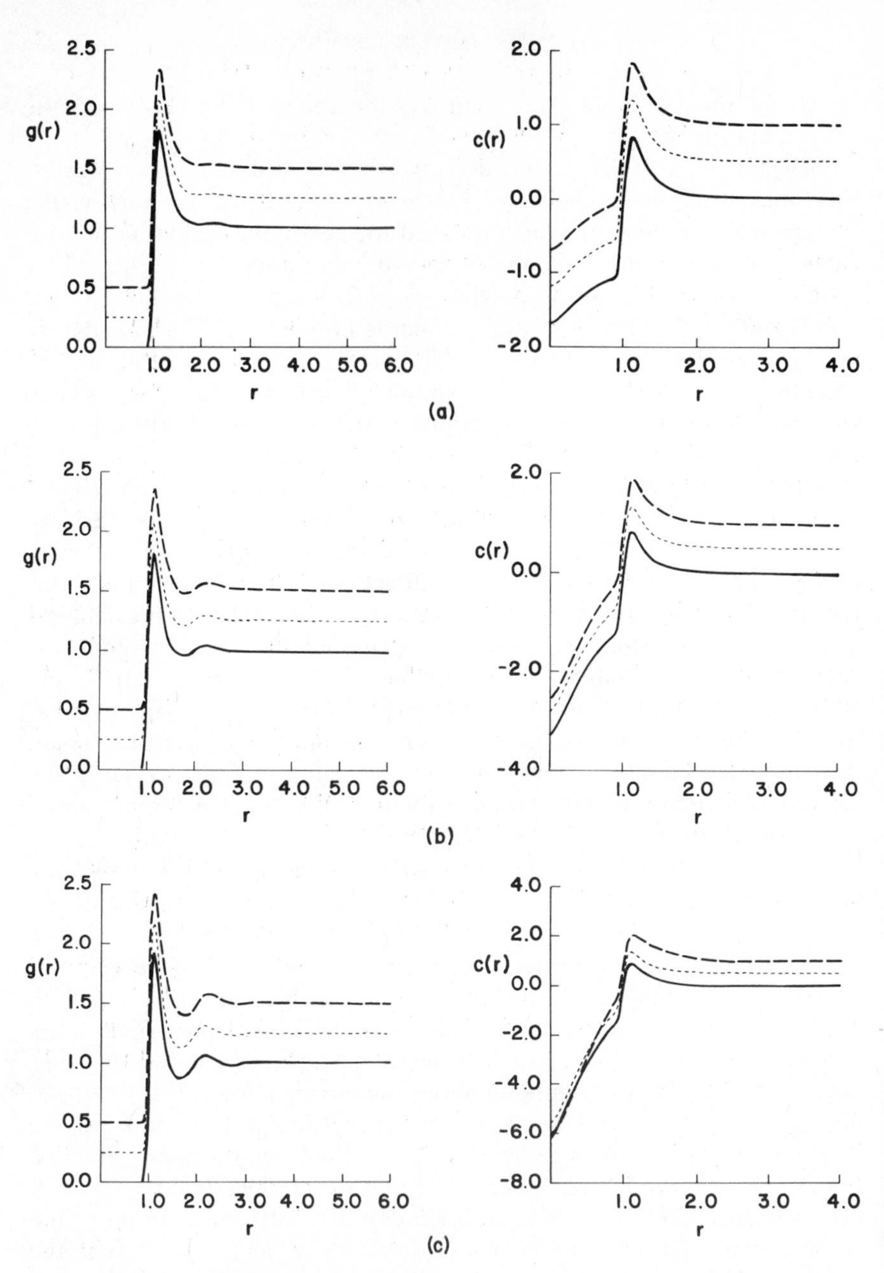

Fig. 2. Pair correlation function $g(r)$ and the direct correlation function $c(r)$ under the PY (—), HNC (---), and BGYK (– –) approximations at temperature $T = 1.65$ and density n equal to 0.10 (a), 0.30 (b), and 0.50 (c). The HNC and BGYK curves have been shifted up by 0.25 and 0.50 for $g(r)$ and by 0.50 and 1.00 for $c(r)$, respectively.

TABLE I
Compressibilities B_c and B_g from the PY, HNC, and BGYK Equations at Different Values of the Cutoff ℓ_1 for $0.5 \leq n \leq 0.65$ and $T = 1.65$

B_c / B_g	$\ell_1 = 6$			$\ell_1 = 12$			$\ell_1 = 18$
n	PY	HNC	BGYK	PY	HNC	BGYK	BGYK
0.05	0.8156	0.8153	0.8135	0.8156	0.8153	0.8135	
	0.8163	0.8159	0.8142	0.8156	0.8153	0.8135	
0.10	0.6686	0.6660	0.6527	0.6686	0.6660	0.6526	
	0.6729	0.6703	0.6573	0.6686	0.6661	0.6526	
0.15	0.5632	0.5551	0.5132	0.5632	0.5551	0.5129	
	0.5755	0.5679	0.5282	0.5634	0.5552	0.5131	
0.20	0.5038	0.4859	0.3931	0.5038	0.4859	0.3926	
	0.5274	0.5112	0.4275	0.5042	0.4864	0.3935	
0.25	0.4955	0.4631	0.2926	0.4955	0.4630	0.2918	
	0.5305	0.5016	0.3546	0.4964	0.4641	0.2954	
0.30	0.5461	0.4931	0.2126	0.5461	0.4931	0.2120	0.2120
	0.5890	0.5422	0.3080	0.5472	0.4947	0.2216	0.2131
0.35	0.6676	0.5869	0.1546	0.6676	0.5869	0.1556	0.1559
	0.7136	0.6415	0.2859	0.6687	0.5886	0.1744	0.1591
0.40	0.8791	0.7615	0.1199	0.8791	0.7615	0.1249	0.1263
	0.9234	0.8163	0.2867	0.8799	0.7629	0.1542	0.1327
0.45	1.2071	1.0405	0.1101	1.2071	1.0405	0.1227	0.1259
	1.2466	1.0912	0.3104	1.2074	1.0413	0.1604	0.1349
0.50	1.6864	1.4522	0.1278	1.6864	1.4522	0.1528	0.1587
	1.7215	1.4975	0.3579	1.6859	1.4523	0.1944	0.1679
0.55	2.3613	2.0288	0.1735	2.3613	2.0288	0.2189	0.2277
	2.3962	2.0715	0.4299	2.3595	2.0279	0.2591	0.2348
0.60	3.2872		0.2409	3.2872	2.8055	0.3919	0.3304
	3.3235		0.5231	3.2822	2.8029	0.3542	0.3349
0.65	4.5327		0.2979	4.5327		0.4399	0.4535
	4.5330		0.6267	4.5208		0.4696	0.4561

do produce a significant decrease in B_g because of the factor of r^2 in the integrand of B_g^{-1}.

An apparently general feature of all three integral equations is that the sensitivity $|\partial y(r)/\partial \ell_1|$ in the range $1 \leq r \leq \ell_1 - 2$ is usually smaller by at least a factor of 0.1 than the sensitivity in the ranges $0 \leq r < 1$ and $\ell_1 - 2 < r \leq \ell_1$. This implies that the short-ranged structure of $c(r)$ is as sensitive to the cutoff of the tail in $g(r)$ as is $g(r)$ itself near the cutoff. This is in accord with the motivation of Ornstein and Zernike in defining $c(r)$, namely, that the short-range structure of the direct correlation function reflect the long-range structure of the pair correlation function.

In the BGYK approximation there is no explicit condition on the behavior of $c(r)$ at large r, unlike both the PY and HNC approximations. One consequence is that the BGYK correlation function, like the BGYK $y(r)$ itself, is sensitive to the cutoff ℓ_1 not only at short distances, but also at large distances $r \lesssim \ell_1$. More importantly, it means that any anomalous behavior in $c(r)$ must be due to the superposition approximation. In particular, the unexpected appearance of negative values in the decaying tail ($r > 2$) of $c(r)$ at intermediate and high densities must be due to the inadequacy of the superposition approximation at these densities. In the superposition approximation the three-point distribution function $g^{(3)}(\mathbf{r}_1, \mathbf{r}_2, \mathbf{r}_3)$ is replaced by the product of three two-point functions:

$$g^{(3)} = g(\mathbf{r}_1, \mathbf{r}_2) g(\mathbf{r}_2, \mathbf{r}_3) g(\mathbf{r}_1, \mathbf{r}_3) . \tag{4.1}$$

If ζ denotes the reduced distance at which the first-neighbor peak occurs in $g(r)$ ($\zeta > 1$), then at high densities the superposition approximation exaggerates the probability of finding a triplet of particles at points ($\mathbf{r}_1, \mathbf{r}_2, \mathbf{r}_3$) all of which lie within a distance ζ of each other.[78,79] This "overcounting" of the number of closely spaced triplets produces a first-neighbor peak in the pair correlation function which is larger than it should be. Moreover, the effect of this large first peak so propagates in the BGYK equation that it produces structure in $g(r)$ at distances r well beyond the first peak position. At low densities the superposition approximation is adequate and the BGYK correlation functions at $T = 1.65$ are comparable to the PY and HNC functions (Fig. 2a). Note from Table I that B_g(BGYK) is within 2% of B_g(PY) or B_g(HNC) for $n < 0.10$. As the density increases the long-range structure in the BGYK $g(r)$ induced by the triplet overcounting becomes significant and B_g(BGYK) quickly drops well below B_g(PY) or B_g(HNC). Of course even at high densities the superposition approximation is not completely without merit, and as the density increases beyond $n = 0.40$, B_g(BGYK) does begin to increase. But the triplet overcounting is responsible for the higher density, $n \cong 0.40$, of the minimum in B_g(BGYK) relative to the density, $n \cong 0.25$, at which B_g(PY) and B_g(HNC) are minimal. The unduly long-range structure in $g(r)$ at high densities due to the inadequacy of the superposition approximation also accounts for the slow rise in B_g(BGYK) as n increases beyond 0.40.

Further insight into the breakdown of the superposition approximation at high density can be gained by examining the sensitivity of the BGYK solution $y(r)$ to the cutoff ℓ_1. It is most useful to contrast the BGYK and HNC equations, since they are alternative approximations for ln $y(r)$. The dependence of the solution $y(r)$ on the cutoff ℓ_1 is determined through the residuals $R[y; r]$ as outlined in Section III. If the linear sensitivity equation (3.5) is written

$$0 = \int_0^{\ell_1} ds\, J(r, s) \frac{\partial y(s)}{\partial \ell_1} + F(r)\,, \tag{4.2}$$

then the inhomogeneous piece $F(r)$ has the form

$$F(r) = \frac{2\pi n \ell_1}{r} \left\{ [g(\ell_1^-) - 1] \int_{\ell_1 - r}^{\ell_1} ds\, s\, c(s) + c(\ell_1^-) \int_{\ell_1 - r}^{\ell_1} ds\, s[g(s) - 1] \right\} \tag{4.3}$$

for the HNC approximation, and the form

$$F(r) = \frac{2\pi n \ell_1}{r} \left\{ [g(\ell_1^-) - 1] \int_{\ell_1 - r}^{\ell_1} ds\, g(s) u'(s)[s^2 - (\ell_1 - r)^2] \right\} \tag{4.4}$$

for the BGYK approximation. In (4.3) g and c refer to the pair and direct correlation functions determined from the HNC solution $y(r)$, while in (4.4) g refers to the pair correlation function computed from the BGYK solution $y(r)$. In the HNC case, $c(\ell_1^-)$ is very small and the second term on the right-hand side of (4.3) can be neglected. It is instructive to compare the inhomogeneous piece at $r = \ell_1$:

$$F_{\mathrm{HNC}}(\ell_1) = 2\pi n [g(\ell_1^-) - 1]_{\mathrm{HNC}} \int_0^{\ell_1} ds\, s\, c_{\mathrm{HNC}}(s) \tag{4.3$'$}$$

and

$$F_{\mathrm{BGYK}}(\ell_1) = 2\pi n [g(\ell_1^-) - 1]_{\mathrm{BGYK}} \int_0^{\ell_1} ds\, s^2 g_{\mathrm{BGYK}}(s) u'(s)\,. \tag{4.4$'$}$$

In both (4.3′) and (4.4′) the factor $[g(\ell_1^-) - 1]$ is most important near the density where B_g is minimal; that is, a minimum in the bulk modulus indicates the presence of a nearby critical point where $g(r)$ is very long-ranged. At densities larger than the critical density, the factor $[g(\ell_1^-) - 1]$ is small when ℓ_1 is reasonably large (e.g., $[g(\ell_1^-) - 1] < 10^{-5}$ for

$\ell_1 \geq 9$), and $F(\ell_1)$ can be significantly different from 0 only if the integral term is large. In the HNC case, the integral over $0 < s < \ell_1$ of $s^2 c_{\text{HNC}}(s)$ is much the same at low, intermediate, and high densities, and so no unusual sensitivity to ℓ_1 arises at high densities. On the other hand, in the BGYK case the relevant integrand is $s^2 g(s)u'(s)$, which depends crucially on the position of the first-neighbor peak in $g(r)$ relative to the pair potential $u(r)$. At high densities we expect too large a first-neighbor peak in $g(r)$ under the superposition approximation, and thus we expect the sensitivity $\partial y(r)/\partial \ell_1$ to be larger at densities greater than the critical density ($n > n_c$) than the sensitivity the same distance below the critical point ($n < n_c$). The change in the BGYK bulk modulus B_c for $\ell_1 = 6$, 12, or 18 seen in Table I at high densities ($n \geq 0.55$) is a result of high-density sensitivity of $y(r)$, and hence $c(r)$, to the cutoff ℓ_1.

The reduced energy per particle e/T (1.5) and the reduced virial pressure P^v (1.4) are plotted against the density n and molecular volume $v = 1/n$, respectively, at $T = 1.65$ in Fig. 3 for all three approximations. We see that the PY and HNC results are in agreement, but both differ from the BGYK results. The differences between the virial pressure P^v and the compressibility pressure P^c are illustrated in Fig. 4. P^c is obtained by integrating the bulk modulus B_c with respect to the density:

$$P^c(n) = T \int_0^n dn' \, B_c(n') \,. \tag{4.5}$$

The effect of changing the potential cutoff ℓ_0 is displayed in Fig. 5 for the PY cause, and similar behavior is found under the HNC or BGYK

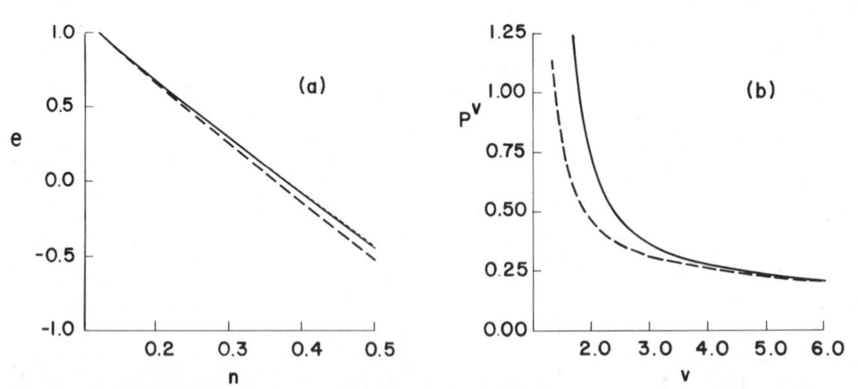

Fig. 3. (a) Reduced energy per particle e as a function of density n and (b) the reduced virial pressure P^v as a function of the molecular volume $v = 1/n$ at $T = 1.65$ under the PY (—), HNC (-----), and BGYK (– –) approximations. The PY and HNC results for P^v are indistinguishable.

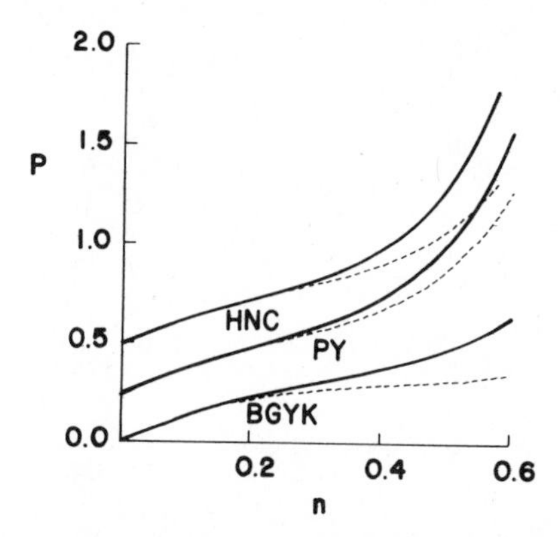

Fig. 4. At $T = 1.65$ the virial pressure P^v (—) and the compressibility pressure P^c (---) are equal at low density, but differ at high density under all three closure approximations. The PY and HNC curves have been shifted up by 0.25 and 0.50, respectively.

closures. The minimum in $B_c(n; \ell_0 = 10)$ is shifted to a slightly lower density and lies above the minimum in $B_c(n; \ell_0 = 4)$. This implies that the critical point (n_c, T_c) is at a higher temperature and lower density when $\ell_0 = 10$ than when $\ell_0 = 4$. Such shifts of the critical temperature and density with a change in the range of the attractive piece of the pair potential are well known in the context of approximate integral equations for the correlation functions,[34,44,47] and have also been documented in a Monte Carlo study of the square-well fluid.[80]

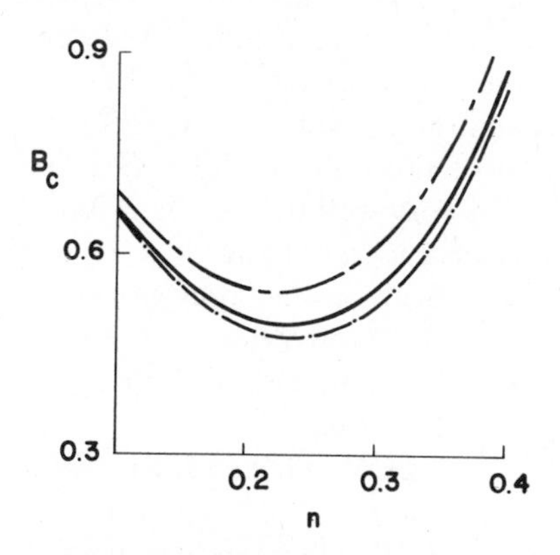

Fig. 5. The minimum in the PY bulk modulus B_c ($T = 1.65$) is shifted to a bit larger value and lower density as the distance ℓ_0 at which the pair potential $u(r)$ is truncated changes from 4 (–··–), to 6 (—), to 10 (–––); the truncated potential is also shifted so that $u(\ell_0) = 0$. Similar results obtain under the HNC or BGYK closure conditions.

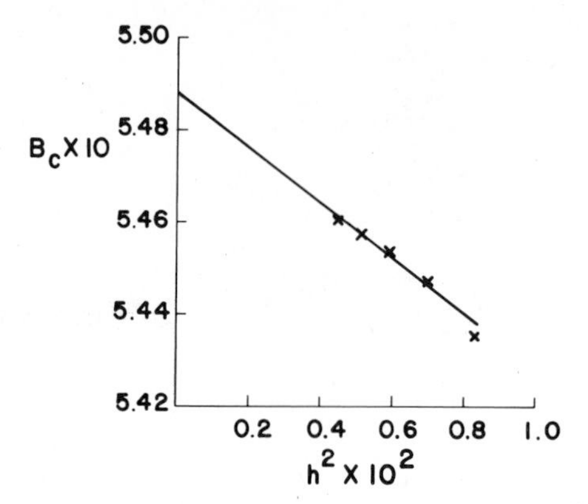

Fig. 6. PY bulk modulus B_c exhibits a linear dependence on h^2, the square of the subdomain size. For $h \leq \frac{1}{15}$, the other thermodynamic quantities display a similar linear dependence on h^2.

To conclude this section we comment on the level of error introduced in our results by the discretization, that is, by the fact that the subdomain size h is greater than 0. Earlier we remarked that the finite-element representation and Swartz–Wendroff approximation errors should be of order h^2. To be more precise, we expect the difference $(y(r) - \mathbf{y} \cdot \mathbf{H}(r))$ between the exact solution and our finite-element solution to be $A(r)h^2 + B(r)h^3 + \cdots$, where $A(r)$, $B(r), \ldots$ are bounded functions of r which decay rapidly to 0 once r is greater than several units. [Once r is greater than several molecular diameters we expect to see a monotonic decay in $y(r)$ which, even if long-ranged, is well represented by piecewise-linear basis functions.] Indeed, we observe that at the common points of a fine mesh of spacing h_1 and a coarser mesh of spacing h_2, where $h_1 < h_2 \leq \frac{1}{10}$, the values of the expansion coefficient $\mathbf{y}$ agree to at least the fourth decimal place. A plot of the bulk modulus B against the square of the subdomain size, h^2, is fit very well by a straight line (Fig. 6). This confirms that the error in the integrand, and hence in $y(r)$ itself, is of order h^2 since the Simpson's rule quadrature would only introduce errors of order h^4. We estimate that (for fixed ℓ_1) our values of the correlation functions and thermodynamic functions have four significant figures.

V. LOW TEMPERATURE

Because the behavior of the low-temperature correlation functions is distinctly different for each of the PY, HNC, and BGYK closure conditions, we present first our PY results, then the HNC results, and finally the BGYK results, comparing the approximations as we proceed.

A. PY Case

At $T = 1.40$ we find a single branch of PY solutions $y(r; n)$ as the density is increased from 0.05 to 0.65. The correlation functions (Fig. 7) at low ($n = 0.10$) and high ($n = 0.50$) densities are comparable to those calculated at $T = 1.65$. At $n = 0.30$, $g(r)$ is longer-ranged at $T = 1.40$ than at $T = 1.65$ since the former temperature is closer to, but still above, T_c(PY). Besides calculations for the truncated and shifted Lennard-Jones (L-J) potential, we made calculations for the truncated but unshifted L-J potential and for a full L-J potential (Table II) in order to better calibrate our algorithm against those reported in the literature. Watts, who used the Baxter transformation[81] to solve the PY equation, published B_c results for the unshifted, truncated L-J potential.[34] At low densities our results agree with his to better than 0.1%, but the difference increases as density increases, with a discrepancy of 4% at $n = 0.60$. When we changed the integral cutoff ℓ_1 from 6 to 12 or decreased the subdomain size h at fixed $\ell_1 = 6$, our bulk modulus B_c was virtually unchanged; so the discrepancy with Watts' results is not an artifact of our numerical approximation. Moreover, the very good agreement between B_c and B_g at high densities, even at $\ell_1 = 6$, indicates that our solutions $g(r)$ and $c(r)$ are reliable. The source of the difference, at high density, between our results and those of Watts is not clear.

In Table II we compare present results on B_c for a full L-J potential with those of Throop and Bearman[29] (which are nearly identical to the results of Levesque[47]). In these calculations the assumptions $g(r) = 1$ and $c(r) = 0$ for $r \geq \ell_1$ were made in solving the PY equation, but the

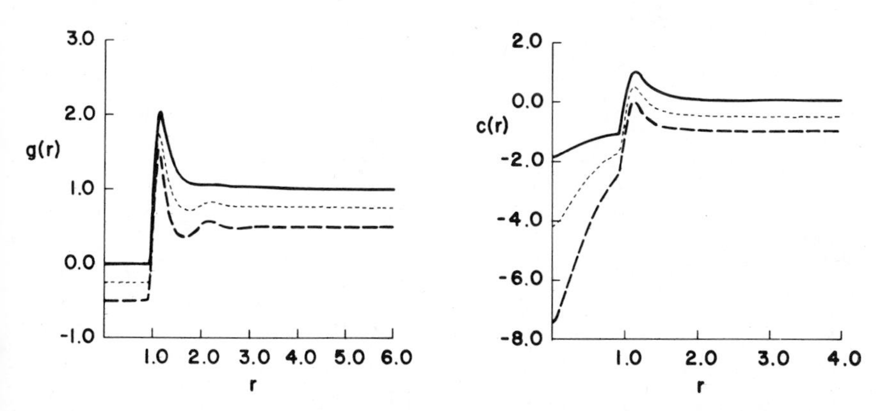

Fig. 7. PY functions $g(r)$ and $c(r)$ at density n equal to 0.10 (—), 0.30 (---, g shifted down by 0.25, c shifted down by 0.50), and 0.50 (– –, g shifted down by 0.50, c shifted down by 1.00) at $T = 1.40$.

TABLE II

PY Compressibilities B_c and B_g for $0.5 \leq n \leq 0.65$ and $T = 1.40$

n	B_c/B_g [a] $\ell_1 = 6$	B_c/B_g [a] $\ell_1 = 12$	B_c/B_g [b] (present)	B_c (ref. 34)	B_c/B_g [c] (present)	B_c (ref. 29)
0.05	0.7370 / 0.7382	0.7370 / 0.7370	0.7342 / 0.7342	0.7345	0.7314 / 0.7343	0.7315
0.10	0.5171 / 0.5262	0.5171 / 0.5172	0.5115 / 0.5125	0.5122	0.5061 / 0.5208	0.5062
0.15	0.3465 / 0.3758	0.3465 / 0.3473	0.3381 / 0.3439	0.3392	0.3300 / 0.3699	0.3303
0.20	0.2310 / 0.2897	0.2308 / 0.2348	0.2199 / 0.2372	0.2215	0.2091 / 0.2835	0.2098
0.25	0.1745 / 0.2612	0.1742 / 0.1837	0.1610 / 0.1925	0.1635	0.1476 / 0.2544	0.1487
0.30	0.1787 / 0.2836	0.1788 / 0.1914	0.1635 / 0.2033	0.1679	0.1469 / 0.2758	0.1486
0.35	0.2497 / 0.3603	0.2500 / 0.2613	0.2328 / 0.2720	0.2405	0.2133 / 0.3505	0.2157
0.40	0.4044 / 0.5092	0.4045 / 0.4119	0.3855 / 0.4173	0.3989	0.3633 / 0.4949	0.3671
0.45	0.6705 / 0.7615	0.6706 / 0.6744	0.6502 / 0.6722	0.6727	0.6254 / 0.7374	0.6311
0.50	1.0864 / 1.1609	1.0864 / 1.0878	1.0653 / 1.0788	1.1026	1.0378 / 1.1137	1.0463
0.55	1.6999 / 1.7617	1.6999 / 1.6995	1.6793 / 1.6872	1.7405	1.6490 / 1.6603	1.6617
0.60	2.5709 / 2.6223	2.5709 / 2.5676	2.5521 / 2.5571	2.652	2.5192 / 2.4011	2.5379
0.65	3.7732 / 3.7775	3.7732 / 3.7631	3.7587 / 3.7597		3.7230 / 3.3158	

[a] A truncated ($\ell_0 = 6$) and shifted Lennard-Jones potential at two different values of ℓ_1.

[b] A truncated ($\ell_0 = 6$) but unshifted Lennard-Jones potential at $\ell_1 = 6$ for which a comparison with Watts[34] can be made.

[c] A full Lennard-Jones potential at $\ell_1 = 6$ for which a comparison with Throop and Bearman[29] can be made.

asymptotic forms $c(r) = e^{-u(r)} - 1$ and $g(r) = e^{-u(r)}$ were used in evaluating B_c and B_g, respectively.[82] In this case the percentage difference between the present and previously published results for B_c has a maximum value of 1.1% at $n = 0.30$. In spite of such excellent agreement, the difference between B_c and B_g is large at intermediate densities, indicating that the cutoff approximation $g(r) = 1$ for $r \geq \ell_1 = 6$ is not really very good.

In addition to calculations at $T = 1.40$, we obtained numerical solutions of the PY equation along the isotherms $T = 1.30, 1.25, 1.20, 1.15$, and 1.10. According to our calculations $T = 1.30$ is quite close to the critical temperature, and the other temperatures are subcritical. Because we are interested in the solutions and the features of the solution space in the coexistence or spinodal regions of the (n, T) parameter plane, we choose the results at $T = 1.15$ to illustrate the typical behavior at subcritical temperatures, well removed from the critical temperature. When $T = 1.15$ the function $y(r; n)$ at low density is much like the low-density $y(r)$ at $T = 1.65$ or $T = 1.40$. As n increases at the subcritical temperature, $y(r)$ becomes longer-ranged and $B_c(n)$ decreases until we reach a (vapor) spinodal density n_{sv} at which $B_c(n_{sv}) = 0$. Continuing on the path of increasing density on the solution branch, past the spinodal point $n = n_{sv}$, we very shortly find a turning point n_{tv} in the density: there the solution branch $y(r; n)$ turns back to lower densities. In fact the parameter $y(r = 0)$ was chosen as the adaptive continuation parameter to get around the turning point in the density (Fig. 8). Once past the spinodal point, the solutions $y(r)$ are very long-ranged, and $B_c(n)$ is roughly constant at a slightly negative value (Fig. 9). The cutoff approximation, $y(r) = 1$ if $r \geq \ell_1$, is undoubtedly a source of substantial quantitative error past the spinodal point, but the qualitative behavior of the solution branch is to be believed; different choices for the values of $\ell_1 \geq 12$ or $h \leq \frac{1}{15}$ slightly altered the exact location of the spinodal and turning-point densities, but did not change the qualitative form of the PY solution branch. The subcritical vapor-phase branch of the PY solution $y(r; n)$ extends up to a spinodal curve where $B_c(n) = 0$, but then turns back in density rather than continuing across the spinodal region to high densities. If we start at a stable $(B_c > 0)$, liquid-phase solution and decrease n, $y(r; n)$ becomes longer-ranged and $B_c(n)$ decreases until we reach a (liquid) spinodal density $n_{s\ell}$ at which $B_c(n_{s\ell}) = 0$. This liquid-phase solution branch continues past $n_{s\ell}$ as density is further decreased all the way to low density, with $B_c(n)$ roughly constant and negative. Again, although quantitatively the solutions $y(r; n)$, $n < n_{s\ell}$, are poor because they are becoming very long-ranged and the cutoff approximation is inadequate, the qualitative behavior of the solution branch is to be believed. The subcritical liquid-

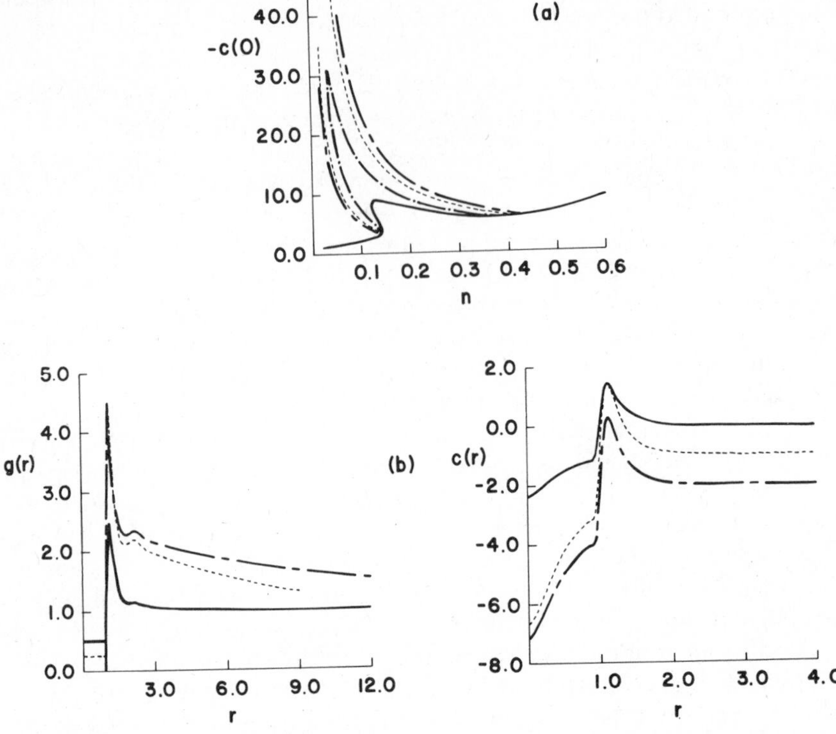

Fig. 8. At the subcritical temperature $T = 1.15$ and fixed potential cutoff $\ell_0 = 6.0$, the turning point in the PY correlation-function space, as mapped into the $(n, c(0))$ plane, depends on the integral cutoff ℓ_1; the progression shown in (a) is ℓ_1 equal to 6 (—), 6.8 (–·–), 9 (---), and 12 (–––). For $n = 0.10$ and $T = 1.15$, (b) shows the stable, vapor-phase g and c solutions (—) which are independent of $\ell_1 \geq 6$, and the unstable-phase g and c at $\ell_1 = 9$ (---, g shifted up by 0.25, c shifted down by 1.00) and at $\ell_1 = 12$ (–––, g shifted up by 0.50, c shifted down by 2.00).

phase branch of the PY solution extends down to the spinodal density $n_{s\ell}$, and then continues below $n_{s\ell}$ to low densities (Fig. 8).

Even though the PY solution branches which start at low and high stable densities appear to be disconnected, they do join at some very low-density, extremely long-ranged solution. The evidence for such a connection of the branches is shown in Fig. 8 where the solutions $y(r; n)$, as projected in the $c(0)$-n plane, are plotted for several values of the cutoff ℓ_1. Figure 8b shows that, at fixed density, the long-ranged solutions do decay; for example, at $n = 0.10$ $g(r = 9; \ell_1 = 12) < g(r = 12, \ell_1 = 12)$. In Fig. 9, we observe that as the cutoff ℓ_1 increases, the vapor spinodal

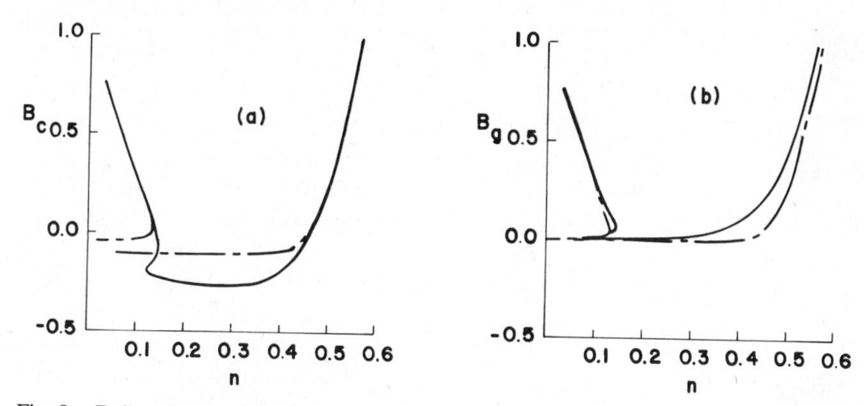

Fig. 9. Bulk moduli B_c (a) and B_g (b) as functions of the density n at $T = 1.15$ for $\ell_1 = 6$ (—) and $\ell_1 = 12$ (---).

density n_{sv} and the vapor turning-point density n_{tv} seem to approach each other. For finite ℓ_1, we find $B_c \lesssim 0$ and $B_g \gtrsim 0$ for solutions that continue past the vapor or liquid spinodal densities, although both B_c and B_g for such solutions decrease as ℓ_1 increases. At all subcritical temperatures $T < T_c \cong 1.30$ for which we made calculations, the PY solution branches are similar to the branches at $T = 1.15$. As the temperature increases from 1.15 to 1.30, the cutoff ℓ_1 must be extended to large distances before the "nearly" disconnected aspect of the vapor and liquid solution branches emerges in place of a smooth, connected, single branch (cf. Fig. 9); as T decreases from 1.15 the nearly disconnected feature becomes apparent even at a cutoff $\ell_1 = 6$.

When following the solution branch past the spinodal densities into the region where $g(r)$ is very long-ranged, we cannot expect B_g and B_c to agree. The pair correlation function $g(r)$ shows the local density correlations in the fluid, and the isothermal compressibility, which is proportional to B_g^{-1}, provides a macroscopic, thermodynamic measure of density correlations based on the microscopic information in $g(r)$. B_g^{-1} is necessarily positive in any exact statistical mechanical calculation. On the other hand, the direct correlation function $c(r)$ indicates the stability of the fluid with respect to local density fluctuations, and the quantity B_c is an index of the macroscopic, thermodynamic stability of the fluid.[49,52] Of course correlations and fluctuations are related, and B_c is usually identified as equal to B_g through the OZ equation (1.1). A necessary condition, however, for B_c to equal B_g is that the integral of $r^2[g(r) - 1]$ over $0 < r < \infty$ converge. As $g(r)$ becomes very long-ranged and B_g^{-1} increases without bound, the condition is violated and B_g^{-1}, a measure of correlation, can no longer be clearly identified with B_c, a measure of stability.

Because our calculations for B_g and B_c are based on approximate correlation functions, the fact that B_g is always found to be positive reflects a degree of reliability in the approximations; the fact that B_c is negative at some densities indicates that the homogeneous fluid is unstable. Indeed, any theory for $g(r)$ which results in $B_g^{-1} < 0$ is unacceptable. This does not mean that any approximate theory which leads to a negative bulk modulus is unacceptable, but only that the negative bulk modulus should be viewed as B_c. For example, at subcritical temperatures the van der Waals equation for the pressure as a function of density implies that the bulk modulus is negative for some values of the density. In analyzing a microscopic model for the van der Waals fluid, van Kampen[83] showed that the bulk modulus calculated is B_c; that is, at a microscopic level the bulk modulus of the van der Waals equation is that calculated through $c(r)$ not $g(r)$. Further analysis shows that even at densities for which B_c is negative, $g(r)$ is defined and decays asymptotically to unity, but decays so slowly that the integral B_g^{-1} is not defined.

In a recent study[37] of the PY critical region, negative compressibilities B_c were reported at subcritical temperatures. The cutoff was fixed at $\ell_1 = 6$, which is tolerable for $T > T_c$, but is inadequate for $T < T_c$. Accordingly, the smooth, regular behavior reported for $B_c(n)$ when $T \lesssim T_c$ is an artifact of the cutoff approximation.[84] Del Rio and Mier y Terán[36] also found negative compressibilities along a subcritical ($T = 0.8$) isotherm when solving the PY equation by a method of orthonormal-series expansion in a Hermite basis. They earlier[33] observed that a Hermite basis does not give correctly the long-tail behavior of the pair correlation function and has an effect similar to the use of a short integration range; hence the regular behavior they encounter is also effectively an artifact of choosing ℓ_1 too small.

The solution space we find for the PY equation, as mapped out in the $c(0)$-T projection (Fig. 8), is different from the solution space determined by Watts[34] for a closely similar pair potential. When $T < T_c$ Watts found the vapor-phase solution branch has a turning point in the density much like the one found here, but he also found a turning point in the density along the liquid-phase solution branch. We suspect that the difference is due to the Baxter transformation[81] which Watts employed. Under the assumptions that $c(r)$ is finite-ranged and that the integral $\int dr\, r^2[g(r) - 1]$ is absolutely convergent, the Baxter transformation reduces the OZ equation, which involves a convolution integral over an infinite domain, to an integral equation that requires only finite domains. But at the vapor and liquid spinodal densities n_{sv} and $n_{s\ell}$, $B_g \cong 0$ and one of the assumptions underlying the transformation is not justified; indeed when Watts' solution branches are continued past the spinodal points, the resulting pair correlation functions do not decay to unity as r increases and do apparently violate the assumption that the integral of $r^2[g(r)-1]$ is absolutely convergent. The appearance of solutions that violate the assumptions connecting the transformed Baxter equation to the original OZ equation suggests that such solutions originate in the transformation rather than being associated with a true solution to the original PY–OZ equation.

Our results differ from those of Baxter[85] who obtained, by solving via the Baxter transformation, PY solutions for the adhesive or sticky hard-sphere pair potential. The sticky-sphere solution space can be mapped out in a λ-n projection, where the dimensionless parameter λ indexing the different PY solutions along a given isotherm is related to $c(0)$ [see Eq. (11) of Baxter's work[85]]. The λ-n plot is much the same as the $c(0)$-n plot of Watts, with turning points in the density on both the vapor- and liquid-phase branches of the subcritical solutions. As the density is decreased along the liquid-phase solution branch at $T < T_c$, a spinodal point $(\lambda_{s\ell}, n_{s\ell})$, at which $B_c(n_{s\ell}) = 0$, is reached before the liquid-branch turning point $(\lambda_{\ell t}, n_{\ell t})$ is reached. Baxter did recognize, however, that the solutions which continue beyond $(\lambda_{s\ell}, n_{s\ell})$ are unphysical and are associated with a breakdown of the Baxter transformation; the solutions beyond the turning point $(\lambda_{\ell t}, n_{\ell t})$ have the additional unphysical feature that $g(r)$ is unbounded at some point r, $0 \leq r < \infty$. As the density increases from 0 at $T < T_c$ along the vapor-phase solution branch, a turning point (λ_{tv}, n_{tv}) is reached at which the pair correlation function becomes unbounded although integrable [i.e., $B(n_{tv}) > 0$]. Because of this unphysical singularity in $g(r)$, the bulk modulus $B_c(n)$ terminates at a square-root branch point on the vapor-phase branch before the spinodal point (λ_{sv}, n_{sv}) is reached.[86] Along no subcritical solution branch did we detect any evidence of $g(r)$ becoming unbounded while $B_c(n)$ was still positive. [Even if $g(r)$ did become unbounded the finite-element analysis could still be applied by introducing singularity basis functions;[73] the Jacobian $J(r, s)$ would be well behaved even in the presence of unbounded but integrable singularities in $y(r)$.] We conclude that there are no singularities in $B(n)$ which occur before the spinodal densities n_{sv} or $n_{s\ell}$ are reached for the PY solutions in the case of the truncated and shifted L-J pair potential; that is, it appears that the spinodal density n_{sv} is always reached before or at the turning-point density n_{tv}. We believe that the same conclusion applies in the case of potentials similar to the truncated and shifted L-J potential (e.g., the full L-J square-well potential), but the sticky-sphere PY results show that n_{tv} may be reached before n_{sv}. The general feature of the subcritical, low-density solution branch of the PY equation is the presence of a spinodal density n_{sv} and a turning-point density n_{tv}. The exact location of n_{sv} relative to n_{tv}, which as noted earlier (Figs. 8 and 9) depends on the cutoff ℓ_1, also has a subtle dependence of the form of the pair potential.[87]

B. HNC Case

Like the PY equation, the HNC equation has been the subject of considerable study.[39–44,47] Most of the numerical studies[39–43,47] are based on a traditional Picard iteration for solving the HNC integral equation, although, as in the PY case, Watts[44] is exceptional in that he used a Newton method on a form of the HNC equation obtained by using Baxter's transformation. Over 20 years ago Green[88] concluded that the HNC equation displays anomalous critical behavior (see also the comments by Fisher and Fishman[93] on the exponent η for the HNC equation), and the numerical studies[42–44] confirmed his conclusion. Again we must emphasize that our interest is in the subcritical coexistence region; the anomalous behavior at the critical point simply points to possible anomalies at subcritical temperatures. We calculated HNC solutions along the isotherms $T = 1.40$, 1.38, 1.35, 1.34, 1.30, and 1.15. We find the critical temperature $T_c \cong 1.35$; we choose the isotherm $T = 1.40$ to illustrate the characteristic behavior along (slightly) supercritical isotherms and the isotherm $T = 1.30$ as a typical example of subcritical isotherms.

As n increased from 0.05 to 0.60 at $T = 1.40$, we find the low-density solution (Fig. 10) connected to the high-density solution by a single,

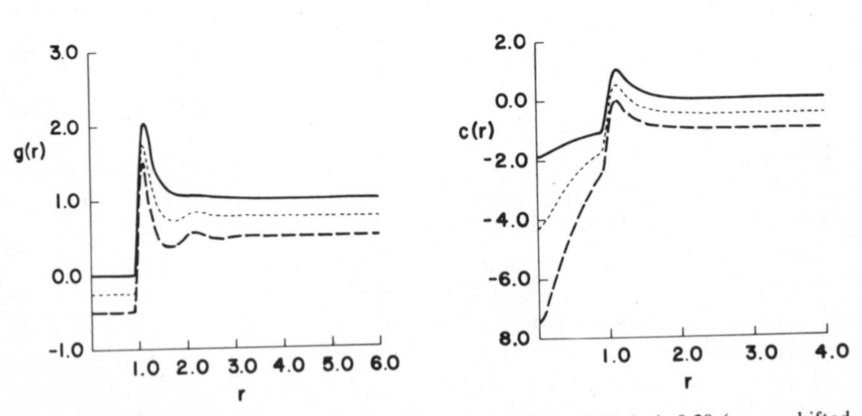

Fig. 10. HNC functions $g(r)$ and $c(r)$ at density n equal to 0.10 (—), 0.30 (---, g shifted down by 0.25, c shifted down by 0.50), and 0.50 (––, g shifted down by 0.50 and c shifted down by 1.00) at $T = 1.40$.

TABLE III
HNC Compressibilities B_c and B_g for $0.05 \le n \le 0.60$ and $T = 1.40$

n	B_c/B_g [a] $\ell_1 = 6$	B_c/B_g [a] $\ell_1 = 12$	B_c/B_g [b] (present)	B_c (ref. 44)	B_c/B_g [c] (present)	B_c (ref. 47)
0.05	0.7364/0.7376	0.7364/0.7364	0.7336/0.7352	0.734	0.7308/0.7318	0.731
0.10	0.5122/0.5217	0.5122/0.5123	0.5065/0.5177	0.507	0.5010/0.5061	0.501
0.15	0.3298/0.3620	0.3293/0.3303	0.3209/0.3570	0.322	0.3124/0.3296	0.312
0.20	0.1942/0.2630	0.1897/0.1962	0.1818/0.2571	0.179	0.1666/0.2099	0.166
0.25	0.1171/0.2223	0.0992/0.1196	0.1015/0.2158	0.077	0.0746/0.1510	0.074
0.30	0.1043/0.2323	0.0899/0.1166	0.0869/0.2254	0.065	0.0618/0.1545	0.062
0.35	0.1505/0.2874	0.1470/0.1691	0.1313/0.2797		0.1116/0.2052	0.112
0.40	0.2652/0.3999	0.2651/0.2803	0.2436/0.3903		0.2247/0.3111	0.225
0.45	0.4772/0.5994	0.4773/0.4859	0.4528/0.5872		0.4331/0.5048	0.433
0.50	0.8249/0.9283	0.8249/0.8289	0.7975/0.9125		0.7775/0.8287	0.778
0.55	1.3500/1.4359	1.3500/1.3513	1.3198/1.4152		1.3007/1.3241	1.302
0.60	2.0961/2.1720	2.0961/2.0948	2.0630/2.1451		2.0463/2.0247	2.049

[a] A truncated ($\ell_0 = 6$) and shifted Lennard-Jones potential at two different values of ℓ_1.
[b] A truncated ($\ell_0 = 6$) but unshifted Lennard-Jones potential at $\ell_1 = 6$ for which a comparison with Watts[44] can be made.
[c] A full Lennard-Jones potential at $\ell_1 = 8$ for which a comparison with Levesque[47] can be made.

smooth solution branch. At intermediate densities $y(r)$ has a long-ranged tail due to the critical-point influence, and it is necessary to fix the cutoff ℓ_1 at a value $\ell_1 > 12$ to produce good agreement between B_c and B_g (see Table III). At low ($n \le 0.20$) and high ($n \ge 0.40$) densities the correlation functions are relatively insensitive to the cutoff and $B_c(\ell_1 = 6)$ is very close to $B_c(\ell_1 = 12)$ at these densities. At intermediate $(0.25 \leq n \leq 0.35)$ densities $y(r)$ is rather more sensitive to the choice of ℓ_1 because of the long-ranged tail in $g(r)$ [see Eqs. (4.3) and (4.3′)]. In Table III we also compare our results for the truncated but unshifted L-J potential with previously published results. The agreement is good on the whole, with the maximum differences occurring at the points where B_c is near its minimum and $g(r)$ is long-ranged. For the truncated but unshifted L-J potential we could improve our agreement with Watts[44] by increasing ℓ_1 from 6; when $B_c > 0$ the Baxter transformation which Watts used, although not exact for the HNC equation, does seem to account partially for the influence of the tail in $g(r)$. For the full L-J potential we set $\ell_1 = 8$ to facilitate the comparison with Levesque.[47] We also made calculations for the full potential with $\ell_1 = 6$ and found excellent agreement with Levesque at low and high densities; at intermediate densities there was a large discrepancy [e.g., $B_c(\ell_1 = 6,\ n = 0.30) = 0.0703$] because the choice $\ell_1 = 6$ cuts off an appreciable portion of the tail in $g(r)$.

At the subcritical temperature $T = 1.30$ there are two separate solution branches of the HNC equation, as indicated in Fig. 11. There are spinodal densities n_{sv} and $n_{s\ell}$ at which $B_c = 0^+$ on the vapor-phase and liquid-phase branches, respectively; on each branch there is also a turning point in the density, n_{tv} or $n_{t\ell}$. If we start at a stable $(B > 0)$, high-density $(n = 0.50)$ subcritical solution $y(r)$ and follow the liquid-phase solution branch as the density is decreased, then the spinodal density $n_{s\ell}$ is reached first. Continuing along this solution branch beyond the spinode, we then reach

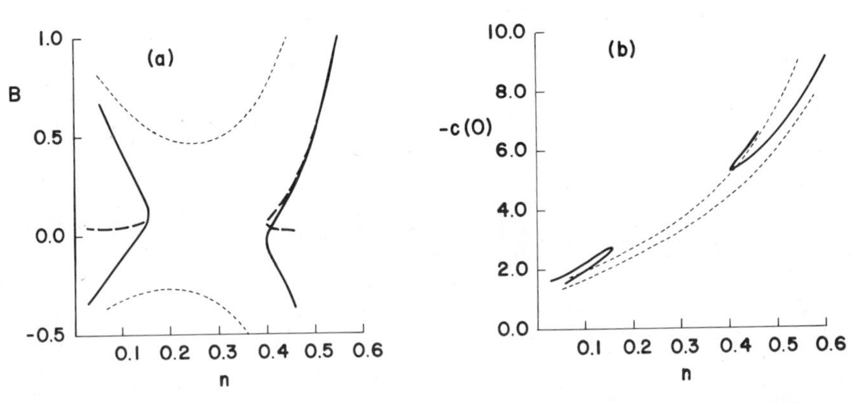

Fig. 11. As shown in (a), both HNC branches of $B_c(n)$ at $T = 1.30$ (—) have positive or stable and negative or unstable portions, although $B_g(n, T = 1.30)$ (-–) is always positive; in contrast, $B_c(n, T = 1.65)$ (---) has a strictly positive branch and a strictly negative branch. The HNC solution space in the $(n, c(0))$ plane is shown in (b) at $T = 1.30$ (—) and $T = 1.65$ (---); the lower $T = 1.65$ curve corresponds to $B_c > 0$ and the upper to $B_c < 0$.

the turning point $n_{t\ell}$. The compressibilities $B_c(n)$ and $B_g(n)$ are in fairly good agreement until the spinode is reached. Beyond the spinode, where $g(r)$ has a long-ranged tail, the discrepancy between B_c and B_g increases because of the long-ranged tail in $g(r)$. Under the HNC approximation $c(r) \cong [g(r)-1]^2$ when $r > \ell_0$ (ℓ_0 is the range of the pair potential); accordingly, $B_c(n)$ takes more negative values as we move along the solution branch past the spinodal point. For all the values of $\ell_1 \leq 12$ at which we made calculations, we found $n_{t\ell} < n_{s\ell}$ although the exact locations of $n_{t\ell}$ and $n_{s\ell}$ did shift slightly with ℓ_1. We believe that in the limit $\ell_1 \to \infty$, the spinodal density $n_{s\ell}$ is reached before the turning-point density $n_{t\ell}$. In contrast, if we start at a stable, low-density ($n = 0.05$) subcritical solution $y(r)$ and follow this vapor-phase solution branch as the density is increased, then the turning-point density n_{tv} is reached before the spinodal density n_{sv}. That is, the bulk modulus $B(n)$ is singular, as a function of n, at a density n_{tv} for which $B_c(n_{tv}) > 0$. This is exactly the situation found for the vapor-phase solution branch of the PY sticky-sphere equation. In the present HNC case, however, the function $y(r)$ is apparently well behaved in the vicinity of the turning point, and the fact that n_{tv} is reached before n_{sv} is an "intrinsic" feature of the HNC solution space for the given potential. (In view of the PY results, it is not likely that this feature is to be expected in general. For some choice of the pair potential it is certainly possible that the spinodal density n_{sv} is reached before the turning-point density n_{tv}.[89]) This "intrinsic" feature is very sensitive to the cutoff ℓ_1; for the choice $\ell_1 = 6$, the spinodal density is reached before the turning-point density (Fig. 12). As ℓ_1 changes, the solutions $y(r)$ undergo minor changes, with the values of $y(r)$ in the ranges $0 < r < 1$ and $\ell_1 - 2 < r < \ell_1$ most sensitive to the choice of ℓ_1. But $B_c(n; \ell_1)$, which is an integral measure of the solution $y(r)$, is highly sensitive to small changes in $y(r)$, particularly in the tail. Thus it is appropriate to view the spinodal density n_{sv}, which is determined by integral measure $B_c(n)$, as shifting relative to the turning-point density n_{tv}, which is fixed by the structure of the HNC equation.

The appearance of two disconnected solution branches of the HNC equation at subcritical temperatures is virtually independent of the cutoff ℓ_1. This is in contrast to the subcritical PY results where two disconnected branches arise as distortions of a single solution branch as ℓ_1 increases (Figs. 8 and 9). With the HNC equation a single solution branch is found when $T < T_c$ if the cutoff ℓ_1 for the range of integration is less than some critical value $\ell_1^c(T)$, but if $\ell_1 > \ell_1^c(T)$ then two disconnected branches are found. There is no continuous crossover between the two types of behavior. Moreover, the critical cutoff $\ell_1^c(T)$ is less than 6, which was the minimum cutoff used, once T is only slightly below T_c. Only at tem-

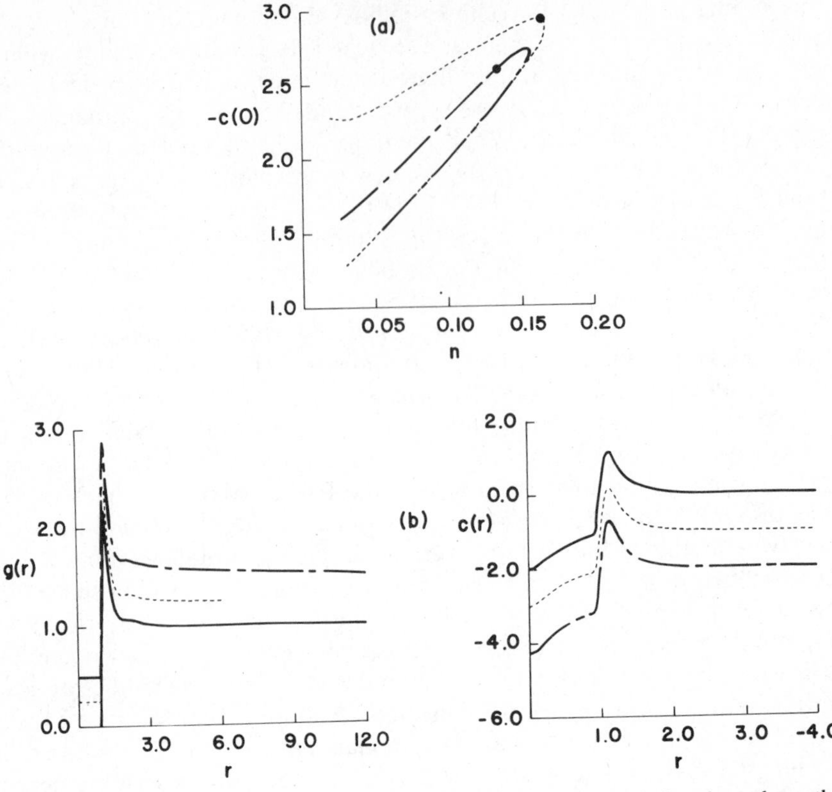

Fig. 12. Location of the turning point on the low-density, HNC solution depends on the cutoff ℓ_1, as exhibited in (a) for $\ell_1 = 6$ (—) and $\ell_1 = 12$ (---); on each curve the spinodal point is marked (●). For $n = 0.10$ and $T = 1.30$, (b) shows the stable, vapor-phase g and c solutions (—) which are independent of $\ell_1 = 6$, and the unstable-phase g and c at $\ell_1 = 6$ (---, g shifted up by 0.25, c shifted down by 1.0) and at $\ell_1 = 12$ (–-–, g shifted up by 0.50, c shifted down by 2.0).

peratures close to $T_c \cong 1.35$ (e.g., $T = 1.34$) did we see just a single solution branch [on which $B_c(n) < 0$ for intermediate densities] at $\ell_1 = 6$ which changed into two disconnected solution branches as ℓ_1 was increased; at $T = 1.30$ two solution branches were found even when $\ell_1 = 6$. Under the HNC approximation both $c(r)$ and $g(r)$ can have long-ranged tails even if $\phi(r)$ is strictly finite-ranged, in contrast to the PY approximation in which only $g(r)$ can have a long-ranged tail. So we expect the HNC equation and its solutions to depend more on the structure of $y(r)$ near $r \cong \ell_1$ and hence to suffer less from artifacts induced by the cutoff for $r > \ell_1$. By this reasoning we expect the HNC critical point to have a more pronounced singular behavior than that of the PY critical point, and this is indeed the case.[42,43]

Using the Baxter transformation, Watts[44] found a solution space for the HNC equation which, as mapped out in the $c(0)$-n plane, is again quite similar to our solution space (Fig. 11). He did not find spinodal points, however, on either the vapor- or liquid-phase solution branches; the bulk modulus $B_c(n)$ reaches a positive minimum but then increases as he follows the solution on past the turning point. We judge that the differences between Watts' results and ours arise from a breakdown of the Baxter transformation in the region where $g(r)$ and $c(r)$ are becoming long-ranged. On the other hand, like Watts, we were able to take an unstable $[B_c(n) < 0]$, low-density HNC solution at a subcritical temperature, continue this solution in temperature at fixed density to a supercritical temperature $T > T_c$, and then track a single unstable solution branch from low to high density at $T > T_c$ (Fig. 11). On this supercritical, unstable solution branch, the function $y(r)$ is very long-ranged and our cutoff approximation $y(r) = 1$ for $r \leq \ell_1$ is extremely poor; nevertheless, the solution $y(r)$ did satisfy the convergence criterion we required. In any case, it seems that for the HNC equation the critical point is, as Watts concluded, a saddle point in the solution space. Above T_c there are two solution branches, each of which runs from low to high densities; the upper branch corresponds to unphysical, extremely long-ranged solutions and the lower branch to physical solutions. Below T_c there are still two solution branches, but now one branch is restricted to low densities while the other branch is restricted to high densities; on each of these branches the upper half corresponds to unphysical solutions whereas the lower half corresponds to the physical solutions. This can be contrasted to the PY equation where the critical point is a point of crossover in the sensitivity of the solution. Above T_c there is one solution branch that is relatively insensitive to the cutoff ℓ_1; below T_c there is also a single solution branch but it is so sensitive to the cutoff ℓ_1 that it appears as two disconnected branches when ℓ_1 is large enough. The pronounced difference between the PY and HNC subcritical results is due to the difference between the PY and the HNC equations in accounting for the long-ranged tail in $y(r)$.

C. BGYK Case

Finally we turn to the BGYK equation (2.7), which is used in combination with the OZ equation (1.1) to obtain the correlation functions. We have made calculations along the isotherms $T = 1.50$, 1.40, 1.35, 1.30, 1.20, and 1.15 which are all subcritical since $T_c \approx 1.55$. Recent analytic work[91] indicates that in three dimensions the BGYK equation probably does not have a true critical point at which the correlation length and isothermal compressibility both diverge. We will return to this point later, but here it should be noted that our basis for classifying a temperature as subcritical is simply a change in the sign of $B_c(n)$ from positive at low density to negative at high density on the same solution branch. As we have already pointed out, the superposition approximation breaks down at high densities, where it induces unphysical direct correlation functions. Because $c(r)$ does not behave well at all densities, we adopted $B_g(n)$ as a primary thermodynamic function of interest, although we still monitored $B_c(n)$.

As a typical subcritical temperature we choose $T = 1.30$. The functions $g(r)$ and $c(r)$ are plotted at low, intermediate, and high densities in Fig. 13. In Fig. 14 we map the solution space in a $c(0)$-n plot and also plot B_c and B_g against n. We find that a single subcritical branch connects the low-density solution to the high-density solution, although there is a pair of density turning points on this curve. In the region of low to intermediate densities at which $B_c(n)$ first takes on negative values, $B_g(n)$ is flat and nearly zero. At high densities B_g increases although B_c remains negative due to the unphysical $c(r)$. We regard the appearance of a density domain over which $B_g \approx 0$ as a feature that further characterizes subcritical isotherms. In such a domain the pair correlation function is not only long-ranged but is distinctly sensitive to the cutoff ℓ_1. When ℓ_1 is increased we find $g(r)$ changes in the neighborhood of the first peak, $0 < r < 2$, and in the tail, $r \lesssim \ell_1$. For a fixed ℓ_1, we can define the vapor and liquid "spinodal" densities n_{sv} and $n_{s\ell}$ as the densities at which $B_g(n)$ flattens out as approached from the stable low- and high-density regions, respectively. Despite a small degree of ambiguity when so defined, n_{sv}

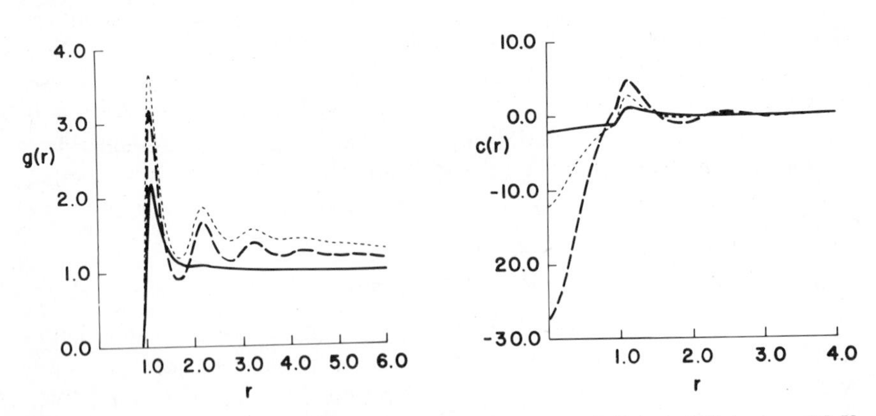

Fig. 13. BGYK functions $g(r)$ and $c(r)$ at density n equal to 0.10 (—), 0.30 (---), and 0.50 (– –) for the subcritical temperature $T = 1.30$.

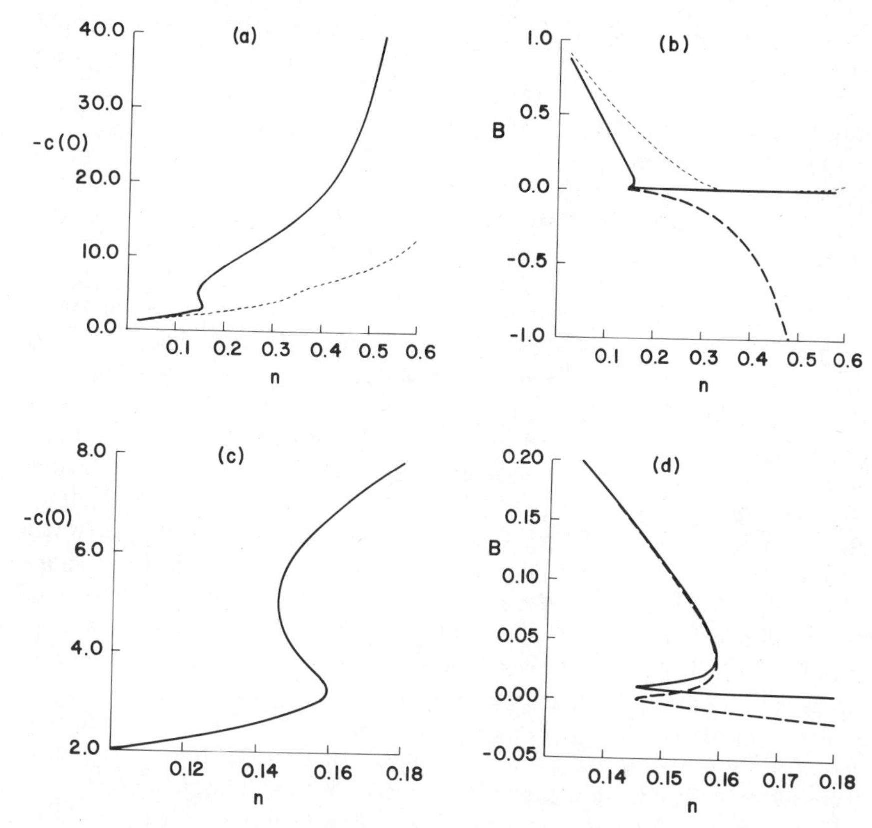

Fig. 14. Single BGYK solution branch, as mapped into the $(n, c(0))$ plane (a) and the (n, B_g) plane (b) is shown at the supercritical temperature $T = 1.65$ (---) and the subcritical temperature $T = 1.30$ (—); in (b) the compressibility $B_c(n, T = 1.30)$ is also plotted (--). An expanded view of (a) and (b) for the subcritical solutions in the neighborhood of the turning point is given in (c) and (d).

always falls in the neighborhood of the density at which $B_c(n)$ changes sign. Although the locations of n_{sv} and $n_{s\ell}$ shift slightly as ℓ_1 is increased, at $T = 1.30$ we find only a single solution branch for all values of ℓ_1 we considered. On this single solution branch there is a pair of turning points at the densities n_{tv} and $\bar{n}_{tv}$. At very low densities ($n \leq 0.05$) we find a solution characteristic of a stable gas phase. As the density is increased solutions $y(r; n)$ are found until $n = n_{tv}$; at this point the solution branch turns back to lower densities, $n < n_{tv}$. As the density now decreases, solutions $y(r; n)$ can again be found until a second turning point in the density is reached at $n = \bar{n}_{tv}$; at $n = \bar{n}_{tv}$ the solution branch turns back to high density. Once past this second turning point, the solution branch

$y(r; n)$ connects smoothly to the stable, liquid solution found at high density. This pair of turning points at densities $(n_{tw}, \bar{n}_{tw})$ develops on the $T = 1.30$ BGYK solution branch as ℓ_1 is increased from 6 to 12, and presumably remains for $\ell_1 \geq 12$ (the pair certainly is present for $\ell_1 \leq 24$). Similar turning-point pairs were observed as ℓ_1 increased, on all the subcritical solution branches we studied, although the further T was below T_c, the less increase required in ℓ_1 before the turning point appeared. This is the same situation observed for the PY equation, and, as in the PY case, the exact locations of n_{tw} and $\bar{n}_{tw}$ depend weakly on ℓ_1.

The behavior we find for $g(r)$ and $B_g(n)$ at subcritical temperatures and $\ell_1 \leq 9$ is like that found by Lincoln, Kozak, and Luks[90] in one of their early studies of the BGYK equation for a subcritical square-well fluid. At low temperatures they found a range of intermediate densities over which $B_g(n) \approx 0$, whereas $B_g(n) > 0$ in the low- and high-density domains that abut this region. Using a bifurcation-like analysis of a simplified, nonlinear integral equation derived from the BGYK equation, they identified the densities at which $B_g \to 0 + \kappa$, $1 \gg \kappa > 0$, as special densities at which the tail of the pair correlation function changes character from short-ranged to long-ranged. With cutoffs $\ell_1 \leq 9$, we, like Lincoln, Kozak, and Luks, find BGYK solutions for which $B_g \approx 0$ connect smoothly to solutions for which $B_g > 0$. They may not have seen a turning point because their cutoff $\ell_1 = 15$ might be inadequate for the square-well potential.

Our calculations show that a subcritical BGYK isotherm has a turning point in the density at n_{tw}. Just as in the PY and HNC cases, this density turning point results in a singularity in the bulk modulus $B(n)$ at $n = n_{tw}$. We expect that there is a unique temperature T_c above which there is no turning point on the isotherms $T > T_c$ even in the limit as $\ell_1 \to \infty$. On the isotherm $T = T_c$ we expect there is a unique density $n = n_c$ at which a turning point is incipient; that is, for $0 < T_c - T' \ll 1$, we would find a turning point developing at $n = n_c$ on the $T = T'$ isotherm in the limit as $\ell_1 \to \infty$. The temperature T_c must lie between $T = 1.65$, where $g(r)$ is short-ranged and $B_g(n) > 0$ $(0.02 \leq n \leq 0.85$ for all $\ell_1 \geq 6.0)$, and $T = 1.30$ where there is a turning point at $n \approx 0.16$ along the curve of $B_g(n)$ versus n; thus the density n_c must lie between 0.02 and 0.84.

Although the point (n_c, T_c) is a special point in the (n, T) parameter space of BGYK solutions, it is not clear that this special point should be identified with a conventional liquid–vapor critical point. At a conventional critical point (n_c, T_c), $g(r)$ has an infinite range, that is, exhibits power-law decay, and $B_g(n_c, T_c) = 0$. In view of the analytic work of Fisher and Fishman[91–93] and the work of the Notre Dame group,[94] we expect that the BGYK special point (n_c, T_c) is not a conventional critical point. They studied the behavior of a BGYK solution $g(r)$ at large r, say

$r > 3\ell_0$ where ℓ_0 is the range of the pair potential, through an asymptotic analysis of the BGYK equation. For large r the BGYK equation (2.7) can be converted, by means of a moment expansion, to a nonlinear differential equation in which two parameters, κ^2 and λ, are relevant to the issue at hand:

$$\lambda\kappa^2 \equiv 1 - \frac{4\pi}{3}\frac{n}{kT}\int_0^{\ell_0} dr\, r^3 \frac{d\phi}{dr} g(r) \tag{5.1a}$$

and

$$\lambda \equiv \frac{2\pi}{15}\frac{n}{kT}\int_0^{\ell_0} dr\, r^5 \frac{d\phi}{dr} g(r)\,. \tag{5.1b}$$

κ^{-1} measures the range of $g(r)$; if, as is expected, $\kappa^{-1} \to \infty$ as a conventional critical point is approached, then $g(r)$ becomes infinitely long-ranged. We note that the quantity $\lambda\kappa^2 - 1$ (5.1a) is simply twice $[P^v/nkT - 1]$; P^v is the virial pressure, and nkT is the ideal-gas pressure at the same density and temperature [see (1.4)]. Fisher and Fishman[91–93] show that:

1. If $g(r)$ is greater than 1 when r is about $3\ell_0$, then κ cannot vanish and there is no true criticality.
2. Conversely, if criticality is achieved, so that κ can vanish, then $g(r) - 1$ must become negative as $r \to \infty$ and $\kappa \to 0$.

[The possibility[95] that $g(r)$ decays as a sum of Bessel functions faster than any power of r as $r \to \infty$ cannot be excluded, but seems unlikely.] The case that $g(r) - 1$ is negative at large r is certainly unphysical, and in all our numerical solutions of the BGYK equation for a truncated and shifted Lennard-Jones potential we have seen only a monotonic decay of positive $[g(r) - 1]$ for $r \geq 9.0$ Moreover, our results imply that κ^2 and λ are positive at temperatures less than 1.55. For example, at $T = 1.30$ and $\ell_1 = 12$, the values of κ^2 and λ are greater than zero in the neighborhood of the turning point at $n_{tw} = 0.16$. On the $T = 1.30$ isotherm κ^2 is less than 0 for $0.44 < n < 0.85$, but this is an artifact of the cutoff of the integral at $\ell_1 = 12$. For densities in the range $0.44 < n < 0.85$ the correlation function $g(r; n, T = 1.30)$ is long-ranged and, as noted earlier, the values of $g(r)$ for $0 < r < 2$ and $\ell_1 - 2 < r < \ell_1$ are most sensitive to the cutoff ℓ_1 [as measured by $\partial g(r)/\partial \ell_1$]. If we increase ℓ_1 to 18, then the short-range structure of $g(r)$ changes enough that κ^2 is now positive except for $0.73 < n < 0.78$; if we increase ℓ_1 to 24, then κ^2 is positive on the entire $T = 1.30$ isotherm (Figs. 15 and 16).

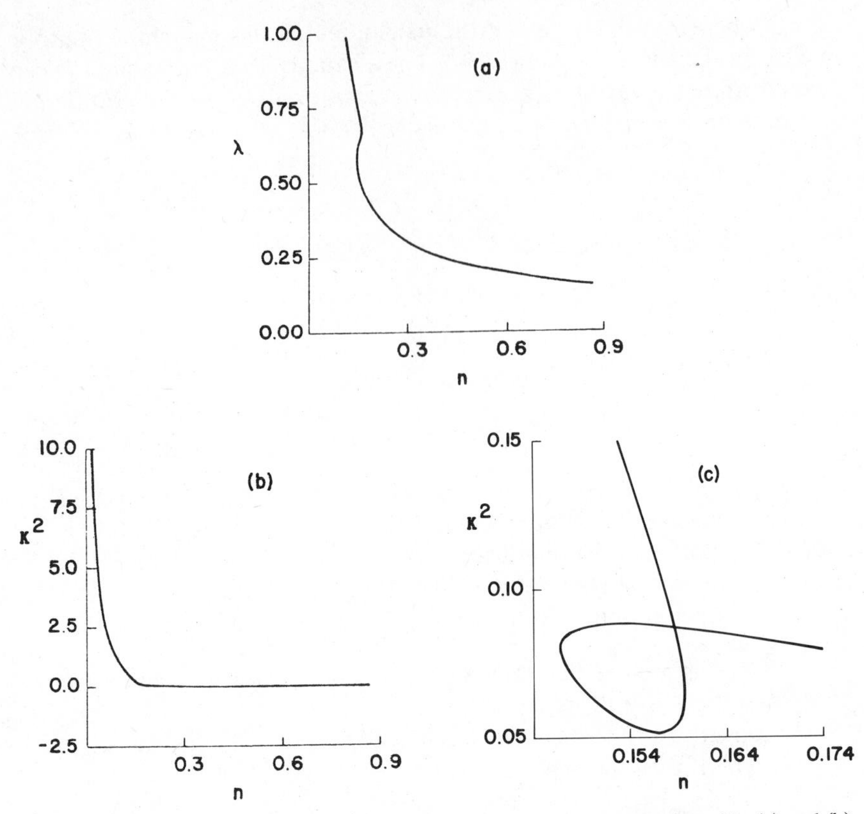

Fig. 15. Moments $\lambda(n)$ and $\kappa^2(n)$ for the BGYK $g(r)$ at $T = 1.30$ are plotted in (a) and (b), respectively. In (c) an expanded view of the moment κ^2 in the neighborhood of the turning point is shown.

Numerical results[94] for the near-critical square-well fluid also support the conclusion that κ^2 is positive for all BGYK solutions. In these numerical studies by Kozak and coworkers, an integral cutoff of $\ell_1 = 100$ was used. A method of iterative substitution was used to find the BGYK solutions and, unfortunately, this scheme did not yield converged solutions when $g(r)$ is very long-ranged. Such convergence difficulties would have prevented these workers from seeing a turning point. Nevertheless, in the near-critical cases where their iteration did converge to a solution, they found κ^2 is positive. [In any case, the numerical work on the square-well fluid points out the advantage of a nonuniform mesh;[75] Jones et al.[94] could afford a cutoff of $\ell_1 = 100$ because nodes or mesh point were located 0.5 units apart in the region $6 < r < 100$ where $g(r)$ is changing very slowly.]

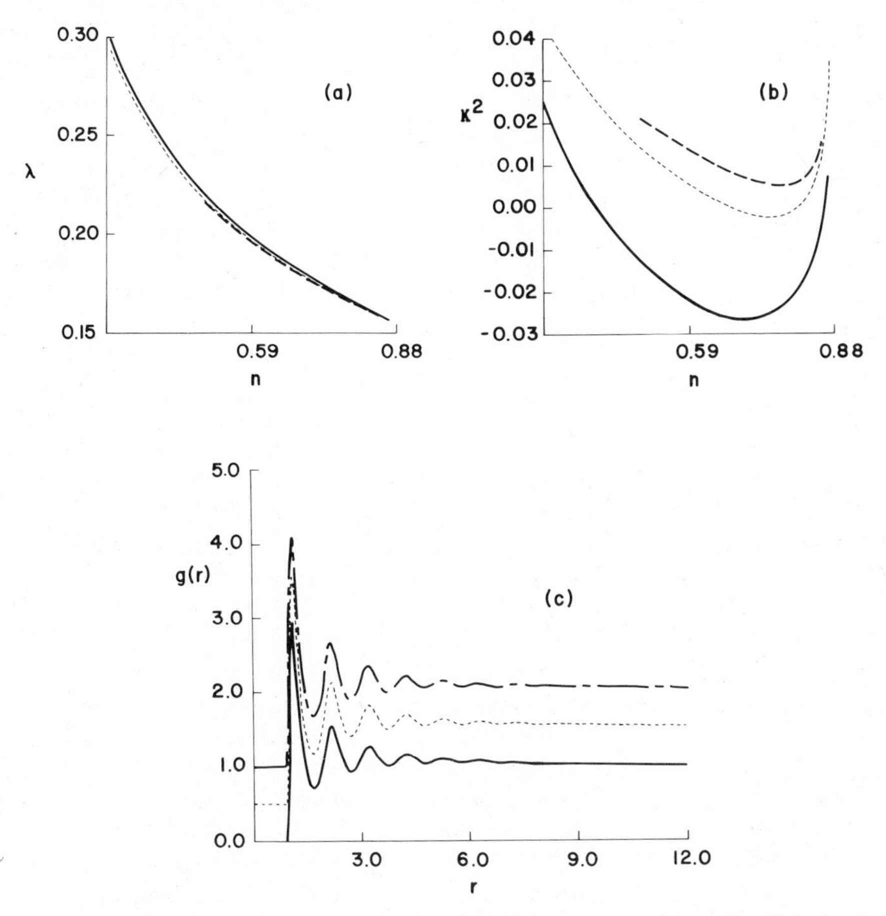

Fig. 16. As the integral cutoff ℓ_1 for the BGYK equation is changed from $\ell_1 = 12$ (—), to $\ell_1 = 18$ (---), to $\ell_1 = 24$ (--), κ^2 increases until it is positive for all $0.0 \leq n < 0.88$ (b); λ remains everywhere positive, but decreases slightly (a). Small shifts of the peaks in the BGYK $g(r; n = 0.73, T = 1.30)$ as ℓ_1 increases can be seen in (c), where ℓ_1 changes from 12 (—), to 18 (---, g shifted up by 1.0), to 24 (– – –, g shifted up by 2.0).

We conclude that our BGYK results, although somewhat limited by the cutoff approximations, that is, $\ell_1 \leq 24$, are consistent with previously published BGYK results. In fact our examination of the dependence of $g(r)$ on the parameters of density and temperature complements the earlier work. Though it bears on the form of the function $g(r)$ in the neighborhood of a turning point in the density, that work cannot disclose the presence of a turning point.

Although the BGYK special point (n_c, T_c) is not a conventional critical point, it could be defined as a pseudo-critical point in the (n, T) parameter plane. For example, we could consider the critical exponent associated with the curve of turning points $n_{tv}(T)$ which terminates at $n_c(T_c)$. The superposition approximation appears to have two paradoxical characteristics: it overestimates fluctuations in such a way as to produce a "critical" turning-point singularity (n_c, T_c) at a higher temperature than either the PY or HNC critical-point singularities, but it also underestimates fluctuations in such a way that the BGYK "critical" point never develops conventional critical-point behavior.

VI. SUMMARY OF RESULTS

The features of the solution space for the correlation function $y(r)$ and the behavior of the bulk modulus $B(n)$ can be summarized as follows. At supercritical temperatures each of the equations, PY, HNC, BGYK, has a single solution branch $y(r; n)$ in the range $0.05 \leq n \leq 0.60$, and this branch is fairly insensitive to the cutoff parameter ℓ_1. At subcritical temperature $T < T_c$(HNC) there are two branches for the HNC solution space. One starts at low densities, follows a path of increasing density until a turning point is reached at an intermediate density n_{tv}, and then turns back to lower densities. The second starts at high density, follows a path of decreasing density until a turning point is reached at an intermediate density $n_{t\ell} > n_{tv}$, and then turns back to higher densities. On each of the vapor and liquid branches, there is a spinodal density, n_{sv} or $n_{s\ell}$, at which $B_c = 0^+$. On the high-density or liquid branch the spinode occurs before the turning point, $n_{s\ell} > n_{t\ell}$, but on the low-density or vapor branch the turning point occurs before the spinode, $n_{tv} > n_{sv}$. For the PY equation at $T < T_c$(PY), there is only one solution branch, but it effectively appears as two disconnected branches. Of these, one PY branch starts at low densities, follows a path of increasing density until a turning point n_{tv} in the density is reached, and then turns back to low densities; the second starts at high densities and continues down in density until it finally does connect, at some very low density, to that portion of the low-density branch which lies beyond the turning point n_{tv}. There are vapor and liquid spinodal densities, n_{sv} and $n_{s\ell}$, with n_{sv} apparently less than, but very close to, n_{tv}. The subcritical solutions $y(r; n)$, which continue beyond the spinodal densities on both the HNC or PY solution branches, are increasingly long-ranged (apparently approaching an infinitely long-ranged behavior) and hence are highly sensitive to the value ℓ_1 of the integral cutoff; for these solutions B_g is positive but small and B_c is negative. In the case of the BGYK equation at $T < T_c$(BGYK) there is only one branch of solutions $y(r; n)$, and it connects the low-density solutions to the high-density solutions; but there is a pair of turning points in the density at n_{tv} and $\bar{n}_{tv}$ on this solution branch, just as in the PY case. As the density is increased from low values on a subcritical BGYK isotherm, the bulk moduli $B_g(n)$ and $B_c(n)$ decrease until, at a vapor spinodal density n_{sv}, they very nearly vanish. This BGYK vapor spinodal density n_{sv} is very close to the turning-point density n_{tv}. [In fact if n_{sv} is specified as the density where the value of $B_c(n)$ changes from positive to negative, then $n_{sv} > n_{tv}$; however, the BGYK direct correlation function exhibits a rather unphysical long-ranged tail when the BGYK pair correlation function $g(r)$ is long-ranged.] There is also a liquid spinodal density $n_{s\ell}$ defined not by B_c, which is notably ill behaved at high densities, but by the condition that $B_g(n) \approx 0$ in the range $n_{tv} < n < n_{s\ell}$. The correlation functions $y(r; n, T, \ell_1)$ and $g(r; n, T, \ell_1)$ are long-ranged for $T < T_c$(BGYK) and densities $n_{sv}(T) \approx n_{tv}(T) < n < n_{s\ell}(T)$, but do not appear to have an infinite range. Thus the BGYK critical point (n_c, T_c) is not associated with conventional critical-point behavior, but, as a limit of turning points $n_{tv}(T)$ at which the function $B_g(n)$ is singular, the point (n_c, T_c) is a special point in the density–temperature parameter plane.

The novel feature in our analysis of the three approximate integral equations for the correlation function $y(r)$ is the dominant role played by the Jacobian $J(r, s)$ [see Eq. (3.4a)]. Using the Jacobian in the Newton iterative method to solve the integral equations, we found quadratic convergence (typically five iterations) to a solution $y(r)$ even in the neighborhood of the coexistence curve, where the traditional method of successive substitution or Picard iteration requires thousands of iterations to achieve convergence. The Jacobian $J(r, s)$ is also used in a strategy of first-order, adaptive continuation to construct a very good initial guess for the solution $y(r)$ at a new set of parameter values from the known solution at an old set of parameter values. Here the solution $y(r)$ depends parametrically on the temperature T, density n, and value ℓ_1 of the integral-cutoff parameter. In addition, $J(r, s)$ is used to determine the stability of a solution with respect to isotropic perturbations and to examine the sensitivity of the solution to the cutoff parameter ℓ_1. In our method of solving the integral equations, we have introduced the use of a finite-element expansion of the function $y(r)$. The finite-element approach provides a convenient framework for systematically decreasing the error incurred by the numerical discretization. And although not used in our calculations, an adaptive finite-element mesh,[75] that makes the integration error approximately the same on each finite-element subdomain, could be employed to great advantage in the integral equations to reduce the number of computer calculations in those regions over which the correlation functions vary slowly.

VII. DISCUSSION

From this summary of the PY, HNC, and BGYK subcritical solutions, we turn to the subjects of metastable states and singularity at the coexistence curve, which were mentioned in the Introduction. We can state nothing definite about the question of a singularity at the coexistence curve, because we did not calculate this curve for any of the three integral equations. Such a calculation would require a (numerically inaccurate) double integration of the bulk modulus with respect to the density to get the free-energy density $f(n)$. Moreover, a specific choice of reference state is required to determine the constants of integration, and although the low-density ideal-gas state provides a well-characterized reference state for the low-density branch of solutions, no such convenient reference state exists at high densities. If the density of free energy $f(n)$ were known, it would still be necessary to apply the double tangent construction to fix the coexisting vapor and liquid densities, n_{cv} and $n_{c\ell}$, respectively. At the coexistence densities n_{cv} and $n_{c\ell}$ the bulk modulus is expected to be positive, and all our solutions for which $B_g \approx B_c > 0$ are well behaved. That is, along a subcritical solution branch we found the solutions $y(r; n)$ to be long-ranged or extremely sensitive to ℓ_1 only in those density regions for which $B_c \approx 0$ and $B_g \approx 0$; these troublesome densities presumably lie between the coexistence densities n_{cv} and $n_{c\ell}$ and hence inside the coexistence curve.

Our results do not appear to indicate any nonanalytic behavior at the coexistence curve, and so it might seem natural to interpret the condition of homogeneous density in the three approximate integral equations as equivalent to the constraint of a restricted ensemble in the molecular theory of metastability.[26] In a classical continuum theory of the liquid–vapor transition, if the volume V occupied by the fluid is imagined to be partitioned into smaller cubical cells of volume V_0, then the restricted ensemble $\mathscr{E}$ is defined to consist of only those configurations for which the number of particles in every cell is within a small, bounded range of values. Using such a restricted ensemble for a Monte Carlo calculation (i.e., an ensemble average over $\mathscr{E}$) on a Lennard-Jones-like fluid, Hansen and Verlet[96] were able to continue a subcritical pressure isotherm from stable vapor phase across the coexistence region into stable liquid phase. As expected in view of the similarity between the restrictions on particle fluctuations and the mean-field assumption behind van der Waals' equation of state, Hansen and Verlet found that the portion of the pressure isotherm inside the coexistence region exhibited a van der Waals loop; in particular, the curve of pressure versus density had a negative slope (i.e., $B < 0$) at densities n greater than a vapor-spinodal density n_{sv} but less than a liquid-spinodal density $n_{s\ell}$. (As remarked above in comments on the correlation functions of the van der Waals model, when the bulk modulus B is negative it should be identified with B_c, the bulk modulus determined from the direct correlation function.) Despite the fact that Hansen and Verlet did not study in detail the way in which the location of the spinodal densities depends on the choice of cell volume V_0, it is significant that, in the restricted ensemble $\mathscr{E}$, they found a smooth, single-branch isotherm connecting the stable vapor and liquid phases at a subcritical temperature. In contrast, for the PY, HNC, and BGYK equations, we found solution branches with turning points in the density at subcritical temperatures. Such turning points imply thermodynamic singularities along the isotherm, namely, in $B(n)$ as a function of n, in contrast to the analytic behavior of $B(n)$ in the case of the van der Waals loops. In fact, on the low-density branch of the subcritical HNC solutions, a density turning point was reached before the spinodal point; that is, a singularity in the bulk modulus B as a function of n was reached at a metastable density (similar features have been found in PY solutions in the case of sticky hard spheres[86]). We conclude that the condition of fixed local density or homogeneity in an *approximate* integral equation for the two-point correlation functions is not *generally* equivalent to the constraint of a restricted ensemble.

If we look back to the origin of the three approximate integral equations for $y(r)$, we can understand why these equations, even under the condition of homogeneous density, do not amount to different approximations for an average in the restricted ensemble $\mathscr{E}$. The set of configurations in which a very large or very small number of particles are clustered in any single partitioning cell is not allowed in $\mathscr{E}$, even though the average of the single-particle density function over this set might be consistent with the condition of a homogeneous density n. That is, not only are configurations associated with an inhomogeneous density $n_I(r)$ and a macroscopic separation of particles into liquid and vapor phases excluded, but also excluded are configurations that contain the nuclei of liquid droplets or gas bubbles associated with the onset of condensation or evaporation, respectively. In spite of such exclusions, when averaging over the restricted ensemble $\mathscr{E}$ all configurations are assigned their full Boltzmann probability factor; that is, the probability of a configuration in $\mathscr{E}$ is determined with respect to all possible configurations, excluding those configurations not in $\mathscr{E}$. In the PY, HNC, and BGYK equations an approximation is made for an exact, full-ensemble, statistical mechanical average [see Eqs. (2.4) and (2.5)–(2.7)], and so there is no outright exclusion of configurations that contain the nuclei for condensation or evaporation, or even droplets of liquid or bubbles of vapor. In a qualitative sense, under the integral-equation approximations all configurations consistent with a homogeneous density n are allowed, but in the statistical average approximate weight factors are assigned, rather than the exact Boltzmann probability factors. [Indeed, we must imagine that the assignment of the approximate weight factors in the ensemble average leading to $g(r; n)$ depends on the quantities $y(r)$ and n themselves.[97]] Consequently, we might expect rather different behavior for the correlation functions found by solving the approximate integral equations than for the correlation functions found by averaging over a restricted ensemble. We contend that the turning-point singularities in $B(n)$ revealed by the PY, HNC, and BGYK results arise from inclusion of configurations that would not be allowed in a restricted ensemble, and are not due simply to the use of incorrect weight factors for ensemble configurations.

For insight into the types of configuration which produce thermodynamic singularities, we turn for a moment to the droplet model,[19,20] where a thermodynamic singularity is predicted at the coexistence curve. To produce this singularity in the droplet model, say at the coexistence vapor density n_{cv}, the key step is to allow not only dilute gas-like clusters but also dense liquid-like clusters. (Recent work[98,99] on the cluster-size distribution in a Lennard-Jones system indicates that the distribution becomes bimodal as the density n_{cv} is approached, in accord with the main physical feature of the droplet model.) The droplet model in a sense amounts to an intuitively appealing and physically reasonable guess about the asymptotic behavior of the Mayer cluster coefficients for large clusters.[20] At n_{cv} the role of dense clusters becomes significant, reflecting the role of critical droplets in the mechanism of condensation, but the only indication of this mechanism apparent in the thermodynamic functions is a weak singularity at n_{cv}. If the approximate integral equations for $y(r)$ are reformulated as functional cluster expansions in the Mayer function and the density,[55] then large clusters are also allowed; that is, summations are over graphs in which an arbitrary number of field points can appear. These graphical clusters cannot be directly identified as physical clusters. For example, ther are four distinct simple graphs with two root points and two field points; the connection between these four graphs and a cluster of four particles is not obvious. Nevertheless, the graphical clusters do require a degree of physical clustering because of the short-ranged nature of the Mayer-function lines in any particular graph. When the root points $\mathbf{r}_1$ and $\mathbf{r}_2$ are far apart, $g(|\mathbf{r}_1 - \mathbf{r}_2|)$ can differ from unity only if there is a significant number of field points connecting $\mathbf{r}_1$ and $\mathbf{r}_2$ such that a bridge of Mayer functions overlaps from $\mathbf{r}_1$ to $\mathbf{r}_2$. The notion that the approximate integral equations do allow contributions from large physical clusters can also be motivated through the exact cluster-like expansion developed in Appendix A. There we show that $g(|\mathbf{r}_1 - \mathbf{r}_2|)$ can be long-ranged only if there is a correlation of the arrangement of particles around $\mathbf{r}_1$ with the arrangement of particles around $\mathbf{r}_2$. In this framework, the most physically plausible mechanism for any long-ranged correlation is large physical clusters of particles. This remains true for the approximate integral equations, even though we cannot define that particular approximation of many-body distribution functions in the exact expansion which would yield the PY, HNC, or BGYK equations.

We have been led to a number of conjectures, however. We propose that the singularities in the HNC bulk modulus $B_c(n)$ at the vapor and liquid turning-point densities, n_{tv} and $n_{t\ell}$, are comparable to the singularities that arise in the droplet model at the condensation and evaporation densities, respectively. The absence of liquid-density turning points on subcritical PY and BGYK isotherms illustrates a deficiency of the two approximations. If a turning point in the density indicates an underlying large-cluster effect associated with a first-order phase transition, then the PY, HNC, and BGYK equations reflect, although as three different distorting mirrors, the liquid–vapor transitions. Since the "goodness" of any one integral-equation approximation depends on the features of the pair potential,[3] we expect the solution space for $y(r)$ to alter as the pair potential is changed; specifically, the location of turning points relative to spinodal points depends on the pair potential, given the integral equation. The difference between the present PY results for the truncated and shifted Lennard-Jones potential and the sticky-sphere PY results[86] demonstrates the dependence of the salient features of the solution space on the pair potential.

Further support for the inference that the turning-point singularities in $B(n)$ are due to large clusters comes from examining the dependence of the features of the solution space on the cutoff parameter ℓ_1. If the pair correlation function $g(r)$, $r = |\mathbf{r}_1 - \mathbf{r}_2|$, is becoming longer-ranged, then large clusters must be playing an increasingly significant role. This implies that if we could suppress the development of a pronounced, long-range tail in $g(r)$, which amounts to suppressing the role of large clusters, then we should find results like those found in a restricted ensemble. And this is indeed the case for the PY and BGYK equations. When we chose the integral cutoff ℓ_1 short enough, we found a single analytic (no density turning points) solution branch connecting the stable, low-density $y(r)$ to the stable high-density $y(r)$ (Fig. 8). Along a portion of this single branch, the potential long-ranged behavior of $g(r)$ is effectively suppressed under the truncation approximation $[g(r)=1, r>\ell_1]$. Arbitrariness in the choice of ℓ_1 (other than it not be too large) is comparable to the arbitrary choice of the cell volume V_0 in the restricted ensemble. Certainly the use of a short cutoff ℓ_1 is a terribly crude way to exclude large-cluster effects, but it does yield a van der Waals loop and a direct correlation function that is reasonable for a homogeneous phase, even if that phase is unstable. Under the HNC approximation it is very hard to suppress turning-point singularities much below the critical temperature because the HNC approximation is better than the PY or BGYK approximation, in the sense that it includes more graphs and hence allows both $c(r)$ and $g(r)$ to develop long-range character; for this better approximation, the use of a short ℓ_1 does not adequately suppress the effect of large clusters.

As for the PY, HNC, and BGYK equations in the limit $\ell_1 \to \infty$, it is not clear what, if any, physical meaning can be given to the very long-ranged solutions that lie either beyond the vapor spinodal or turning-point density, n_{sv} or n_{tv}, or beyond the liquid spinodal or (in the HNC case) turning-point density, $n_{s\ell}$ or $n_{t\ell}$. If these long-ranged subcritical solutions refer to a disordered fluid, then the long tail in $g(r)$ implies that all scales of density fluctuations in this fluid are important, just as in a fluid near the liquid–vapor critical point. Alternatively, these solutions might describe an orientationally averaged, ordered fluid; the orderly arrangement of fluid particles would produce a long-range correlation and the average over orientations could lead to a monotonic decay. The most likely possibility, however, is that these solutions have no physical meaning and instead result from inadequacy or breakdown of the approximations underlying the integral equations, or from the constraint of homogeneous fluid enforced in our solutions.

The role of homogeneous correlation functions at densities inside the coexistence curve in the theory of inhomogeneous fluid [cf. Eqs. (1.9a)–(1.9e) and the remarks following them] warrents further comment. Those theories that replace the correlation function of an inhomogeneous fluid by the correlation function of a homogeneous fluid at some local density $n_I(\mathbf{r})$ implicitly assumed that there is a single branch of correlation functions which extends smoothly across the coexistence region. Our results (for large ℓ_1 and in three dimensions) show that the notion of a smooth single branch must be reexamined. We argued above that by suppressing large clusters we could obtain such a single branch. Accordingly, the functions $c(r)$ or $g(r)$ at a local density $n_I(\mathbf{r})$ which appear in theories of inhomogeneous fluid must be viewed as the correlation functions of a homogeneous fluid in which the role of large clusters has been suppressed. (Note that the method of suppressing clusters as we move from a stable one-phase region into the coexistence region is not equivalent to the method of analytic continuation into the two-phase region from the one-phase region.) From this viewpoint we can better appreciate the usefulness of perturbation theories in the study of inhomogeneous fluids. In perturbation theories, the pair potential is typically split into a repulsive piece and an attractive piece.[3,100] At high densities any one fluid particle is most often a partner in a repulsive interaction with any neighbor because of crowding. Thus, at least at high densities, it is natural to define a reference system using only a repulsive piece of the pair potential, and to consider the attractive piece of the pair potential as a small perturbation to this reference system. Similar perturbation theories can be applied in nonuniform fluids.[18] Again the inhomogeneous correlation functions are replaced by the correlation functions of a homogeneous fluid at a local density $n_I(\mathbf{r})$, but now the correlation functions at $n_I(\mathbf{r})$ are determined in a perturbative approach. In particular, the basic structure of the homogeneous correlation function is fixed by the repulsive reference system. The perturbation due to the attractive piece of the potential does not appreciably alter the basic structure. Since the repulsive reference fluid is stable over the entire range of fluid densities, the correlation functions never develop the pronounced long-range structure found in the PY and HNC correlation functions. The choice of a repulsive reference system automatically suppresses the role of large clusters. The modified van der Waals theory of inhomogeneous fluid developed by Bongiorno and Davis[50] is in fact built on a repulsive reference state in which the correlation functions of homogeneous fluid are well behaved.

While this study of the PY, HNC, and BGYK integral equations for the correlation functions of a subcritical, homogeneous fluid has produced some insight into the questions of thermodynamic singularities and metastability in the coexistence region, there are several interesting points not addressed which deserve further study. We introduced the droplet model as a convenient reference model to illustrate the way in which thermodynamic singularities could arise. It is not clear, however, whether the turning-point singularities of the integral equations and the essential singularities of the droplet model are manifestations of the same underlying physical singularities and differ in form only because of the approximations involved. That there might be distinct physical singularities is suggested by the following argument. Suppose that the approximate integral equation for $y(r)$ represents an ensemble average in which configurations with many particles found in a dense, compact cluster are given zero weight and all other configurations are given some nonzero, approximate weight. Then if the turning-point singularity in the bulk modulus $B(n)$ is present, it must be due to the influence of large, ramified clusters. In contrast, the essential singularity found in a droplet-like model is due to the influence of all large clusters, compact and ramified, and hence would probably occur at a different density than the turning-point singularity.

An examination of higher-order approximations for $y(r)$, for example, PY II,[101–104] modified HNC,[105,106] or improved BGYK,[48,58] might lead to a clearer picture of subcritical thermodynamic singularities. In the approximate integral equations the effects of dimensionality on the correlation functions and thermodynamic singularities could also be examined. Specifically, it would be interesting to see if the subcritical solution space for $y(r)$ has one or two branches in four and higher dimensions. The dimensionality d could be treated as a formal parameter on which the correlation functions depend; that is, $y = y(r; d)$. The Jacobian $J(r, s)$ could be used to examine the sensitivity of $y(r)$ to d or, perhaps more practically, used in a scheme of first-order continuation in the parameter d to devise good initial estimates of $y(r)$ in dimensions other than three. From a study of the stability of solutions $y(r)$ for homogeneous fluid with respect to inhomogeneous density perturbations there might emerge a deeper understanding of what inhomogeneity in a fluid entails. Toward this goal of understanding inhomogeneous fluid, we are undertaking numerical calculations to solve approximate integral equations for the correlation function $y_I(\mathbf{r}', \mathbf{r}+\mathbf{r}')$ in an inhomogeneous fluid with density $n_I(\mathbf{r})$.

Acknowledgments

This work was supported in part by a grant from the U.S. Department of Energy and the Petroleum Research Fund. We also acknowledge grants for computing from the University of Minnesota Computer Center and Control Data Corporation.

APPENDIX A

In Section II we defined the normalized covariance $K(r)$ as

$$K(r) = K(|\mathbf{r}-\mathbf{r}'|) = \frac{\langle e^{-\beta\psi(\mathbf{r})} e^{-\beta\psi(\mathbf{r}')}\rangle - \langle e^{-\beta\psi(\mathbf{r})}\rangle\langle e^{-\beta\psi(\mathbf{r}')}\rangle}{\langle e^{-\beta\psi(\mathbf{r})} e^{-\beta\psi(\mathbf{r}')}\rangle}, \tag{A.1}$$

where the angular brackets denote an ensemble average and

$$\psi(r) = \sum_{i=1}^{N} u(\mathbf{r}, \mathbf{r}_i). \tag{A.2}$$

$\psi(r)$ is the energy of interaction of an imaginary test particle inserted at position $\mathbf{r}$ with the N real particles in the fluid. We commented briefly in the text on the role of $K(r)$ as a measure of fluctuations and correlations in the fluid. In this appendix we support those remarks by deriving a formal expansion that clearly shows the effect of correlations in $K(r)$. We adopt an expansion that blends a hierarchy of distribution functions with the well-known Mayer-function cluster expansions.

Let us first examine the term $\langle e^{-\beta\psi(\mathbf{r})}\rangle$ which appears in $K(r)$. By virtue of the indistinguishability of particles in a grand canonical ensemble we find

$$\langle e^{-\beta\psi(\mathbf{r})}\rangle = \left\langle \prod_{i=1}^{N} [f(\mathbf{r}, \mathbf{r}_i) + 1] \right\rangle$$

$$= \sum_{i=0}^{\infty} \frac{1}{i!} \int d\mathbf{r}_1 \cdots \int d\mathbf{r}_i \prod_{j=1}^{i} f(\mathbf{r}, \mathbf{r}_j) n_i(\mathbf{r}_1, \mathbf{r}_2, \ldots, \mathbf{r}_i), \tag{A.3}$$

where $f(\mathbf{r}_1, \mathbf{r}_j)$ is the Mayer function for the pair potential $u(\mathbf{r}, \mathbf{r}_j)$,

$$f(\mathbf{r}, \mathbf{r}_j) = \exp[-\beta u(\mathbf{r}, \mathbf{r}_j)] - 1, \tag{A.4}$$

and $n_i(\mathbf{r}_1, \mathbf{r}_2, \ldots, \mathbf{r}_i)$ is the i-particle distribution function in the grand canonical ensemble. For a short-ranged potential u of the type we wish to consider (i.e., the Lennard-Jones potential and its variants under truncation and shift), $f(\mathbf{r}, \mathbf{r}_j)$ is very near to 0 for $|\mathbf{r} - \mathbf{r}_j| \geq \ell_z$, where ℓ_z is two or three particle diameters (i.e., $\ell_z = 2\sigma$ or 3σ for a Lennard-Jones potential). For this reason the integrand in (A.3) is appreciable only for those locations $\mathbf{r}_j$ of particles j, $j = 1, 2, \ldots, i$, within a distance ℓ_z of $\mathbf{r}$. At the same time the distribution function $n_i(\mathbf{r}_1, \ldots, \mathbf{r}_i)$ for a set of i particles, each within a distance ℓ_z of $\mathbf{r}$, vanishes if i is large (say larger than 12 or so) as a consequence of the repulsive-core interactions between fluid particles. Therefore the value of $\langle e^{-\beta\psi(\mathbf{r})}\rangle$ is determined by the average arrangement of a limited number of particles in a shell around a central test particle at $\mathbf{r}$.

After similar manipulations for the correlation factor $\langle e^{-\beta\psi(\mathbf{r})} e^{-\beta\psi(\mathbf{r}')}\rangle$ we find

$$\langle e^{-\beta\psi(\mathbf{r})} e^{-\beta\psi(\mathbf{r}')}\rangle = \sum_{m=0}^{\infty} \sum_{\ell=0}^{\infty} \sum_{k=0}^{\infty} \int d\mathbf{r}_1 \cdots \int d\mathbf{r}_{m+\ell+k} n_{m+\ell+k}(\mathbf{r}_1, \ldots, \mathbf{r}_{m+\ell+k})$$
$$\times \frac{\prod_{j=1}^{m} f(\mathbf{r}, \mathbf{r}_j) \prod_{j=m+1}^{m+\ell} f(\mathbf{r}', \mathbf{r}_j) \prod_{j=m+\ell+1}^{m+\ell+k} f(\mathbf{r}, \mathbf{r}_j) f(\mathbf{r}', \mathbf{r}_j)}{m!\,\ell!\,k!}. \tag{A.5}$$

The integrand for this equation requires the $(m + \ell + k)$-particle distribution function in the case where m particles interact with the test particle at $\mathbf{r}$, ℓ particles interact with the test particle at $\mathbf{r}'$, and k particles interact with both test particles.

By combining Eqs. (A.3) and (A.4) we have

$$\langle e^{-\beta\psi(\mathbf{r})} e^{-\beta\psi(\mathbf{r}')}\rangle - \langle e^{-\beta\psi(\mathbf{r})}\rangle\langle e^{-\beta\psi(\mathbf{r}')}\rangle = \sum_{m=0}^{\infty}\sum_{\ell=0}^{\infty}\sum_{k=1}^{\infty}\int d\mathbf{r}_1 \cdots \int d\mathbf{r}_{m+\ell+k}\,\rho_{m+\ell+k}$$
$$\times \frac{\prod_{j=1}^{m} f(\mathbf{r}, \mathbf{r}_j) \prod_{j=m+1}^{m+\ell} f(\mathbf{r}', \mathbf{r}_j) \prod_{j=m+\ell+1}^{m+\ell+k} f(\mathbf{r}, \mathbf{r}_j) f(\mathbf{r}', \mathbf{r}_j)}{m!\,\ell!\,k!}$$
$$+ \sum_{m=1}\sum_{\ell=1}\int d\mathbf{r}_1 \cdots \int d\mathbf{r}_{m+\ell}[n_{m+\ell} - n_m n_\ell]\,\frac{\prod_{j=1}^{m+\ell} f(\mathbf{r}, \mathbf{r}_j) \prod_{j=m+1}^{m+\ell} f(\mathbf{r}', \mathbf{r}_j)}{m!\,\ell!} \tag{A.6}$$

Expression (A.6), although cumbersome, highlights the physics of interest. The triple sum on the right-hand side of (A.6) includes all terms where at least one fluid particle j interacts with both test particles. Because the product $f(\mathbf{r}, \mathbf{r}_j)f(\mathbf{r}', \mathbf{r}_j)$ is very near 0 if $|\mathbf{r} - \mathbf{r}_j|$ or $|\mathbf{r}' - \mathbf{r}_j|$ is greater than ℓ_z, this sum contributes to $K(r)$ only if $|\mathbf{r} - \mathbf{r}'| \leq 2\ell_z$; that is, this sum contributes only to the short-range structure of $K(r)$. The magnitude of the contribution depends through $n_{m+k+\ell}$ on the structure or correlation in the arrangement of the set of $m + \ell + k$ particles around the test particles at $\mathbf{r}$ and $\mathbf{r}'$. The second term on the right-hand side of (A.6), the double sum, requires an integration over the positions $\mathbf{r}_j$ for a set of $\ell + m$ particles. Because of the factors $f(\mathbf{r}, \mathbf{r}_j)$, $j = 1, \ldots, m$, and $f(\mathbf{r}', \mathbf{r}_j)$, $j = m + 1, \ldots, m + \ell$, we need only consider those arrangements for which the positions of the set of m particles are within a distance ℓ_z of $\mathbf{r}$ and the positions of the set of ℓ particles are within a distance ℓ_z of $\mathbf{r}'$; again, because of repulsive-core effects in $n_{m+\ell}$, n_m, and n_ℓ, the relevant values of m and ℓ are limited. This second sum is physically interpreted as a measure of how much the simultaneous arrangement of particles in the pair of shells around the test particles at $\mathbf{r}$ and $\mathbf{r}'$ differs, on average, from the average arrangement of two single shells at $\mathbf{r}'$ and $\mathbf{r}$.

If the two shells overlap (i.e., if $|\mathbf{r} - \mathbf{r}'| < 2\ell_z$), the arrangement of particles near $\mathbf{r}$ certainly influences the arrangement of particles near $\mathbf{r}'$; then $n_{m+\ell} \neq n_m n_\ell$ and the double sum contributes to the short-range structure of $K(r)$. On the other hand, when $|\mathbf{r} - \mathbf{r}'| > 2\ell_z$, the double sum is nonzero only if $n_{m+\ell}$ is still different from $n_m n_\ell$. That is, in a homogeneous phase at a given density n and temperature T, $K(r)$ can be long-ranged only if the average simultaneous arrangement of paired shells, measured by $n_{m+\ell}$, is different from the product of the average single-shell arrangement, measured by $n_m n_\ell$.

Although Eq. (A.6) explicitly illustrates the role of correlations in determining $K(r)$, we have glossed over a mathematical detail in its derivation: we have made several rearrangements of infinite sums without ensuring that these sums are absolutely convergent. This is reasonable on physical grounds because we expect that $\langle e^{-\beta\psi(\mathbf{r})}\rangle$ and $\langle e^{-\beta\psi(\mathbf{r})}e^{-\beta\psi(\mathbf{r}')}\rangle$ are bounded and well behaved, but no mathematically rigorous justification is at hand. In a fluid constrained to have overall density n and temperature T, if large clusters or extended fluctuations are important, then we would expect $n_{m+\ell} \neq n_m n_\ell$ even for $|\mathbf{r}-\mathbf{r}'| > 2\ell_z$, and hence we would expect a long-ranged tail in $K(r)$. But it may well be that those points (n, T) at which $K(r)$ becomes long-ranged mark the radius of convergence for one of the infinite sums which we have rearranged.

APPENDIX B

A detailed description of our numerical algorithm for solution of the OZ equation was given in Section III, and in Section IV the relative reliability of the algorithm was evidenced by comparison of our results with those from the literature. In this appendix we illustrate the absolute accuracy of the algorithm against an exact solution to an OZ-type convolution equation.

If

$$f(r) = e^{-\alpha r^2/2} \tag{B.1a}$$

and

$$n = \left(\frac{2\alpha}{\pi}\right)^{3/2}, \tag{B.1b}$$

then the integral equation

$$y(r) = \frac{2\pi n}{r}\int_0^\infty ds\, s\, f(s)y(s)\int_{r-s}^{r+s} dt\, t\, f(t)y(t) \tag{B.2}$$

for the unknown function $y(r)$ has a form very much like the OZ equation under the PY approximation [see Eq. (2.5)]; when $\alpha = 1$ the known function $f(r)$, (B.1a), is short-ranged, not unlike the Mayer f function. An exact solution to this equation is

$$y^{\text{ex}}(r) = e^{-\alpha r^2/2} \tag{B.3}$$

(which admittedly has a simpler structure than the correlation function $y(r)$ calculated under the PY, HNC, or BGYK approximations). With the truncation approximation $y(r) = 0$ for $r \geq \ell_1$ we obtained a numerical solution to (B.2) using the same algorithm of collocation, finite-element expansion, and convergence criteria given in Section III. The first four moments M_i of $y(r)$,

$$M_i = \int_0^\infty dr f^i y(r), \qquad i = 0, 1, 2, 3, \tag{B.4}$$

which can be evaluated analytically, were also calculated for the numerical solution $y(r)$ using the same scheme of interpolation and integration by Simpson's rule outlined in Section III.

The h^2 dependence of the difference $y(r) - y^{\text{ex}}(r)$ between the numerical solution [as in Eq. (3.13)] and the exact solution [as in Eq. (B.3)] is given in Table B.I for $r = 0$, 1.0, 2.0, and 5.0 when $\alpha = 1.0$ and $\ell_1 = 6.0$. For $h \leq \frac{1}{15}$ and $y(r) \geq 10^{-4}$ the difference $y(r) - y^{\text{ex}}(r)$ is less than 0.5%. For $10^{-7} \leq y(r) \leq 10^{-4}$ the difference is larger, but typically less than 2%. We have also made calculations for $\ell_1 = 9.0$, 12.0, and 18.0 at $\alpha = 0.5$, 2.0, and

TABLE B.I
h^2 Dependence of the Difference $y(r) - y^{\text{ex}}(r)$ Between the Numerical and Exact Solutions of Eq. (B.2)

	$[y(r) - y^{\text{ex}}(r)] \times 10^n$			
	$r = 0.0$	1.0	2.0	5.0
h^2	$n = 4$	$n = 4$	$n = 4$	$n = 8$
$(\frac{1}{10})^2$	25.04	10.11	9.04	15.72
$(\frac{1}{12})^2$	17.38	7.02	6.27	10.88
$(\frac{1}{15})^2$	11.12	4.49	4.01	6.94
$(\frac{1}{20})^2$	6.25	2.53	2.26	3.89
$(\frac{1}{24})^2$	2.78	1.76	1.57	2.70
$(\frac{1}{30})^2$	4.34	1.12	1.00	1.73
Intercept	−0.01	0.00	0.00	−0.04
Correlation coefficient	1.00	1.00	1.00	1.00

Table B.II
h^2 Dependence of the Difference $M_i = M_i^{ex}$ Between the Numerical and Analytic Evaluations of the First Four Moments, $i = 0, 1, 2, 3$, of $y(r)$

h^2	$[(M_i - M_i^{ex})/M_i^{ex}] \times 10^4$ $i = 0$	1	2	3
$(\frac{1}{10})^2$	17.37	33.42	50.12	66.86
$(\frac{1}{12})^2$	11.98	23.19	34.78	46.39
$(\frac{1}{15})^2$	7.61	14.83	22.25	29.67
$(\frac{1}{20})^2$	4.25	8.34	12.51	16.68
$(\frac{1}{24})^2$	2.94	5.79	8.68	11.58
$(\frac{1}{30})^2$	1.88	3.70	5.56	7.41
Intercept	−0.09	−0.02	−0.02	−0.04
Correlation coefficient	1.00	1.00	1.00	1.00

4.0 and find that even for $10^{-11} < y(r) < 10^{-7}$ results for $y(r)$ and $y^{ex}(r)$ agree in the first significant figure; for $y(r) < 10^{-11}$ we are essentially limited by an accumulation of roundoff error. The calculated correlation coefficients of 1.00 indicate that $y(r; h^2)$ is a linear function of h^2. This is just the dependence on h^2 we expect when using linear basis functions in a finite-element expansion. The difference between the linear intercepts and 0 is so small as to be insignificant.

Results for the normalized moments, again for the case $\alpha = 1.0$ and $\ell_1 = 6.0$, are reported in Table B.II. The calculated correlation coefficient of 1.00 for a linear regression analysis of the difference between the calculated and exact moments, $M_i - M_i^{ex}$, versus h^2 for $i = 0, 1, 2, 3$ is expected because of the h^2 error due to $y(r)$ in the integrand. The difference of the linear intercepts from 0 is still small enough to be negligible (the actual magnitude of the difference probably reflects errors of order h^4 due to interpolation and numerical integration).

References

1. T. L. Hill, *Statistical Mechanics*. McGraw-Hill, New York, 1956.
2. A. Munster, *Statistical Thermodynamics*, Vol. I. Springer-Verlag, Berlin, 1969; Vol. II, Springer-Verlag, Berlin, 1974.
3. J. A. Barker and D. Henderson, *Rev. Mod. Phys.* **48**, 587 (1967).
4. R. O. Watts, in *Statistical Mechanics*, Vol. I, a specialist Periodical Report of the London Chemical Society, K. Singer (ed.). Billings and Sons Ltd., London, 1973, pp. 1–70.

5. L. S. Ornstein and F. Zernike, *Proc. Akad. Sci. (Amsterdam)* **17**, 793 (1914); reprinted in *The Equilibrium Theory of Classical Fluids*, H. L. Frish and J. L. Lebowitz (eds.). W. A. Benjamin, New York, 1964, pp. III 2–16.
6. J. K. Percus and G. J. Yevick, *Phys. Rev.* **110**, 1 (1958).
7. J. M. J. van Leeuwen, J. Groenveld, and J. De Boer, *Physica* **25**, 792 (1959).
8. E. Meeron, *J. Math. Phys.* **1**, 192 (1960).
9. M. S. Green, *J. Chem. Phys.* **33**, 1403 (1960).
10. T. Morita and K. Hiroike, *Prog. Theor. Phys.* **23**, 1003 (1960).
11. L. Verlet, *Nuovo Cimento* **18**, 77 (1960).
12. G. S. Rushbrooke, *Physica* **26**, 259 (1960).
13. J. G. Kirkwood, *J. Chem. Phys.* **3**, 300 (1935).
14. J. Yvon, *Actualitiés Scientifiques et Industriel.* Herman et Cie, Paris, 1935.
15. M. Born and H. S. Green, *A General Kinetic Theory of Liquids.* Cambridge, London, 1949.
16. See ref. 2, Vol. I, p. 333. In ref. 1, pp. 203–206, Hill also discusses the conversion of an integrodifferential equation to an integral equation, but in the context of the Kirkwood coupling hierarchy for the pair correlation function rather than the BGYK hierarchy.
17. For references and a recent summary of results see *Phase Transitions, Cargèse 1980*, M. Lévy, J. C. Le Guillon, and J. Zinn-Justin (eds.). Plenum Press, New York, 1980.
18. F. F. Abraham, *Phys. Rep.* **53**, 93 (1979).
19. M. E. Fisher, *Physics* (*N.Y.*) **3**, 255 (1967).
20. J. S. Langer, *Ann. Phys.* (*N.Y.*) **41**, 108 (1967).
21. K. Binder, *Ann. Phys.* (*N.Y.*) **98**, 390 (1976).
22. W. Klein, *Phys. Rev. B* **21**, 5254 (1980).
23. R. J. Baxter and I. G. Enting, *J. Stat. Phys.* **21**, 103 (1979).
24. G. A. Baker and D. Kim, *J. Phys. A* **13**, L103 (1980).
25. C. Domb, *J. Phys. A* **9**, 283 (1976).
26. O. Penrose and J. L. Lebowitz, in *Fluctuation Phenomena*, E. W. Montroll and J. L. Lebowitz (eds.). North-Holland, Amsterdam, 1979, pp. 293–340.
27. A. A. Broyles, *J. Chem. Phys.* **35**, 493 (1961).
28. A. A. Broyles, S. U. Chung, and H. L. Sahlin, *J. Chem. Phys.* **37**, 2462 (1962).
29. J. Throop and R. J. Bearman, *Physica* **32**, 1298 (1966).
30. F. Mandel, R. J. Bearman, and M. Y. Bearman, *J. Chem. Phys.* **52**, 3315 (1970).
31. D. J. Henderson, J. A. Barker, and R. O. Watts, *IBM J. Res. Dev.*, 668 (1970).
32. D. Henderson and R. D. Murphy, *Phys. Rev. A* **6**, 1224 (1972).
33. L. Mier y Terán and F. del Rio, *J. Chem. Phys.* **72**, 1044 (1980).
34. R. O. Watts, *J. Chem. Phys.* **48**, 50 (1968).
35. C. Ebner, W. F. Saam, and D. Stroud, *Phys. Rev. A* **14**, 2264 (1976).
36. F. del Rio and L. Mier y Terán, *J. Chem. Phys.* **72**, 5776 (1980).
37. J. J. Brey, A. Santos, and F. Romero, *J. Chem. Phys.* **77**, 5058 (1982).
38. L. Mier y Terán, A. H. Falls, L. E. Scriven, and H. T. Davis, in *Proceedings of the Eighth Symposium on Thermophysical Properties*, Vol. I, J. V. Sengers (ed.). American Society of Mechanical Engineers, New York, 1982, pp. 45–56.

39. L. Verlet and D. Levesque, *Physica* **28**, 1124 (1962).
40. M. Klein and M. S. Green, *J. Chem. Phys.* **39**, 1367 (1963).
41. J. de Boer, J. M. J. van Leeuwen, and J. Groenveld, *Physica* **30**, 2265 (1964).
42. M. I. Guerrero, G. Saville, and J. S. Rowlinson, *Mol. Phys.* **29**, 1941 (1975).
43. S. M. Foiles and N. W. Ashcroft, *Phys. Rev. A* **24**, 424 (1981).
44. R. O. Watts, *J. Chem. Phys.* **50**, 1358 (1968).
45. J. G. Kirkwood, V. A. Lewinson, and B. J. Alder, *J. Chem. Phys.* **20**, 929 (1952).
46. A. A. Broyles, *J. Chem. Phys.* **33**, 456 (1960); **34**, 359 (1961).
47. D. Levesque, *Physica* **32** 1985 (1966).
48. A. D. J. Haymet, S. A. Rice, and W. G. Madden, *J. Chem. Phys.* **75**, 4696 (1981).
49. H. T. Davis and L. E. Scriven, *Adv. Chem. Phys.* **49**, 357 (1982).
50. V. Bongiorno and H. T. Davis, *Phys. Rev. A* **12**, 2213 (1975).
51. A. J. M. Yang, P. D. Fleming, and J. H. Gibbs, *J. Chem. Phys.* **64**, 3732 (1976).
52. R. Evans, *Adv. Phys.* **28**, 143 (1979).
53. W. F. Saam and C. Ebner, *Phys. Rev. A* **15**, 2566 (1967).
54. J. K. Percus, in *The Equilibrium Theory of Classical Fluids*, H. L. Frish and J. L. Lebowitz (eds.). W. A. Benjamin, New York, 1976, pp. II 33–70.
55. G. Stell, *ibid.*, pp. 171–266.
56. E. Meeron, *J. Chem. Phys.* **27**, 1238 (1957).
57. E. E. Saltpeter, *Ann. Phys. (N.Y.)* **5**, 183 (1958).
58. A. D. J. Haymet, S. A. Rice, and W. G. Madden, *J. Chem. Phys.* **74**, 3033 (1981).
59. B. Widom, *J. Chem. Phys.* **39**, 2808 (1963).
60. J. L. Jackson and L. S. Klein, *Phys. Fluids* **7**, 279 (1964).
61. B. Widom, *J. Phys. Chem.* **86**, 869 (1982).
62. M. E. Fisher, *J. Math. Phys.* **5**, 944 (1964).
63. V. Volterra, *Theory of Functionals.* Dover Publications, New York, 1959. For a more recent discussion of the functional or Fréchet derivative see the articles by L. Collatz, R. A. Tapia, and M. Z. Nashed, in *Nonlinear Functional Analysis and Applications*, L. B. Rall (ed.). Academic Press, New York, 1971, pp. 1–43, pp. 45–102, and pp. 103–309, respectively.
64. J. M. Ortega and W. C. Rheinbolt, *Iterative Solution of Nonlinear Equations in Several Variables.* Academic Press, New York, 1970.
65. E. Riks, *J. Appl. Mech.* **39**, 1060 (1972).
66. H. B. Keller, in *Applications of Bifurcation Theory*, P. Rabinowitz (ed.). Academic Press, New York, 1977, pp. 359–384.
67. J. P. Abbott, *J. Comp. Appl. Math.* **4**, 19 (1978).
68. W. C. Rheinboldt, *SIAM J. Numer. Anal.* **17**, 221 (1980).
69. A. Fulinski, *Acta Phys. Polonica* **A59**, 707 (1981).
70. R. A. Lovett, C. Y. Mou, and F. P. Buff, *J. Chem. Phys.* **65**, 570 (1976).
71. M. S. Wertheim, *J. Chem. Phys.* **65**, 2377 (1976).
72. C. T. H. Baker, *The Numerical Treatment of Integral Equations.* Clarendon Press, Oxford, 1977.
73. G. Strang and G. J. Fix, *An Analysis of the Finite Element Method.* Prentice-Hall, Englewood Cliffs, N.J., 1973.

74. B. Swartz and B. Wendroff, *Math. Comput.* **23**, 37 (1969).
75. R. E. Benner, Jr., H. T. Davis, and L. E. Scriven, *SIAM J. Sci. Stat. Comput.*, in press (1985).
76. For a review of the numerical methods traditionally used in solving the correlation-function integral equations see ref. 4.
77. M. J. Gillan, *Mol. Phys.* **38**, 1781 (1979).
78. A. Rahman, *Phys. Rev. Lett.* **12**, 575 (1964).
79. S. Wang and J. A. Krumhansl, *J. Chem. Phys.* **56**, 4287 (1972).
80. D. Henderson, O. H. Scalise, and W. R. Smith, *J. Chem. Phys.* **72**, 2431 (1980).
81. R. J. Baxter, *Phys. Rev.* **154**, 170 (1967).
82. Although the attractive tail in $u(r)$ is small for $\ell_1 \geq 6$, it still makes an appreciable contribution to B_c. See the comments by F. Mandel and R. J. Bearman, *J. Chem. Phys.* **50**, 4121 (1969) and the subsequent reply by R. O. Watts, *ibid.*, **50**, 4122 (1969).
83. N. G. van Kampen, *Phys. Rev.* **135**, 362 (1964).
84. In a recent letter, *J. Chem. Phys.* **79**, 4652 (1983), Brey and Santos recognize the deficiency of the cutoff $\ell_1 = 6$, but they incorrectly suggest that the cutoff effect cannot be important for finite-ranged potentials.
85. R. J. Baxter, *J. Chem. Phys.* **49**, 2770 (1968).
86. S. Fishman and M. E. Fisher, *Physica* **108A**, 1 (1981).
87. That n_{tv} is reached before n_{sv} in the case of the sticky-sphere potential is a peculiar, but not necessarily pathological, consequence of the unrealistic form of the potential. Similar remarks about the peculiar behavior associated with the sticky-spere potential have been made by P. T. Cummings and G. Stell, *J. Chem. Phys.* **78**, 1917 (1983).
88. M. S. Green, *J. Chem. Phys.* **33**, 1403 (1960).
89. Foiles and Ashcroft, ref. 43, report a vapor spinode at $n_{sv} = 0.183$ and $T = 1.360$ for an exponential pair potential which very well approximates the Lennard-Jones 6–12 form. Since they use a standard iterative algorithm to solve the HNC equation it is clear that the turning point n_{tv} must lie beyond n_{sv} or else there would have been convergence problems at n_{tv}. While we suspect that they reached n_{sv} before n_{tv} because of the cutoff approximation for ℓ_1, it indeed may be that for the exponential potential n_{tv} lies beyond n_{sv}.
90. W. W. Lincoln, J. J. Kozak, and K. D. Luks, *J. Chem. Phys.* **62**, 2171 (1975).
91. M. E. Fisher and S. Fishman, *J. Chem. Phys.* **78**, 4227 (1983).
92. S. Fishman, *Physica* **109A**, 382 (1981).
93. M. E. Fisher and S. Fishman, *Phys. Rev. Lett.* **47**, 421 (1981).
94. G. L. Jones, E. K. Lee, and J. J. Kozak, *J. Chem. Phys.* **79**, 459 (1983), and references therein, especially K. A. Green, K. D. Luks, G. L. Jones, E. Lee, and J. J. Kozak, *Phys. Rev.* **25**, 1060 (1982). The latter reference is a first report on numerical evidence (fully converged) for the absence of true critical behavior in the BGYK approximation for a square-well fluid. The recent interest in the critical behavior of the BGYK equation was stimulated by two papers: K. A. Green, K. D. Luks, and J. J. Kozak, *Phys. Rev. Lett.* **42**, 985 (1979) and K. A. Green, K. D. Luks, E. Lee, and J. J. Kozak, *Phys. Rev. A* **21**, 356 (1980).
95. M. Alexanin, *Phys. Rev. A* **25**, 582 (1982).
96. J. P. Hansen and L. Verlet, *J. Chem. Phys.* **184**, 151 (1969).

97. It is rather imprecise and possibly even incorrect to describe the PY, HNC, and BGYK equations in these terms. For example, the approximation of $\langle e^{-\beta\psi(\mathbf{r})}\rangle$ by $e^{-\beta\langle\psi(\mathbf{r})\rangle}$ cannot easily be described as an approximation of the exact Boltzmann factors by some inexact form. Nevertheless, our qualitative picture adequately portrays the spirit of the PY, HNC, and BGYK approximations.

98. J. H. Gibbs, B. Bagchi, and U. Mohanty, *Phys. Rev. B* **24**, 2893 (1981).
99. A. J. Yang, Z. Pavlin, and J. H. Gibbs, *J. Chem. Phys.* **78**, 7005 (1983).
100. H. C. Anderson, D. Chandler, and J. D. Weeks, *Adv. Chem. Phys.* **34**, 105 (1976).
101. L. Verlet, *Physica* **30**, 95 (1964); **31**, 959 (1965); **32**, 304 (1966).
102. L. Verlet and D. Levesque, *Physica* **36**, 254 (1967).
103. M. S. Wertheim, *J. Math. Phys.* **8**, 927 (1967).
104. R. J. Baxter, *Ann. Phys.* **46**, 509 (1968).
105. Y. Rosenfeld and N. Ashcroft, *Phys. Rev. A* **20**, 1208 (1979).
106. F. Lado, S. M. Foiles, and N. W. Ashcroft, *Phys. Rev. A* **28**, 2374 (1983).

AUTHOR INDEX

Numbers in parentheses are reference numbers and indicate that the author's work is referred to although his name is not mentioned in the text. Numbers in *italics* show the pages on which the complete references are listed.

Aartsma, T. J., 37(72, 86, 87), *44*
Abbott, J. P., 228(67), *277*
Abella, I. D., 31(58), 35(58), 36(58), *43*
Åberg, T., 116(11), 122(11), 125(11), *154*
Abraham, F. F., 218(18), 269(18), *276*
Abram, I. I., 5(35), 6(35), 22(35), *43*
Adam, M. Y., 141(135), 142(135), *158*
Admuir, T., 55(11), 110
Ågren, H., 117(40), 118(56, 60, 61), 127(40), 141(40), 142(40), 151(40), 152(40), *155, 156*
Aksela, H., 118(64), *156*
Aksela, S., 118(64), *156*
Albrecht, O., 100(100), 101(100), *112*
Alder, B. J., 219(45), 234(45), *277*
Alexanin, M., 259(95) *278*
Allan, C. J., 117(18), 133(114), 135(114), 136(114), *154, 157*
Allan, J. D., Jr., 133(106), 139(128), *157, 158*
Allen, L., 3(19), 4(19), *42*
Allison, D. A., 117(18), 133(114), 135(114), 136(114), 141(132), 142(132), 147(153), 154, *157–159*
Al'shits, E. I., 31(52), *43*
Anderson, P. W., 5(32), *42*
Andersen, H. C., 1(3), *41*, 269(100), *279*
Arneberg, R., 117(40, 41), 123 (41), 127(40), 141(40), 142(40), 147(41), 151(40), 152(40), *155*
Arnold, R., 99(97), *112*
Arnold, V. I., 190(14), 197(14), *213*
Asbach, G. C., 80(58), 109(58), *111*
Åsbrink, L., 117(17), 129(98), 131(95, 98), 136(17), 139(95, 98), 141(17, 98), 143(98, 140), *154, 157, 158*
Ashcroft, N. W., 219(43), 234(43), 250(43), 253(89), 255(43), 270(105, 106), *277–279*

Baer, Y., 115–118(2), *154*
Bagchi, B., 266(98), *279*
Bagus, P. S., 116(13), 121(73), *154, 156*
Baker, A. D., 115(4), 116(4), 121(4), 129(4), 131(4), 133(4), 136(4), 139(4), 141(4), 143(4), *154*
Baker, C., 115(4), 116(4), 121(4), 129(4), 131(4), 133(4), 136(4), 139(4), 141(4), 143(4), *154*
Baker, C. T. H., 230(72), *277*
Baker, G. A., 218(24), *276*
Baker, J., 117(36), *155*
Balle, T., 118(50), 126(50), 148(50), *155*
Bancroft, G. M., 133(109), *157*
Banna, M. S., 129(125), 139(125), 140)(125), *158*
Barker, J. A., 215(3), 217(3), 219(31), 221(3), 234(31), 267(3), 269(3), *275, 276*
Barnes, J. D., 56(14), 72(45), *110, 111*
Basilier, E., 129(100), 131(100), 139(124), 143(100), *157, 158*
Bastian, H., 80(58), 109(58), *111*
Bawn, C. E. H., 145(150), *159*
Baxter, R. J., 218(23), 243(81), 248(81), 249(85), 270(104), *276, 278, 279*
Bearman, M. Y., 219(30), 234(30), *276*
Bearman, R. J., 219(29, 30), 234(29, 30), 243(29), 244(29), *276*
Bellemans, A., 63–65(26), *110*
Benner, R. E., Jr., 233(75), 260(75), 263(75), *278*
Bergman, T., 115(1, 2), 116(1, 2), 117(2), 118(2), 122(1), *154*
Berkowitz, J., 115(5), 116(5), 121(5), 129(105), 133(105), *154, 157*
Berndtsson, A., 139(124), *158*
Berry, M. V., 162(3), 171(3), 197(3f), 198(3), 199(3), *213*
Beunker, R. J., 143(142), *158*
Bieri, G., 118(49), 120(49), 129(98), 131(98), 139(98), 141(98), 143(98), *155, 157*

Bigelow, R. W., 118(47), 126(47), 148(47), *155*
Bigotto, A., 99(95), *112*
Binder, K., 218(21), *276*
Birang, B., 31(55), 35(55), 36(55), *43*
Bizau, J. M., 117(20), 148(20), *154*
Blasenbrey, S., 51
Bloom, B., 162(7), 198(7), *213*
Bogges, G. W., 133(106), *157*
Boistelle, R., 69(32), *110*
Bonart, R., 108(113), *113*
Bongiorno, V., 220(50), 269(50), *277*
Borges da Costa, J. A., 57(17), 62(17), *110*
Born, G., 118(70), 126(70), *156*
Born, M., 150(161), *159*, 217(15), 221(15), *276*
Boyer, R., 97(84, 85), *112*
Bradshaw, A. M., 141(133), 142(133), 148(133, 158), 149(133), *158*, *159*
Brandrup, J., 98(93), *112*
Brewer, R. G., 2(9), *42*
Brey, J. J., 219(37), 234(37), 248(37), *276*
Brezinskii, W. L., 92(74), *112*
Brion, C. E., 115(6), 116(6), 117(16, 21, 22, 25), 118(66), 127(93), 129(93), 131(101), 133–136(93), *154*, *156*, *157*
Bristow, D. J., 133(109), *157*
Broadhurst, M. G., 69(33), *110*
Broyles, A. A., 219(27, 28, 46), 234(27, 28, 46), *276*, *277*
Bruna, P. J., 143(142), *158*
Brundage, R. T., 31(61), *43*
Brundle, C. R., 115(4), 116(4), 121(4), 129(4), 131(4), 133(4), 136(4), 139(4), 141(4), 143(4), *154*
Buenker, R. J., 139(129), 147(152), *158*, *159*
Buff, F. P., 230(70), *277*
Bunn, C. W., 100(99), *112*
Burke, F. P., 31(48, 49), 32–34(48), *43*
Bürkle, K. R., 99(97), *112*
Burns, M. J., 2(11), 3(11), 21(11), 22(11), *42*
Burum, D. P., 37(73), *44*

Cacelli, I., 117(35), 123(83), 124(83), 143(145), *155*, *156*, *159*
Caille, A., 100(101), *112*
Calviello, J. A., 31(54), 35(54), 36(54), *43*
Camilloni, R., 117(23), 123(80), *154*, *156*
Camli, R., 117(27), 123(27), 131(102), *155*, *157*
Cantow, H. J., 100(98), *112*
Caplan, A., 139(127), *158*
Caprace, G., 133(122), *157*
Carlson, T. A., 124(85), 133(116), 136(116), 139(128), *156–158*
Carravetta, V., 117(35), 123(83), 124(83), 143(145), *155*, *156*, *159*
Caudano, R., 129(100), 131(100), 143(100), *157*
Cavell, R. G., 141(132), 142(132), 147(153), *158*, *159*
Cederbaum, L. S., 116(12, 14), 117(14–17, 28–32, 38), 118(28, 32, 48, 49), 120(49), 122(31), 123(12, 14, 15, 38), 124(12, 28), 125(28), 126(28, 29, 32, 92), 127(32, 38, 92, 93), 129(93), 131(48), 133(93, 104), 134–136(93), 137(48, 119), 139(48, 129), 140(48), 141(48, 133, 137), 142(48, 133), 143(48), 147(12), 148(104, 133, 155, 158), 149(133), 151(162), 152(14, 38, 163, 164), 153(154), *154*, *155*, *157–159*
Chandler, D., 269(100), *279*
Chapman, D., 102(107), *112*
Chen, J. C. Y., 162(3), 171(3), 198(3), 199(3), *213*
Chester, C., 198(21), *214*
Chong, D. P., 118(46), *155*
Chraplyvy, A. R., 31(62), *43*
Chung, S. U., 219(28), 234(28), *276*
Ciullo, C., 117(27), 123(27), *155*
Coatworth, L. L., 133(109), *157*
Cocksey, B. G., 133(107), *157*
Cohen-Tannoudji, C., 5(34), *43*
Collatz, L., 226(63), *277*
Colle, R., 143(143), *159*
Collet, A., 71(40), *111*
Collin, J. E., 133(111), *157*
Connor, J. N. L., 162(3), 171(3), 197(3e), 198(3), 199(3), *213*
Cook, J. P. D., 117(21, 25), 127(93), 129(93), 133–136(93), *154*, *157*
Cooper, D. E., 1(5), *41*
Cornelius, P. A., 5(29), 6(29), 26(29), 27(29), 30(29), *42*
Courant, R., 165(9), *213*
Craievich, A. F., 71(39–42), 72(41), *111*
Cruickshank, D. W. J., 33(64), *44*
Csanak, G., 150(160), *159*
Cuillo, G., 131(102), *157*
Cummings, P. T., 250(87), *278*

Dale, J., 97(90), *112*
Danby, C. J., 133(107), *157*
Davidson, E. R., 139(130), 140(130), *158*
Davis, H. T., 219(38, 49), 220(49, 50), 233(75), 234(38), 247(49), 260(75), 263(75), 269(50), *276–278*
Davis, M. J., 162(3), 171(3), 198(3), 199(3), *213*
Davydov, A. S., 1(1), *41*
De Boer, J., 217(7), 219(41), 221(7), 234(41), 250(41), *276*, *277*
de Bree, P., 5(30), 6(30), 27(30), 30(30), 39(30), 49(30), *42*
de Gennes, P. G., 77(52), 79(52), *111*
Dehmer, J. L., 125(88), 141(134), 148(156), 150(149), *157–159*
Delos, J. B., 162(7), 198(7), *213*
del Rio, F., 219(33, 36), 234(33, 36), 248(33, 36), *276*
Delwiche, J., 133(111, 112), 146(151), *157*, *159*
Demekhin, V. F., 118(58), *156*
Demekhina, L. A., 118(58), *156*
Demicolo, I., 71(39–42), 72(41), *111*
Denny, L. R., 97(84), *112*
Dettenmaier, M., 76(50), *111*
Deverteuil, F., 100(101), *112*
DeVoe, R. G., 2(9), *42*
deVries, H., 4(21), *42*
Dey, S., 143(138), *158*
DiBartolo, B., 2(8), 5(8), 31(55), 35(55), 36(8, 55), *42*, *43*
Dicker, A. I. M., 37(29, 84), *44*
Diercksen, G. H. F., 118(48, 49), 120(49), 137(48), 139(129, 155), 140–143(48), 148(155), *155*, *158*, *159*
Dill, D., 125(88), 148(156), 150(149), *157*, *159*
Di Martino, V., 117(27), 123(27), *155*
DiMarzio, E. A., 84(64), 85(64), *111*
Dinse, K. P., 37(76) *44*
Dirlikov, S., 99(95), *112*
Dixon, A. J., 139(126), 140(126), 141(136), 142(136), 143(138), *158*
Dobkowski, J., 37(79), *44*
Dollhopf, W., 80(57), 109(57), *111*
Domb, C., 219(25), *276*
Domcke, W., 116(14), 117(14–17, 28, 30, 31, 38), 118(28, 48), 122(31), 123(14, 15, 38), 124–126(28), 127(38, 93), 129(93), 131(48), 133–136(93), 137(48, 119), 139(48), 140(48), 141(48, 133), 142(48, 133), 143(48), 148(133, 158), 149(133), 151(162), 152(163), 153(165), *154*, *155*, *157–159*
Doniach, S., 77–79(53), 86(53), 102(106), (76, 77), *111*, *112*
Doucet, J., 71(39–42), 72(41), *111*
Douglas, T. B., 35(67), *44*
Duppen, K., 37(75), *44*

Eastwood, E., 145(150), *159*
Eberhardt, W., 141(133), 142(133), 148(133), 149(133), *158*
Eberly, J. H., 3(19), 4(19), *42*
Ebner, C., 219(35), 220(35), 234(35), *276*, *277*
Ederer, D., 117(20), 141(134), 148(20), *154*, *158*
Edqvist, O., 143(140), *158*
Eland, J. H. D., 133(107) *157*
Engel, A., 80(57), 109(57), *111*
Enting, I. G., 218(23), *276*
Evans, R., 220(52), 247(52), *277*
Ewen, B., 73(46, 47), 75(46, 47), *111*

Fahlman, A., 115(1), 116(1), 122(1), 131(99), 143(99), *154*, *157*
Falls, A. H., 219(38), 234(38), *276*
Fantoni, R., 117(23, 27), 123(27), 131(102), *154*, *155*, *157*
Fayer, M. D., 1(3, 5), 2(12), 37(73), *41*, *42*, *44*
Fedoriuk, M. V., 162(6), 163(6), 173(6), *213*
Fetter, A. L., 8(44), 9(44), *43*, 122(76), 126(76), *156*
Fischer, E. W., 73(46, 47), 75(46, 47), 76(50), 109, *111*
Fisher, E. W., 31(54), 35(54), 36(54), *43*
Fisher, M. E., 218(19), 222(62), 249(86), 250(93), 255(91), 258(91, 93), 259(91, 93), 265(86), 266(19), 267(86), *276–278*
Fishman, S., 249(86), 250(93), 255(91), 258(91–93), 259(91–93), *278*
Fitchen, D. B., 17(46), *43*
Fix, G. J., 230(73), 231(73), 233(73), 249(73), *277*
Fleming, P. D., 220(51), *277*
Flory, P., 57(18), 70(37), 83(18), *110*, *111*
Foiles, S. M., 219(43), 234(43), 250(43), 253(89), 255(43), 270(106), *277–279*
Franconik, B. M., 56(14), *110*

Freed, K. F., 117(33), 118(69), 152(33), *155*, *156*
Freiborg, A., 37(80, 88) , *44*
Freund, H.-J., 118(45), *155*
Frey, R., 133(115), *157*
Friedel, J., 105(110), *113*
Friedman, B., 198(21), *214*
Fridh, C., 131(95), 139(95), *157*
Frost, D. C., 118(46), *155*
Fukuda, Y., 31(57), 35(57), 36(57), *43*
Fulinski, A., 229(69), *277*
Furukawa, J. T., 35(67), *44*
Fuss, I., 117(24), *154*

Garner, W. E., 69(35), 70(35), *111*
Gelius, U., 115(2), 116(2), 117(2, 18, 40), 118(2, 52), 127(40), 129(100), 131(99, 100), 133(114), 135(114), 136(114), 139(52, 124), 140(52), 141(40), 142(40), 143(99, 100), 151(40), 152(40), *154*, *155*, *157*, *158*
Genack, A. Z., 37(83), *44*
George, T. F., 162(3), 171(3), 198(3), 199(3), *213*
Giardini-Guidoni, A., 117(23, 26, 27, 80), 123(27), 131(102), *154–157*
Gibbs, J. H., 220(51), 266(48, 99), *277*, *279*
Gillan, M. J., 234(77), *278*
Ginnings, D. C., 35(67), *44*
Glasbeek, M., 37(82), *44*
Gleiter, H., 109
Godyaev, E. D., 31(52), *43*
Goldstein, H., 165(8), 201(8), *213*
Gordon, R. G., 7(43) *43*
Gorokhovski, A. A., 37(77), *44*
Goscinski, O., 122(75), *156*
Gotchev, B., 133(115), *157*
Gottfried, K., 123(79), *156*
Green, A. K., 118(68), *156*
Green, K. A., 258(94), 260(94), *278*
Green, H. S., 217(15), 221(15), *276*
Green, M. S., 217(9), 219(40), 221(9), 234(40), 250(40, 88), *276–278*
Green, T. J., 102(107), *112*
Grimm, F. A., 133(116), 136(116), 139(128), *157*, *158*
Groenveld, J., 217(7), 219(41), 221(7), 234(41), 250(41), *276*, *277*
Grossman, H. P., 98(92), 99(92, 97), *112*
Groth, P., 97(89), *112*
Gruler, H., 100(100), 101(100), *112*
Guerrero, M. I., 219(42), 234(42), 250(42), 255(42), *277*
Gussoni, M., 99(95), *112*

Ha, T.-K., 118(51), *155*
Hägele, P. C., 49(5), 108(114), *110*, *113*
Hall, C. K., 62(24), *110*
Halperin, B., 5(27), 36(68, 69), *42*, *44*, 46(2), *110*
Hammet, A., 115(6), 116(6), 131(101), *154*, *157*
Hanrin, K., 115(1, 2), 116(1, 2), 117(2), 118(2), 131(99), 143(99), *154*, *157*
Hansen, J. P., 264(96), *278*
Harris, C. B., 5(29), 6(29), 26(29), 27(29), 30(29), *42*
Hartmann, S. R., 31(58), 35(58), 36(58), *43*
Hashi, T., 31(57), 35(57), 36(57), *43*
Haymet, A. D. J., 219(48), 221(58), 234(48), 270(48, 58), *277*
Heading, J., 162(1), *212*
Heden, P. F., 115–118(2), *154*
Hedman, J., 115(1, 2), 116(1, 2), 117(1, 2), 118(2), 122(1), 139(124), *154*, *158*
Hegarty, J., 31(61), *43*
Heilbronner, E., 129(97), 131(97), *157*
Helfand, E., 62(20–23), 65(23), *110*
Helrich, W., 77(54), 78(54), *111*
Heller, E., 162(3), 171(3), 198(3), 199(3), *213*
Heller, Z. H., 31(54), 35(54), 36(54), *43*
Henderson, D., 215(3), 217(3), 219(31, 32), 221(3), 234(31, 32), 241(80), 267(3), 269(3), *275*, *276*, *278*
Herman, M. F., 117(33), 118(69), 126(69), 152(33), *155*, *156*
Hesselink, W. H., 2(10), 3(10), 37(72, 85), *42*, *44*
Hessler, J. P., 31(61), *43*
Hilbert, D., 165(9), *213*
Hill, T. L., 215(1), 217(1), 230(1), *275*
Hiroike, K., 217(10), 221(10), *276*
Hirth, J. P., 105(11), *113*
Hobbler, J. A., 118(50), 126(50), 148(50), *155*
Hochstrasser, R. M., 37(74), *44*
Höcker, H., 97(91), 100(98), *112*
Hoffman, J. D., 90(71), 109(115a), *112*, *113*
Hohlneicher, G., 118(45), 141(137), *155*, *158*
Hohne, G. W., 76(49), *111*
Holland, D. M. P., 141(134), *158*
Holland, V. F., 108(112), *113*

Holland-Moritz, K., 99(95), *112*
Holstein, T., 2(7), *42*
Holz, A., 57(17), 62(17), 80(59), 84(59), 68(67), 87(68), 92(75), 95(80), 96(81), 103(109), 109, *110–113*
Honjou, N., 117(37), *155*
Hood, S. T., 117(22), 131(101), 139(126), 140(126), *154*, *157*, *158*
Hotokka, M., 118(64), *156*
Howells, E. R., 100(99), *112*'
Huang, K., 150(161), *159*
Hubermann, B. A., 77–79(53), 86(53), (76, 77), *111*, *112*
Hubin-Franski, M. J., 133(112), 146(151), *157*, *159*
Huggins, M. L., 70(36), *111*
Hsu, D., 6(37–40), 7(41), 8(37), 9(37), 12(37), 13(37, 38), 14(40), 15(38), 16(38), 18(37), 19(38), 20(38), 22(38), 23(38, 39), 24(39), 25(39), 30(39), 31–33(40), 35(40), 36(38), 37(40) 38(39, 40), 40(39), *43*

Imbusch, G. F., 31(63), *43*
Immergut, E. H., 98(93), *112*
Imre, D., 11850), 126(50), 148(50), *155*
Ingram, M. G., 139(120), *158*
Ishinabe, T., 91(73), *112*
Ito, T., 109(115b), *113*

Jackson, J. L., 221(60), *277*
Jahnig, F., 102(104), 103(104), *112*
Jäntti, M., 118(64), *156*
Jennison, D. R., 118(63), *156*
Johansson, G., 115(1, 2), 116(1, 2), 117(2, 18), 118(2), 122(1), 131(99), 133(114), 135(114), 136(114), 143(99), *154*, *157*
Johnson, L. W., 37(84), *44*
Jones, G. L., 258(94), 260(94), *278*
Jones, K. E., 5(26), 9(26), 13(26), 15(26), 21(26), 26(26), 27(26), *42*
Jones, T. B., 129(97), 131(97), *157*

Kaliteevskii, M. Yu., 37(70), 38(70), 41, *44*
Karlson, S. E., 115(1), 116(1), 122(1), *154*
Karlsson, L., 115–117(3), 136(117), *154*, *157*
Karplus, M., 48(3, 4), *110*
Karwasz, G., 150(148), *159*
Katrib, A., 143(139), *158*
Kay, H. F., 97(86, 87), *112*
Keith, H. D., 80(61), 108(61), *111*
Kelber, J. A., 118(63), *156*
Keller, J. B., 162(3), 171(3), 198(3), 199(3), *213*
Keller, H. B., (66) *277*
Keller, P. R., 139(128), *158*
Kern, C. W., 48(4), *110*
Khan, I., 148(158), *159*
Kikuchi, M., 31(56), 35(56), 36(56), *43*
Kim, D., 218(24), *276*
Kim, Q., 31(59), *43*
King, A. M., 69(35), 70(35), *111*
Kirkwood, J. G., 217(13), 219(45), 221(13), 234(45), *276*, *277*
Kitaigorodskii, A. J., 68(30), *110*
Kitaigorodskii, S. I., 33(66), *44*
Kizel', V. A., 31(50), *43*
Klasson, M., 139(124), *158*
Klein, L. S., 221(60), *277*
Klein, M., 219(40), 234(40), 250(40), *277*
Klein, W., 218(22), *276*
Kloplenstein, C. E., 118(50), 126(50), 148(50), *155*
Kluge, G., 118(43), *155*
Knudson, S. K., 162(7), 198(7), 212, *213*
Kobayashi, M., 91(72), *112*
Koch, E. E., 115(8), *154*
Koenig, T., 118(50), 126(50), 148(50), *155*
Kondratenko, A. V., 118(57), *156*
Koni, Yu. Ya., 37(88), *44*
Koningstein, J. A., 36(69), *44*
Koopmans, T., 120(72), *156*
Köppel, H., 151(162), 152(163, 164), 153(164), *159*
Korotaev, O. N., 37(70), 38(70), 41, *44*
Kosterlitz, J. M., 46(1), 86(1), 88(1), 92(1), *110*
Kovač, B., 133(110), *157*
Kovac, J., 109
Kovacs, A., 90(69), *111*
Kozak, J. J., 258(90, 94), 260(94), *278*
Kraemer, W., 118(48), 131(48), 137(48), 139–142(48), 148(155), *155*, *159*
Kraemers, H. A., 64(27), *110*
Krause, M. O., 133(116), 136(116), 139(128), 147(154), *157–159*
Krivoglaz, M. A., 5(25), 9(25), 13(25), *42*
Krüger, J. K., 76(49), 80(58), 100(98), 109(58), *111*, *112*
Kruizinga, B., 1(6), *42*
Krumhansl, J. A., 238(79), *278*

Krummacher, S., 117(19, 20), 148(19, 20), 152(19), *154*
Kubo, R., 5(31), 8(45), 9(45), *42*, *43*
Kunz, C., 115(7), *154*
Kuppermann, A., 139(121), *158*
Kurnit, N. A., 31(58), 35(58), 36(58), *43*
Kushida, T., 31(56), 35(56), 36(56), *43*
Kvalheim, O. M., 118(62), *156*

Lablanquie, P., 141(135), 142(135), *158*
Lado, F., 270(106), *279*
Lamberg, W. R., 37(82), *44*
Landau, L. D., 60(19), *110*, 162(2), 171(2), *212*
Langer, J. S., 218(20), 266(20), *276*
Langhoff, P. W., 117(38), 123(38), 127(38), 150(160), 152(38), *155*, *159*
Langhoff, S. R., 117(38), 123(38), 127(38), 152(38), *155*
Larkin, A. I., 94(79), *112*
Larkins, F. P., 117(26), *155*
La Roe, R. R., 118(68), *156*
Lau, W. M., 118(46), *155*
Lax, M., 7(42), *43*
Leal, E. P., 150(160), *159*
Lebowitz, J. L., 219(26), 264(26), *276*
Leclerc, B., 146(151), *159*
Lee, E. K., 258(94), 260(94), *278*
Lee, H. W. H., 1(5), 37(73), *41*, *44*
Lee, Y. T., 139(122), *158*
Levesque, D., 219(39, 47), 234(39, 47), 241(47), 243(47), 244(47), 250(39, 47), 251(47), 252(47), *277*, *279*
Levine, Z. H., 125(89), 148(89), *157*
Levinson, H. J., 141(133), 142(133), 148(133), 149(133), *158*
Levinson, V. A., 219(45), 234(45), *277*
Liegener, C.-M., 118(65), *156*
Lifschitz, E. M., 162(2), 171(2), *212*
Lifshiftz, E. M., 60(19), *110*
Lifson, S., 55(9, 10), *110*
Lin, C. D., 126(90), *157*
Lincoln, W. W., 258(90), *278*
Lindberg, B., 115(1), 116(1), 122(1), *154*
Lindenmeyer, P. H., 108(112), *113*
Lindgren, I., 115(1), 116(1), 122(1), *154*
Lindholm, E., 131(95), 139(95), 143(140), *157*, *158*
Liu, W. K., 2(11), 3(11), 21(11), 22(11), *42*
Lochte-Holtgreven, W., 145(150), *159*
Longmire, M. S., 55(11), *110*
Lonsdale, K., 68(28b), *110*
Lothe, J., 105(111), *113*
Lovett, R. A., 230(70), *277*
Loring, R., 1(3), *41*
Lublin, D. M., 77–79(53), 86(53), *111*
Lucchese, R. R., 124(85), 148(157), *156*, *159*
Luks, K. D., 258(90, 94), 260(94), *278*
Lundqvist, S., 118(53), *155*
Lynch, M. G., 150(149), *159*
Lyo, S. K., 2(7), *42*

McCarthy, I. E., 116(9, 10), 117(27), 123(10, 27, 80), 141(136), 142(136), 143(138), *154–156*, *158*
McClure, D. W., 90(70), *112*
McCoeskey, R. E., 35(67), *44*
McCullough, R. L., 55(12), *110*
McCumker, D. E., 5(23, 24), 9(23, 24), 13(23, 24), 15(23), 31(23, 63), 34(23), 35(23), 36(23), *42*, *43*
McDowell, C. A., 118(46), *155*
Macfarlane, R. M., 2(9), 37(73, 78, 83), *42*, *44*
Machado, L. E., 150(160) *159*
Maciag, K., 150(148), *159*
McKoy, B. V., 150(160), *159*
McKoy, V., 124(85), 148(157), *156*, *159*
McMillan, W. L., 78(56), 79(56), *111*
Madden, W. G., 219(48), 221(58), 234(48), 270(48, 58), *277*
Magni, R., 99(95), *112*
Maier, J. P., 129(97), 131(97), *157*
Malmqvist, P.-Å., 117(40), 127(40), 129(100), 131(100), 141(40, 135), 142(40, 135), 143(100), 151(40), 152(40), *155*, *157*, *158*
Malzahn, K., 99(96), *112*
Mandel, F., 219(30), 234(30), *276*
Manne, R., 115(2), 116(2), 117(2, 40, 41), 118(2), 123(41), 127(40), 141(40), 142(40), 147(41), 151(40), 152(40), *154*, *155*
Maradudin, A. A., 2(14), 5(14), 21(14), *42*
Marcelja, S., 101(103), *112*
Marcus, R. A., 162(3), 171(3), 198(3), 199(3), *213*
Maripuu, R., 136(117), *157*
Marks, S., 5(29), 6(29), 26(29), 27(29), 30(29), *42*
Maroncelli, M., 56(15), 57(15), 73(15), 74(15), 76(15), 93(15), *110*
Martin, R. L., 123(81), 139(130), 140(130), *156*, *158*

Maslov, V. P., 162(4, 6), 163(6), 173(6), *213*
Mason, E. A., 55(11), *110*
Massih, A. R., 102(105), *112*
Mattsson, L., 136(117), *157*
Mazalov, L. N., 118(57), *156*
Medeiros, J. F. N., 86(67), *111*
Meeron, E., 217(8), 221(8, 56), *276, 277*
Mehaffy, D., 139(128), *158*
Meldner, H. W., 148(159), *159*
Messiah, A., 162(2), 171(2), *212*
Meyer, W., 131(103), *157*
Mier y Terán, L., 219(33, 36, 38), 234(33, 36, 38), 248(33, 36), *276*
Migdal, A. B., 122(77), 139(127), *156, 158*
Miller, W. H., 162(3), 171(3), 198(3), 199(3), *213*
Minchington, A., 117(24, 26), *154, 155*
Mintz, D. M., 139(121), *158*
Mishra, M., 117(34), 147(34), *155*
Mnyukh, Y. V., 69(31), *110*
Moccia, R., 117(35), 123(83), 124(83), 143(145), *155, 156, 159*
Moerner, W. E., 31(62), *43*
Mohanty, U., 266(98), *279*
Molenkamp, L. W., 37(71, 75), 39(71), 40(71), *44*
Montroll, E. W., 80(62), 83(62), *111*
Moore, J. H., 139(127), *158*
Morin, P., 141(135), 142(135), *158*
Morita, T., 217(10), 221(10), *276*
Morsink, J., 37(86), *44*
Morsink, J. B. W., 1(6), 37(75), *42, 44*
Moscardó, F., 143(144), *159*
Mostoller, M., 31(59), *43*
Mou, C. Y., 230(70), *277*
Müller, A., 67(28), 68(28b), 73(28), *110*
Müller, J., 117(39–41), 123(41), 126(39), 127(40), 141(40), 142(40), 147(41), 151(40), 152(40), *155*
Müller, M., 100(98), *112*
Munster, A., 215(2), 217(2), 230(2), *275*
Muramoto, T., 31(57), 35(57), 36(57), *43*
Murphy, R. D., 219(32), 234(32), *276*
Müser, H., 109

Nabarro, F. R. N., 85(66), 105(66), *111*
Naghizadeh, J., 57(16, 17), 62(16, 17), 80(59), 84(59), 87(68), 95(78), 102(105), *110–112*
Nagle, J. F., 100(102), 102(108), *112*
Naiman, C. S., 31(55), 35(55), 36(55), *43*
Nakama, K., 109(115b), *113*
Nakatsuji, H., 117(42), 126(42), 133(42), 136(42), 139(131), 140(131), *155, 158*
Nashed, M. Z., 226(63), *277*
Natalis, P., 133(111, 112), *157*
Neiman, D. M., 118(57), *156*
Nelson, D. R., 46(2), 77(55), 78(55), *110, 111*
Nenner, I., 141(135), 142(135), *158*
Newman, B. A., 97(86, 87), *112*
Nicollin, D., 36(69), *44*
Nilsson, R., 139(124), *158*
Noach, F., 72(44), *111*
Nohre, C., 136(117), *157*
Noid, D. W., 162(3), 171(3), 198(3), 199(3), *213*
Nordberg, R., 115(1), 116(1), 122(1), *154*
Nordgren, J., 118(56), *156*
Nordling, C., 115(1, 2), 116(1, 2), 117(2), 118(2, 56), 122(1), 131(99), 139(124), 143(99), *154, 156–158*
Norell, K. E., 136(117), *157*
Nozières, P., 122(78), *156*
Nyi, C. A., 37(74), *44*

Ohno, M., 118(54), 147(54), *155*
Ohrn, Y., 117(34), 118(70), 126(70), 147(34), *155, 156*
Olson, R. W., 1(5), 37(73), *41, 44*
Orbach, R., 2(7), *42*
Orchard, A. F., 133(108), *157*
Orlowski, T. E., 37(89), *44*
Ornstein, L. S., 216(5), *275*
Ortega, J. M., 266(64), *277*
Osad'ko, I. S., 2(15), 4(15), 5(36), 6(36), 12(36), 13(15), 17(15), 18(36), 22(36), 25(15), 31(15), 37(15, 36), 38(15, 36), *42, 43*

Paniagua, M., 143(144), *159*
Parks, C. C., 118(68), *156*
Parr, A. C., 141(134), 148(156), *158, 159*
Parr, R. G., 48(3), *110*
Pars, L. A., 201(19), *214*
Passaglia, E., 80(61), 108(61), *111*
Patterson, F. G., 1(5), 37(73), *41, 44*
Pavlin, Z., 266(99), *279*
Pawley, G. S., 33(65), *44*
Peatman, W. B., 133(115), *157*
Pechhold, W., 49(5), 51, 80(57), 109(57), *110, 111*
Pechukas, P., 162(5), 212, *213*

Peetz, L., 76(49), *111*
Penrose, O., 219(26), 264(26), *276*
Percival, I., 162(5), *213*
Percus, J. K., 217(6), 221(6, 54), *276*, 277
Peredecki, P., 80(60), 108(60), *111*
Perez, J. D., 148(159), *159*
Perry, J. W., 37(82), *44*
Personov, R. I., 31(51, 52), *43*
Petersson, J., 109
Peyerimhoff, S. D., 139(129), 143(142), 147(152), *158, 159*
Pickup, B. T., 122(75), 123(82), *156*
Picszek, W., 73(46, 47), 75(46, 47), 99(95), *111, 112*
Pietralla, M., 76(49), 80(58), 109(58), *111*
Pikin, S. A., 94(79), *112*
Pines, D., 122(78), *156*
Pink, D., 100(101), 102(107), *112*
Pireaux, J. J., 129(100), 131(100), 143(100), *157*
Pitzer, R. M., 48(4), *110*
Pollak, H., 133(115), *157*
Pollard, J. E., 139(122), *158*
Poston, T., 211(20), *214*
Potts, A. W., 131(96), 133(113), 136(113), 143(96, 141), *157*, *158*
Potts, W. A., 139(123), 141(123), *158*
Poulin, A., 146(151), *159*
Powell, R. C., 31(55, 59), 35(55), 36(55), *43*
Price, W. C., 131(96), 143(96), *157*
Prigogine, I., 162(5), *213*

Qi, S. P., 56(15), 57(15), 73(15), 74(15), 76(15), 93(15), *110*

Rabalais, J. W., 115(5), 116(5), 121(5), 143(139), *154, 158*
Rahman, A., 238(78), *278*
Ramaker, D. E., 118(59, 67), *156*
Rand, S. C., 2(9), *42*
Rebane, K. K., 2(16), 4(16), *42*
Rebane, L. A., 37(77, 80, 81, 88), *44*
Redfield, A. G., 5(33), *43*
Rehn, V., 118(68), *156*
Reineck, I., 136(117), *157*
Rheinbolt, W. C., 226(64), 228(68), *277*
Rice, S. A., 31(53), *43*, 162(5), *213*, 219(48), 222(58), 234(48), 270(48, 58), *277*
Richards, J. L., 31(53), *43*
Richardson, N. V., 133(108), 148(158), *157, 159*
Richartz, A., 143(142), *158*
Riebel, K., 97(91), *112*
Riks, E., 227(65), *277*
Robin, M. B., 150(147), *159*
Romero, F., 219(37), 234(37), 248(37), *276*
Root, L., 1(2), *41*
Rosenberg, R. A., 118(68), *156*
Rosenfeld, Y., 270(105), *279*
Rowlinson, J. S., 219(42), 234(42), 250(42), 255(42), *277*
Roy, D., 133(112), 146(151), *157, 159*
Rubin, R. J., 84(64, 65), 85(64), *111*
Rubinow, S. I., 162(3), 171(3), 198(3), 199(3), *213*
Rublein, G., 212
Rushbrooke, G. S., 217(12), 221(12), *276*
Ryckaert, J. P., 63–65(26), *110*
Rye, R. R., 118(63), *156*

Saam, W. F., 219(35), 220(35), 234(35), *276, 277*
Sackmann, E., 100(100), 101(100), *112*
Saddei, D., 118(45), *155*
Sahlin, H. L., 219(28), 234(28), *276*
Salem, S., 51(6), 54(6), *110*
Saltpeter, E. E., 221(57), *277*
Salvetti, O., 143(143), *159*
Sambe, H., 118(59), *156*
Samulski, E. T., 97(83), *112*
San-Fabian, E., 143(144), *159*
Santos, A., 219(37), 234(37), 248(37), *276*
Sapozhnikov, M. N., 2(17), 31(50), *42, 43*
Sasajima, T., 117(37), *155*
Sasaki, F., 117(37), *155*
Saville, G., 219(42), 234(42), 250(42), 255(42), *277*
Scalise, O. H., 241(80), *278*
Schachtschneider, J. H., 56(13), *110*
Schaefer, H. F., 118(71), 123(71), *156*
Schawlow, A. L., 31(60, 63), *43*
Scherer, C., 57(16, 17), 62(16, 17), *110*
Schlag, E. W., 133(115), *157*
Schirmer, J., 116(14), 117(15–17, 29–32, 38), 118(32, 48, 49), 120(49), 122(31), 123(14, 15, 38), 126(29, 32, 92), 127(32, 38, 92, 93), 129(93, 94), 131(48), 133–136(93), 137(48, 119), 139(48), 140–143(48), 148(155), 152(14, 38, 94, 163), *154, 155, 157–159*
Schmidt, J., 5(29), 6(29), 26(29), 27(29), 30(29), *42*
Schmidt, V., 117(19, 20), 148(19, 20),

152(19), *154*
Scholz, M., 118(43, 44), 126(44), 148(44), *155*
Schwartz, M. E., 121(74), *156*
Schweig, A., 118(44), 126(44), 148(44), *155*
Schweitzer, G. K., 133(106), *157*
Schwickert, H., 99(94), 100(94), *112*
Sckurda, J., 109(115b), *113*
Scott, W. C., 31(60), *43*
Scriven, L. E., 219(38, 49), 220(49), 233(75), 234(38), 247(49), 260(75), 263(75), *276–278*
Selander, L., 118(56), *156*
Sergio, A., 131(102), *157*
Sgamellotti, A., 117(27), 123(27), 131(102), *155, 157*
Shatzki, T. F., 63(25), *110*
Shearer, H. M., 68(29), *110*
Shelby, R. M., 2(9), 5(29), 6(29), 26(29), 27(29), 30(29), 37(73), *42, 44*
Sh'ih, S. K., 139(129), *158*
Shirley, D. A., 123(81), 129(125), 139(122, 125), 140(125), *156, 158*
Shoemaker, R. L., 2(9), *42*
Shpol'skii, E. V., 31(51), *43*
Siegbahn, H., 115(3), 116(3), 117(3, 18), 118(60), 133(114), 135(114), 136(114), *154, 156, 157*
Siegbahn, K., 115(1, 2), 116(1, 2), 117(2, 18, 56), 118(2), 122(1), 129(100), 131(99, 100), 133(114), 135(114), 136(114, 117), 143(99, 100), *154, 156, 157*
Siegel, J., 150(149), *159*
Sievers, A. J., 31(62), *43*
Silbey, R., 3(18), 13(18), *42*
Silsbee, R. H., 5(22), *42*
Simons, J., 126(91), *157*
Skinner, J. L., 1(2), 3(20), 6(37–40), 7(41), 8(37), 9(20, 37), 12(37), 13(37, 38), 14(40), 15(38), 16(38), 18(20, 37), 19(38), 20(38), 22(38), 23(38, 39), 24(39), 25(39), 30(39), 31–33(40), 35(40), 36(38), 37(40), 38(20, 39, 40), 40(39), *41–43*
Skolnick, J., 62(23), 65(23), *110*
Small, G. J., 5(28), 6(28), 21(28), 26(28), 27(28), 31(48, 49), 32–34(48), 41, *42, 43*
Smith, D. D., 37(82), *44*
Smith, W. D., 126(91), *157*
Smith, W. R., 241(80), *278*
Snell, W., 118(50), 126(50), 148(50), *155*
Spivak, M., 211(22), *214*
Southworth, S., 118(50), 126(50), 148(50), *155*
Soven, P., 125(89), 148(89), *157*
Sovers, O. J., 48(4), *110*
Stamm, M., 76(50), *111*
Starace, A. F., 125(87), 147(87), *157*
Statton, W. O., 80(60), 108(60), *111*
Stefani, G., 117(23), 123(80), *154, 156*
Steidle, N., 76(50), *111*
Stell, G., 221(55), 250(87), 266(55), *277, 278*
Stewart, I., 211(20), *214*
Stockbauer, R., 139(120), *158*
Stohrer, M., 72(44), *111*
Strang, G., 230(75), 231(73), 233(73), 249(73), *277*
Straupe, J. F. N., 90(69), *111*
Strauss, H., 56(15), 73(15), 74(15), 76(15), 93(15), *110*
Streets, D. G., 139(123), 141(123), 143(141), *158*
Strobl, G., 54(7), 55(7), 71(43), 73(46, 47), 75(7, 46–48), 76(7, 50), 94(7), 99(94, 96), 100(94), 109(43), 109, *110–112*
Stroud, D., 219(35), 234(35), *276*
Sturge, M. D., 5(23), 9(23), 13(23), 15(23), 31(23, 63), 34–36(23), 41, *42, 43*
Sukhorukov, V. L., 118(58), *156*
Suzuki, I. H., 117(22), *154*
Svensson, S., 117(40), 127(40), 129(100), 131(100), 139(124), 141(40, 135), 142(40, 135), 143(100), 151(40), 152(40), *155, 157, 158*
Swartz, B., 232(74), *278*
Synder, R. G., 56(13–15), 57(15), 73(15), 74(15), 76(15), 77(15), 93(15), 97(15), *110–111*
Szmytkovski, C., 150(148), *159*

Tan, K. H., 117(16), *154*
Tapia, R. A., 226(62), *277*
Tarantelli, F., 117(27), 123(27), 131(102), *155, 157*
Tau, K. T., 117(21), *154*
Taylor, J. W., 139(128), *158*
ter Haar, D., 172(12), 173(12), *213*
Thom, R., 197(18), *214*
Thoules, D. J., 46(1), 86(1), 88(1), 92(1), *110*
Throop, J., 219(29), 234(29), 243(29), 244(29), *276*
Thunemann, K. H., 139(129), *158*
Timoshevskaya, V. V., 118(58), *156*
Tiribelli, R., 117(23), *154*
Tomita, K., 5(31), *42*

Trommsdorff, H. P., 37(75, 83), *44*
Toner, J., 77(55), 78(55), *111*
Tossell, J. A., 139(127), *158*
Trevor, D. J., 139(122), *158*
Trzebiatowski, T., 97(88), 99(94), 100(94), *112*
Turner, D. W., 114(4), 116(4), 121(4), 129(4), 131(4), 133(4), 136(4), 139(4), 141(4), 143(4), *154*

Unger, G., 71(38), *111*
Unroh, H. G., 109
Unwin, R., 148(158), *159*
Ursell, F., 198(21), *214*

Van Bibber, K., 69(35), 70(35), *111*
Vand, V., 68(29), *110*
van der Waals, J. H., 37(78, 83, 84), *44*
van der Wiel, M. J., 118(66), *156*
van Kamper, N. G., 248(83), *278*
van Leeuwen, J. M. J., 217(7), 219(41), 221(7), 234(41), 250(41), *276, 277*
van't Hof, C. A., 5(29), 6(29), 26(29), 27(29), 30(29), *42*
Veenhuizen, H., 136(117), *157*
Verlet, L., 217(11), 219(39), 221(11), 234(39), 250(39), 264(96), 270(101, 102), *276–279*
Vezzetti, D. J., 162(3), 171(3), 198(3), 199(3), *213*
Vigren, D. T., 84(63), 87(68), 103(109), *111, 113*
Viinikka, W.-K., 116(13), *154*
Völker, S., 37(78, 79, 83, 84), *44*
Volterra, V., 226(63), *277*
von Niessen, W., 116(4), 117(14–17, 30–32, 38), 118(32, 48, 49), 120(49), 122(31), 123(14, 15, 38), 126(32), 127(32, 38, 93), 129(93), 131(48), 133(93, 104), 134–136(93), 137(48, 119), 139(48, 129), 140(48), 141(48, 137), 142(48), 143(48), 148(104, 155), 152(14, 38), *154, 155, 157–159*
Vrij, A., 70(37), *111*

Wäckerle, G., 37(76), *44*
Walecka, J. D., 8(44), 9(44), *43*, 122(76), 126(76), *156*
Wallace, S., 125(88), *157*
Walter, O., 126(92), 127(92), 129(94), 152(94), *157*
Wang, C. S., 162(3), 171(3), 198(3), 199(3), *213*
Wang, S., 238(79), *278*
Wannberg, B., 136(117), *157*
Warren, W. S., 1(4), *41*
Warshel, A., 55(9, 10), *110*
Wasserman, Z. R., 62(22), *110*
Watson, K. M., 162(3), 171(3), 198(3), 199(3), *213*
Watts, R. O., 215(4), 217(4), 219(31, 34, 44), 221(4), 228(34, 44), 230(4), 234(31, 34, 44), 241(34, 44), 243(34), 244(34), 248(34), 250–252(44), 255(44) *275–277*
Weber, T. A., 26(22), *110*
Weeks, J. D., 269(100), *279*
Weigold, E., 116(9, 10), 117(22, 24, 26), 123(10), 139(126), 140(126), 141(136), 142(136), 143(138), *154, 155, 158*
Weir, C. E., 109(115a), *113*
Wendin, G., 118(53–55), 147(54, 55), *155*
Wendroff, B., 232(74), *278*
Werme, L. O., 115–118(2), *154*
Wertheim, M. S., 230(71), 270(103), *277, 279*
Wertheimer, R., 3(18), 13(18), *42*
West, J. B., 141(134), *158*
White, M. G., 127(93), 129(93), 133–136(93), *157*
Widom, B., 221(59, 61), *277*
Wiersma, D. A., 1(6), 2(10, 13), 3(10), 4(21), 5(30), 6(30), 27(30), 30(30), 37(71–73, 75, 85–87), 39(30, 71), 40(30, 71), 41, *42, 44*
Wieselek, R. A., 118(50), 126(50), 148(50), *155*
Wildner, W., 76(49), *111*
Williams, D. E., 54(8), *110*
Williams, G. R. J., 117(22), 139(126), 140(126), 141(136), 142(136), *154, 158*
Williams, T. A., 131(96), 133(113), 136(113), 143(96), *157*
Wilson, R. M., 117(26), *155*
Wilson, T. M., 31(59), *43*
Wilson, W. L., 37(73), *44*
Wobser, G., 108(114), *113*
Wokaun, A., 2(9), *42*
Wright, G. R., 118(66), *156*
Wuilleumier, F., 117(19, 20), 147(154), 148(19, 20), 152(19), *154, 159*
Wunderlich, B., 49(82), 69(34), 71(34), 97(82), *110, 112*

Yang, A. J., 266(99), *279*
Yang, A. J. M., 220(51), *277*
Yannoni, C. S., 2(9), *42*
Yeager, D. L., 117(33), 118(69), 126(69), 152(33), *155, 156*
Yen, W. M., 31(60, 63), *43*
Yevick, G. J., 217(6), 221(6), *276*
Yonezawa, T., 117(43), 126(42), 133(42), 136(42), *155*
Yousif, M., 118(59), *156*
Yvon, J., 217(14), 221(14), *276*
Zerbi, G., 99(95), *112*
Zernike, F., 216(5), *276*
Zewail, A. H., 1(4), 2(11), 3(11), 5(26), 9(26), 13(26), 15(26), 21(11, 26), 22(11), 26(26), 27(26), 37(82, 89), *41, 42, 44*
Zhdavov, S. A., 5(36), 6(36), 12(36), 18(36), 22(36), 37(36), 38(36), *43*
Zimmermann, H., 37(76), *44*
Zittlau, W., 118(44), 126(44), 148(44), *155*
Zuckermann, M. J., 100(101), 103(109), *112, 113*

SUBJECT INDEX

Absorption, 37, 39, 41
Absorption line shape, 3, 7–8
Absorption spectrum, 31
Acetylene:
  extended 2ph-TDA, 150–152
  photoelectron spectra, 148–151
Acoustic phonons, 4–5, 7, 23, 31–36, 40
  coupling to, 14–19
  Debye model of, 5, 14
  weak coupling theory, 21
Adhesive hard-sphere potential, 249–267
  thermodynamic singularity, 249
  unphysical PY solutions, 249
Adiabatic approximation, 150
Airy function, 173, 186, 198
Ammonia, ionization, 117
Angle-resolved photoelectron spectrum, hydrocarbons, 139
Anharmonic phonos, 30, 40–41
Approximation:
  adiabatic, 150
  asymptotic, 171–172
  BGYK, 217
  HNC, 217
  JWKB, 161
  Kirkwood superposition, 238, 256
  PY, 217
  semiclassical, 161
  stationary phase, 187–190
  sudden, 125
  Swartz Wendroff, 232, 242
  uniform, 172, 198
Arrhenius temperature dependence, 4–5, 25, 37
Asymmetry parameters, 148–153
Asymptotic approximations, 171–172
  formal, 200
  global, 185–198
Asymptotic expansion, formal, of ψ, 208–210
Auger effect, 118
Autoionization, 118, 125

Baxter transformation, 248–249, 254
BGYK equation, 255–262
  critical behavior:
    analytical, 258
    numerical, 260
  definition, 217
  direct correlation function, 238
  high-density breakdown, 238
  HNC comparison, 239
  integral equation, 223
  low temperature, 255
  singularity behavior, 258
  solution space structure, 262, 263, 268
  spinodal density, 256–257
  turning point location, 256
Binary (e,2e) spectroscopy, 116, 123
  carbonyl sulfide, 133
Binding energy, 119
Bloch equations, 3–4
  optical, 3
Born-Green-Yvon hierarchy, *see* BGYK equation
Born-Oppenheimer approximation, 150
Borrow intensity, 134–136
Breakdown of molecular orbital picture of ionization, 117–153
  hydrocarbons, 137–143
  methane, 129–139
Breakdown phenomenon, 117
Brillouin spectroscopy, 76, 109
Brownian bistable oscillator, 67
Buckingham potential, 48
Bulk modulus, isothermal, 216, 264–265
  alternate microscopic expressions, 216–247
  discretization, 242
  negative values, 247
  van der Waals loop, 248
Burgers circuit, 105–106

Cadmium dichloride, 133
Canonical chart, 191

Carbon dioxide, ionization, 117
Carbon disulfide, ionization, 117
Carbon monoxide, 148
  ionization, 116
Carbonyl sulfide, 133–136
  binary (e,2e) spectroscopy, 133
  correlation satellites, 136
  dipole (e,2e) spectroscopy, 133
  photoelectron spectrum, 133
  satellite lines, 134
  shake-up satellites, 134–136
Catastrophes, 197–198
Caustics, 167, 197–198
  cusps, 198
  folds, 198
  stable, 198
Chain axis slip, 108
Chain folding, 86–90
Chapeau functions, 231
Characteristic function, 167
Charts, 211
  canonical, 191
  singular, 191
Classical density, 170–171
  mixed spaces, 179
Cluster expansion, 223–224, 266
  closure comparison, 223–224
  covariance factor, 271
  potential distribution theory, 271
Clusters, physical, 224, 265, 267
  cutoff suppression, 267–268
  origin of singularity, 267–270
C_2N_2, 152
Coexistence curve, 218, 266
  analytic continuation, 219, 269
  perturbation theories, 269
  singularity, 218, 264
Configuration-interaction, 126–127, 139
  hydrocarbons, 139–141
Configurations:
  electron, 120–121
  single-hole, 147
Conformational transitions, 47, 61–67
  crank shaft, 63
  type 1, 63
  type 2, 63
  type 3, 63
Conformations:
  all trans, 50
  gauche, 50
  helical, 50
  jog, 50
  kinks, 50
  trans, 50
Continuation, parametric, 227–228, 245, 263
Continuum states, 129,
Correlation effects, 116–153
  final-state, 147
  ionization, 116–153
Correlation function, 11
  direct, 216
  inhomogeneous, 219
  pair, 215–216
  subcritical fluids, 215–275
Correlation length, 224
Correlation satellites, 122
  carbonyl sulfide, 136
  hydrocarbons, 140–141
  initial-state, 122, 152
  methane, 131–136
  zinc dichloride, 132
COS, 152
Coupling:
  electron phonon, 3
  interchannel, 123, 125, 150
  vibrionic, 151–152
Coupling theory, weak, 13–15, 21, 23, 35
Covariance, potential distribution theory,
    222, 271
Covariance factor, cluster expansion, 271
Critical behavior, BGYK, 258–260
Critical point, 218, 224
  BGYK, 258, 262–263
  closure effect, 235
  HNC, 250, 254–255
  pair potential range, 240–241
  Crystals, 1–41
  heat capacity, 36
Crystal structure:
  herring bone, 68, 75
  hexagonal, 68, 71
  monoclinic, 68
  orthorhombic, 68
  pseudohexagonal, 68, 71
  triclinic, 68
Cumulant average, 8
Cycloparaffins, 53, 99–100

Debye model, 14, 36, 40
Debye temperature, 32, 35

Defects, 78–90
  chain folding, 86–90
  classification, 78
  dislocation loops, 83–86
  paraffins, 78–90
Density:
  classical, 170–171
  local, inhomogeneous fluid, 229
Density function, in mixed spaces, 206–208
Density functional formalism, 220
Dephasing:
  optical, 1–41. *See also* Optical dephasing.
  pure, 2, 31
Diagrammatic perturbation theory, 9
1,3-Diazaazulene, in naphthalene, 31–34
3,4,6,7-Dibenzopyrene, in N-octane, 37–39
Diffeomorphism, 176, 211
Dimethylisoindene, 147
Dipalmtoyl phosphatidylcholin (DPPC), 100
Dipole (e,2e) spectroscopy, 116, 123
  carbonyl sulfide, 133
Direct correlation function, 216
  BGYK equation, 238
  inhomogeneous fluid, 220, 230
  physical structure, 216, 223
Direct process, 36
Disclination loops, 78
Dislocation loops, 83–86
Dislocations, 103–108
  edge, 106–107
  movement, 108–109
Droplet model, 218, 266

Edge dislocations, 103–108
Einstein, model, 19, 22
Elastic constants, 108–109
  paraffins, 108–109
Electron configurations, 120–121
Electron-phonon coupling, 3
Electron-phonon interactions, 2–3
Elementary catastrophes, 197
Energy:
  binding, 119
  ionization, 119
  orbital, 120
  orientational, 54–55
  thermodynamic, 217, 225, 240
Equation-of-motion method, 126
Equations:
  BGYK, 217, 223, 238–239, 255–262
  Bloch, 3–4
  HNC, 217, 222–223, 250–255
  motion of chain, 57–62
    generalized coordination, 57–59
    mass tenor, 59–61
    normal modes, 61–62
    torsional specific heat, 61–62
  OZ, 216, 230, 265
  PY, 217, 222, 243–250
  Schroedinger, 171–172
  transport, 170–171
ESCA spectroscopy, 131, 139, 147
  hydrocarbons, 139
Ethane, 143
Ethylene, 139
Exchange theory, 5, 27–30
Extended 2ph-TDA, 152
  acetylene, 150–152
  hydrocarbons, 139–141
  methane, 129–131
  zinc dichloride, 131

Fatty acids, 46
Final state correlation, effects, 147
Finite element method, 230–232
  adaptive mesh, 233, 263–264
  convergence, 233, 263
  discretizations, 232, 242, 274
  Jacobian, 233
  Swartz-Wendroff approximation, 232, 242
Fluids, inhomogeneous, *see* Inhomogeneous fluids
Fluorescence, 37
Fluorescence spectrum, 13
Focal defects, 79
Focus, 200
Fold catastrophe, 197
Fourier transform, 180
  importance of, 185
Frenkel exciton states, 1

Generating function, 177–178
Green's function method, 116–117, 126, 131, 139,
  hydrocarbons, 141–143

Hamilton characteristic function, 167
Hamilton-Jacobi equation, 165

Hamilton-Jacobi equation (*Continued*)
  mixed space, 179, 204–206
  solution to, 201–202
Hartree-Fock particles, 120
High-density breakdown, BGYK, 238
Hindered rotation potential, 49
HNC equation, 250–255
  BGYK comparison, 239
  definition, 217
  integral equation, 222–223
  low temperature, 250
  multiple solution branches, 253–255
  solution space structure, 262, 268
  spinodal density, 252
  turning point location, 252–253
Hole burning, 4, 37, 41
Homogeneous linewidths, 3–4, 37
Horizontal dislocation loops, 78
Hydrocarbons:
  angle-resolved photoelectron spectrum, 139
  breakdown of ionization, 137–143
  configuration-interactions, 139–141
  correlation satellites, 140–141
  ESCA spectra, 139
  extended 2ph-TDA, 139–141
  initial-state correlation satellites, 139–141
  inner valence orbitals, 137–140
  main lines, 137
  photoelectron spectrum, 140–141
  satellite lines, 137–141
  spectra, 137–143
  spectroscopic factor, 139–142
Hypernetted chain (HNC) approximation, *see* HNC equation

Independent particle picture, 120
Infrared spectra, 56
  paraffins, 56
Inhomogeneous broadening, 3–4, 31
Inhomogeneous correlation functions, 219
Inhomogeneous fluids, 219
  direct correlation function, 220, 230
  hard-sphere perturbation, 269
  local density, 229
  OZ equation, 230
  pair correlation function, 230
Initial-state correlation satellites, 122, 152
  hydrocarbons, 139–141
  methane, 130
Inner valence orbital, 116, 122, 143
  hydrocarbons, 137–140
  methane, 130–1347
  nitrogen, 152
  zinc dichloride, 132–133
Integrals, complete, 165–166
Interchannel coupling, 123, 125, 150
Interferences, 149
Integrals, complete, 165
Intralayer melting transition, 79–90
  chain folding, 86–90
  dislocation loops, 83–86
Ionization:
  ammonia, 117
  auto-, 118, 125
  carbon dioxide, 117
  carbon disulfide, 117
  carbon monoxide, 116
  computational aspects, 126–129
  correlation effects, 116–153
  cross section, 123
  nitrogen, 116
  water, 116
Ionization cross section, 123
Ionization energy, 119
Ionization potentials, 126
Ionization spectrum, 119–126
Ising theory, 80–83
Isothermal compressibility, 216

Jacobian, 170, 226, 229
  mixed spaces, 179
JWKB approximation, 161

Kinks, 78
Kirkwood superposition approximation, 238, 256
Koopmans' theorem, 120–121, 134

Lagrange manifold, 165, 167–170, 202–204, 211
  definition, 175–176
  diffeomorphic, 176, 178
  essential properties of, 175–178
  generating function, 177–178
Lagrangian coordinate planes, 176, 211
Lagrangian plane, 211
Lennard-Jones potential, 225, 267, 271
  HNC, 252
  PY, 243
  truncation, 243–245

Linewidth, temperature behavior of, 17–18, 31
Lipids, 100–103. *See also* Phospholipids
Liquid-liquid transition, 76–77
Local mode, 7, 23
Local mode lifetime, 24–25

Main bands, 116
  methane, 131
Main lines, 117, 122, 127, 146–147, 150
  hydrocarbons, 137
Manifold, 211
  Lagrangian, 165, 167–170
Mapping, 211
Maslov index, 190–193
  μ, 191
  ν, 192
Mass tensor, 59–61
Mayer function, 267, 271
Melting transitions, 69–72
  intralayer, 73–74, 79–90
Metastability, 218, 264
Methane:
  breakdown of ionization, 129–139
  correlation satellites, 131–136
  extended 2ph-TDA, 129–131
  initial-state correlation satellites, 130
  inner valence orbitals, 130–137
  main bands, 131
  relaxation effects, 130–134
  satellite lines, 130–137
  shake-up satellites, 131–134
  spectroscopic factor, 129–132
Mixed spaces, 170–175
  classical density, 179
  density function, 206–208
  Hamilton-Jacobi equation, 204–206
  Jacobian, 179
  wavefunctions, 180–183
Mixing orbitals, 120
Molecular orbit picture, breakdown, 116–153
Momentum space, 172–175
Motion of chain equation, 55–67
  conformational transitions, 62–67
  infrared spectra, 56
Movement dislocations, 108–109

Newton method, 226
Nickel dichloride, 147
p-Nitroaniline, 147
Nitrogen, 148
  inner-valence orbital, 152
  ionization, 116
Nonadiabatic effects, 151–153
Nonorthogonality correction term, 124, 150
Nonpertubative theory, 6–14
Normal modes, 6, 11, 19, 23, 59–61

One-particle-one hole (1p-1h) configuration, 120
Optical Bloch equations, 3
Optical dephasing, 1–41
Optical phonons, 19–23
Orbital energy, 120
Orbitals, mixing, 120
Orientational energy, 54–55
Orientational potential, 54–55
Ornstein-Zernike (OZ) equation, *see* OZ equation
OZ equation, 216
  homogeneity constraint, 265
  inhomogeneous fluid, 230
  numerical calibration, 273

Pair correlation function, 215–216
  asymptotic behavior, 222
  inhomogeneous fluid, 219, 230
  long-ranged solutions, 268
  physical structure, 216
  potential distribution theory, 221
Pair distribution theory, pair correlation function, 221
Paraffins:
  cyclic, 46
  defects, 78–90
  elastic constants, 108–109
  infrared spectra, 56
  motion of chain, 55–67
    conformational transitions, 62–67
    equations, 57–62
    infrared spectra, 56
  normal 46,
  potential, 47–55
  single chain, 48–50
  two-dimensional phase transitions, 45–109
Partial-channel ionization cross section, 119, 123, 126, 148–150
Partial distribution theory, pair correlation function, 221
PE, *see* Photoelectron spectroscopy
Pearcy function, 198
Pentacene, in benzoic acid, 39–40

Percus-Yevick approximation, *see* PY equation
n-Perfluoroalkanes, 100
Perturbation theories, 13–14
  coexistence curve, 269
  diagrammatic, 9
Phase transitions, 45–109
  classification, 95-97
  conformational, 47, 62–67
  intramellar melting, 95
  Kosterlitz-Thouless, 46
  lamellar smectic A, 95
  liquid-crystal, 47
  liquid-liquid, 76–77
  melting, 46, 69–72
    intralayer, 73–74, 79–90
  nematic-isotropic, 95
  premature transition, 46
  rotation transition, 46, 95
  rotator, theory, 90–92
  smectic, 77
  smectic A, 95–97
  smectic-nematic, 95–97
  smectic-smectic, 47
  structural, 47
  surface melting, 93, 95
  surface roughening, 95–96
  theory, 77-97
  Transition III II, 76
  Transition IV II, 94–95
  Transition IV III, 75–76
  Transition V IV, 75
  Transition V IV γ, 93–94
  two-dimensional, 45–109
Phonon-assisted transfer, 1–2
Phonon correlation time, 8, 18–19
Phonon echo, 4, 37, 39
Phonons, 1–41
  acoustic, 4–5, 7, 14–19, 23, 31–36, 40
  anharmonic, 30, 40–41
  local mode lifetime, 24–25
  optical, 19–23
  phonon correlation time, 8, 18–19
  pseudolocal, 4–5, 7, 23–30, 37–40
    local mode lifetime, 23–25
  sidebands, 4–5, 31, 41
Phonon sideband, 5
Phospholipids, 99, 100–103
  membrane monolayer, 46
Photodissociation, 118
Photoelectron spectroscopy (PE), 116
  acetylene, 148–151
  angle-resolved, hydrocarbon, 139
  carbonyl sulfide, 133
  hydrocarbons, 140–141
Photoionization, 123
Photon stimulation desorption, 188
2ph-TDA, *see* Two-particle-hole-Tamm-Dancoff approximation
Picard iteration, 234, 250
Pole strength, 119
  zinc dichloride, 131
Population relaxation, 2
Potential, 47–55
  adhesive hardsphere, 249, 267
  Buckingham, 48
  energy surfaces, 151
  hindered rotation, 49
  of interaction, 47–55
  ionization, 126
  Lennard-Jones, 225, 267, 271
  orientational, 54–55
  paraffins, 47–55
  Salem, 51–54
  single chain, 48–50
  valence, 49
Potential distribution theory, 221–222
  clustered expansion, 221
  covariance, 222, 271
Potential energy surfaces, 151
Pressure:
  compressibility, 216, 240
  virial, 216, 240, 259
Propiolic acid, 120
Pseudolocal phonons, 4–5, 7, 37–40
  coupling to, 23–30
Pseudo-state, 119
Pure dephasing, 2, 31
PY equation, 243–250
  definition, 217
  integral equation, 222
  singularity behavior, 297
  solution space structure, 262, 268
  spinodal density, 245
  turning point location, 245, 248

Quantum mechanical wavefunction, 261–262
"Quasi-particle picture," 122

Radiation, synchrotron, 142, 148, 152
Raman spectroscopy, 76
Range of influence, 215

Rayleigh depolarizatioin, 76
Regular/Singular, 211
Relaxation, vibrational, 34
Relaxation effects, 116, 122
  carbonyl sulfide, 134
  methane, 130–134
Relaxation process, 2
Resonances, shape, 148
Restricted ensemble, 219, 264, 266
Rotator phase, melting transition, 71
Rotator transition, theory, 90–92
Ruby, 34–36
Rydberg states, 129, 145

Salem potential, 51–54
Satellite bands, 116, 152
Satellite lines, 117, 121, 122, 147, 149, 152
  carbonyl sulfides, 134
  hydrocarbons, 137–141
  methane, 130–137
  zinc dichloride, 131
Scattering:
  Lennard-Jones, 163
  two-dimensional, 163
Schroedinger equation, 171–172
  formal asymptotic solution of, 180–183
Screw dislocation, 103–108
Semiclassical approximation:
  independence of coordinates, 197
  primitive, 172
  two or three dimensions, 162
  validity, 183–185
Sensitivity of fluid structure:
  cut off dependence, 226–238, 140
  perturbations, 229
Separated-channel approximation, 150
Shake-up bands, 116
Shake-up satellites, 122
  carbonyl sulfide, 134–136
  methane, 131–134
  zinc dichloride, 132
Shape resonances, 125, 145, 148
Sharing of intensity, 127
Sidebands, 4–5, 31, 41
Single chain paraffins, 48–50
Single chain potential, 48–50
Single-channel approximation, 124–125
Single-hole (1h) configuration, 120, 144, 147
Single-particle picture, 116
Singlet-triplet splitting, 145
Singular chart, 191
Singularity, coexistence curve, 218, 264
Singularity behavior, BGYK, 258
Singular points, 191
Smooth, 211
Solution, asymptotic, 171–172
Solution space structure:
  BGYK, 262–263, 268
  HNC, 262, 268
  PY, 262, 268
Spaces:
  mixed, 170–175
  momentum, 170–175
Spectroscopic factor, 119, 123, 125, 148
  hydrocarbons, 139–142
  methane, 129–132
  zinc dichloride, 132
Spectroscopy:
  binary (e,2e), 116, 123
  dipole (e,2e), 116, 123
  ESCA, 131, 139
  photoelectron (PE), 116
Spinodal curve, 218
Spinodal density:
  BGYK, 256–257
  HNC, 252
  PY, 245
Square-well fluid, 242, 258
Stationary phase approximation, 187–190
Strain field, 14
Strong coupling limit, 27
Subcritical fluid, correlation functions, 215, 275
Successive substitution, 234
Sudden approximation, 125
Sudden limit, 149
Surface melting transition, 93
Swartz-Wendroff approximation, 232, 242
Switching functions, 187
Synchrotron radiation, 142, 148, 152

Temperature dependencies, 18
Thermodynamic consistency, 217
  cutoff dependence, 226
  numerical calculation, 234
Thermodynamic energy, 217, 225, 240
Tilt transition, 103
Torsional specific heat, 61–62
Trajectories:
  classical, 165–167
  families of, 167–169
Transformation, Baxter, 248–249, 254

Transition amplitude, 119, 121, 125–126
Transitions:
  III II, 76
  IV II, 94
  IV III, 75–76
  V IV, 75
  V IV γ, 93–94
Transport equation, 170–171
Triacetylene, 147
Tunneling, 200
Turning point location, 245, 267
  BGYK, 256
  dimensionality effect, 270
  HNC, 252–253
  pair potential dependence, 249–250, 253, 267
  PY, 245, 248
Two-dimensional scattering, 163
Two-hole-one-particle (2h-1p) configuration, 120–121, 133, 136, 144–145
Two-particle-hole-Tamm-Dancoff approximation, 117, 127
  extended, 127. *See also* Extended 2pH-TDA
Two-particle-two-hole (2p-2h) configuration, 120–121

Ultraviolet PE spectroscopy (UPS), 116
Uniform approximations, 172, 187, 190, 198
  strictly, 198

Vacancies, 78
Valence potential, 49
Valence-shell binding energy spectrum, 133
Vertical dislocation loops, 78
Vibrational relaxations, 34
Vibronic coupling, 151–152
  multistate, 152
Vibronic spectrum, 31
Virial pressure, 216, 240, 259
Vortex loops, 78

Water, ionization, 116
Wavefunctions:
  calculation of, 162–212
  canonical formula, 194–197
  mixed spaces, 180–183
  momentum spaces, 180–183
  primitive semiclassical forms, 185–186
  quantum mechanical, 161–212
Waves:
  incoming, 186
  outgoing, 186
Weak coupling theory, 13–15, 21, 25, 35

Xenon, 147
X-ray emission, 118
X-ray induced PE spectroscopy (XPS), 116

Zero-phonon line (ZPL), 4, 31, 37
Zinc dichloride, 131–134
  correlation satellites, 132
  extended 2pH-TDA, 131
  inner-valence orbitals, 132–133
  pole-strength, 131
  satellite lines, 131
  shake-up satellites, 132
  spectroscopic factor, 132
ZPL, *see* Zero-phonon line